Corporate Finance Under Climate Crisis

"As someone who navigates both the open waters as a competitive swimmer and the complexities of modern business leadership, I found this book to be a profound guide. It masterfully equips business students and future corporate leaders with the tools and insights needed to turn the tide on climate challenges and drive impactful sustainability transitions."

—Jordi Díaz Martín, *Dean of EADA Business School*

"This textbook provides a comprehensive framework linking corporate finance with climate challenges, offering insights into the financial motivations that drive climate solutions. It equips finance and business students with climate literacy while fostering financial literacy for environmental management students, empowering them to navigate risks, seize opportunities, and shape strategic responses to the pressing realities of climate change."

—Junjie Zhang, *Professor, Nicholas School of the Environment, Duke University*

"This book is a tour de force in sustainability-focused corporate finance. Packed with innovative tools, practical strategies, and a vision for the future, it's indispensable reading for anyone shaping the financial systems of tomorrow."

—Alex Lazarow, *Venture Capitalist and Author of Out-Innovate*

"Emphasizing sustainability as central to long-term success, this essential textbook equips students and professionals with strategies to assess, manage, and maximize value amid climate challenges. A vital resource for driving resilient, forward-thinking financial decisions in the pursuit of a Net-Zero future."

—William Hua Wang, *Dean, emlyon business school Asia*

"As the largest French-speaking business school in North America, ESG UQAM champions the integration of sustainability into business education. This book provides an unparalleled synthesis of climate science, policy, and corporate finance, equipping readers with actionable frameworks and practical applications to lead in a net-zero economy. A must-read for those shaping tomorrow's sustainable business practices."

—Komlan Sedzro, *Dean, École des sciences de la gestion, Université du Québec à Montréal*

"This timely textbook reflects the growing consensus that economic growth at the expense of the environment is unsustainable. It integrates ESG factors into corporate finance, covering investment decisions, asset pricing, and risk management, and offers a solid foundation for sustainable planning. Ideal for students and leaders, it bridges theory and practice to address urgent global challenges."

—LIU Feng, *Chief Economist, International Institute of Green Finance, Central University of Finance and Economics*

"This critically important text will have readers wishing they had read it sooner. It provides a solid orientation to a set of issues that every modern executive will confront in their careers. Corporate Finance under Climate Crisis should quickly become part of the finance canon."

—Sara Olsen, *Founder, SVT Group*

"Climate risk is financial risk. Addressing climate issues is no longer solely the domain of science and policy; it has become increasingly critical for business leaders and financial investors. This book provides a comprehensive overview of how climate impacts finance and offers actionable insights for businesses to navigate these challenges effectively."

—GUO Peiyuan, *Chairman of SynTao Green Finance and China Sustainable Investment Forum*

Sandra Dow · Yuwei Shi

Corporate Finance Under Climate Crisis

Sustainability Transitions in Theory and Practice

Sandra Dow
Middlebury Institute of International
Studies at Monterey
Monterey, CA, USA

Yuwei Shi
University of California at Santa Cruz
Santa Cruz, CA, USA

ISBN 978-3-031-83486-8   ISBN 978-3-031-83487-5   (eBook)
https://doi.org/10.1007/978-3-031-83487-5

This Palgrave Macmillan imprint is published by the registered company Springer Nature Switzerland
AG
The registered company address is: Gewerbestrasse 11, 6330 Cham, Switzerland

If disposing of this product, please recycle the paper.

For Jack and Will—With love and the hope that tomorrow's challenges find solutions in today's choices.

For Jiesheng Shi, who held an unshakable belief that books improve the world.

Preface

The financial world stands at an unprecedented juncture, where the growing reality of climate change demands a fundamental transformation in corporate finance. No longer a distant environmental concern, climate change has evolved into a financial crisis affecting every industry, market, and region. Projections suggest that unchecked climate risks could cause global financial losses exceeding $23 trillion by 2050. The evidence is already stark: devastating hurricanes, such as Milton and Helene in 2024, not only caused unprecedented destruction but highlighted vulnerabilities in insurance markets, real estate, and critical infrastructure. Elsewhere, Europe's heat waves exposed the fragility of energy systems, while catastrophic flooding in Pakistan underscored the disproportionate burden on developing economies. These interconnected crises reveal the inescapable influence of climate dynamics on economic stability and corporate decision-making.

This textbook, *Corporate Finance Under Climate Crisis: Sustainability Transitions in Theory and Practice*, is designed to prepare finance leaders for this era of complexity and uncertainty. Unlike traditional approaches, this book integrates corporate finance with sustainability strategy, reflecting the critical need to balance financial objectives with long-term environmental and social resilience. By equipping readers with adaptive, systems-based thinking and innovative frameworks, it empowers them to manage risks, seize emerging opportunities, and drive sustainable transformation. Finance is no longer just about profits; it is about shaping the future.

Thank you for embarking on this journey. May it inspire you to drive change through finance.

Monterey, USA
Santa Cruz, USA
January 2025

Sandra Dow, Ph.D.
Professor Emeritus, Middlebury
Institute of International Studies

Yuwei Shi, Ph.D.
Professor Emeritus, Middlebury
Institute of International Studies

Acknowledgments

This book is the culmination of years of teaching, research, and exploration into how corporate finance intersects with one of the defining issues of our time: climate change. It all began at the Middlebury Institute of International Studies (MIIS) in 2010, where we set out to create a frontier course in ESG—Environmental, Social, and Governance principles. Back then, this was uncharted territory, and we aimed to go beyond the traditional boundaries of siloed business disciplines by integrating strategic thinking, scientific understanding, impact investing, and development economics into one cohesive framework.

It was a genuine smorgasbord of disciplines, each essential to understanding how corporate strategy and finance must adapt to a world facing unprecedented environmental and social challenges. Our thanks go to the colleagues and students at MIIS who contributed ideas, challenged assumptions, and shared insights that deepened and enriched this work. A special acknowledgment goes to the students who participated in our ESG course spinoffs. These raw case studies, which tackle sustainable strategy and finance as tools to solve complex social and environmental problems, require a level of dedication and commitment far surpassing even the most demanding MBA courses.

Special thanks to those early advocates and contributors who helped shape this vision, as well as to the ever-growing community of academics, practitioners, and students who continue to push the boundaries of business in the climate crisis. Together, we are forging new pathways in finance that not only seek returns but also consider the legacy we leave behind.

We extend our heartfelt gratitude to the anonymous referees whose insightful feedback illuminated paths we might not have otherwise explored, shaping the foundation of this book from its inception. We are equally indebted to our editors for their unwavering support, boundless patience, and the thoughtful guidance that made this journey both meaningful and achievable. Their encouragement and steady hand have been invaluable companions throughout this process.

About This Book

Corporate Finance Under Climate Crisis: Sustainability Transitions in Theory and Practice is a pioneering textbook that bridges the domains of corporate finance and sustainability strategy. It addresses the urgent challenges posed by climate change while equipping students and professionals with the tools and insights needed to lead transformative change. With climate risks reshaping the global economy, this book offers a roadmap for navigating uncertainty, managing risks, and fostering innovation through a sustainability lens.

The textbook is organized into four key sections, each addressing a critical aspect of climate-conscious corporate finance:

1. **Foundations and Context**
   The opening Chapters 1–6 establish the foundational understanding of corporate finance within the context of climate change. Topics include an introduction to climate finance, scientific principles of climate change, and the implications for central banks and financial institutions. It also delves into non-financial disclosures, greenhouse gas measurement frameworks, and strategic tools like Integrated Assessment Models (IAMs) and Shared Socioeconomic Pathways (SSPs) for setting firm-level goals aligned with global climate targets.
2. **Risks and Opportunities in Climate Change and Financial Implications**
   Chapters 7–10 explore how climate risks—both physical and transitional—impact financial decision-making. From assessing the financial implications of extreme weather to understanding transition pathways like net-zero financing, this section integrates advanced discussions on green technology adoption and its financial ramifications. Readers will learn to analyze the risks and opportunities presented by the changing climate landscape and their effects on capital costs and firm valuation.
3. **Practical Tools and Applications**
   Chapters 11 and 12 focus on advanced techniques for project valuation and capital budgeting, with applications in sustainable and traditional asset investments. This section also introduces the role of carbon markets as critical tools for providing price signals and incentivizing emissions reductions. The section concludes with a comparative case study in Chapter 13 of ExxonMobil and

Royal Dutch Shell, illustrating how two major oil companies are navigating the Net-Zero challenge and the strategic implications for their long-term value.

4. **Governance and the Future of Finance**

The final Chapters (14 and 15) examine the evolving nature of corporate governance in addressing climate change. They highlight the shift from shareholder primacy to stakeholder capitalism, emphasizing long-term value creation for all stakeholders. This section encourages critical thinking about the systemic shifts required in capitalism, offering thought experiments and strategic insights to envision a sustainable financial future.

Throughout the book, case studies and real-world examples provide vivid illustrations of how businesses across industries and geographies are adapting to climate challenges. Readers will find actionable tools for evaluating green technology investments, integrating ESG factors into capital strategies, and navigating carbon markets. The global perspective ensures that the lessons are applicable to both developed and emerging economies, fostering an understanding of the universal nature of climate finance.

Corporate Finance Under Climate Crisis is more than a textbook—it is a guide for the future of finance. By merging rigorous analysis with strategic foresight, it equips readers to tackle the greatest challenge of our time and turn it into an opportunity for innovation, resilience, and meaningful change.

Contents

About the Authors

Dr. Sandra Dow is Professor Emeritus at the Middlebury Institute of International Studies. She holds a PhD in International Finance from Concordia University in Montreal, a Masters in Economics from Dalhousie University, and a Bachelor of Economics from Mount Allison University. Sandra began developing an advanced corporate finance class in ESG Risk Assessment in 2010 at the Middlebury Institute of International Studies. In the decade that followed the course evolved to its latest iteration as Corporate Finance in Climate Change and was the inspiration for this book. Professor Dow, along with colleague Yuwei Shi, was the recipient of the prestigious Best Paper Award in Graduate Management Education from the Academy of Management. The award recognized their work in devising a new pedagogical approach, the raw case method, to address complex problems. In 2017 she received the Middlebury Institute Faculty Excellence Award in recognition of teaching excellence. She has written extensively on corporate governance, in addition to research focusing directly on environmental and social risks and reviews regularly for several finance and pedagogical journals.

Dr. Yuwei Shi is Professor Emeritus at the Middlebury Institute, where he served as Dean of the Graduate School of International Policy and Management. He is also Adjunct Professor at the University of California Santa Cruz and Academic Director of the Blue Pioneers Accelerator, a global leadership development program for ocean and climate innovators. Dr. Shi's current research and practice interests center on impact management and sustainability transitions. He serves as Associate Editor of Elementa: Science of the Anthropocene, a scholarly journal by the University of California Press. Dr. Shi holds PhD in Strategic Management from University of Texas at Dallas, Master of Comparative Law from Southern Methodist University Dedman School of Law, and Bachelor of Engineering from Shanghai Jiaotong University.

List of Figures

List of Tables

Part I

Foundations and Context

Introduction to Corporate Finance Under Climate Crisis

1

Introduction

Imagine a world where finance is not just about managing assets and maximizing shareholder value but about navigating the shifting tides of a planet in flux. The practice of corporate finance is undergoing a profound transformation as climate change forces firms to confront new types of risks—and to explore emerging opportunities. A warming planet brings volatility, not just in weather patterns but in financial markets and corporate strategies. Adapting to these changes requires more than an expansion of traditional finance knowledge; it demands an understanding of how a company's financial decisions intersect with climate science, policy shifts, and stakeholder expectations.

Throughout this book, we use the terms "global warming" and "climate change" interchangeably. However, it is important to distinguish that global warming—primarily the result of increasing greenhouse gas emissions—drives climate change. Climate change encompasses the broader consequences of this warming, including shifts in weather patterns, rising sea levels, and increased frequency of extreme events. Understanding this causal link is critical to framing corporate finance's role in addressing both risks and opportunities.

In the realm of corporate finance, the challenges of climate change reshape the basics. Consider that finance professionals now need to keep pace with climate science and develop fluency in reading climate projections—not as abstract forecasts but as strategic determinants that could impact a company's future revenues, asset valuations, and even its very survival. Finance teams are following policies around the world as governments intensify regulations and shift incentives, altering the landscape of financial risks and rewards across borders. This is not business as usual; this is corporate finance on the frontlines of climate change.

Climate change introduces both physical and transitional risks that challenge more than just corporate profitability. These risks also have the potential to disrupt supply chains, damage reputations, alter employee and consumer behavior, and even lead to regulatory or legal repercussions. As we explore in later chapters, these risks are interrelated and often compound each other in ways that force businesses to reconsider their traditional financial and operational models.

To measure a company's greenhouse gas emissions or assess its exposure to climate-related risks, finance professionals now work with data that goes beyond traditional financial statements. They assess physical risks—like those posed by wildfires, floods, and storms—to a company's assets and operations. They also evaluate transition risks, which emerge as policies and markets adapt to a low-carbon future, affecting a firm's cost structures, competitive landscape, and even reputation.

With the risks come new opportunities, and climate change has opened a fresh frontier for profitable investments. Renewable energy, clean transportation, energy efficiency, and carbon capture and storage are not just buzzwords; they are the building blocks of tomorrow's economy. But these climate-aligned investments require fresh valuation approaches. For example, estimating the NPV of a carbon capture project involves calculating not only the operational costs but also the value of captured carbon in a regulatory landscape that is still evolving. In fossil fuels, what used to be a straightforward investment in oil field expansion now involves factoring in potential emissions liabilities and a fast-approaching horizon where high-carbon assets might become stranded.

In the face of these risks and opportunities, companies are increasingly realizing that addressing climate-related challenges requires more than just financial adjustments. It requires a rethinking of business models themselves, impacting everything from supply chain design to customer engagement strategies. The resilience of a business, therefore, depends not only on financial flexibility but on its ability to innovate and adapt to a rapidly changing world. In subsequent chapters, we will see how businesses are integrating these considerations into their long-term strategic planning, from sustainability initiatives to operational overhauls.

In response to these growing risks, financial markets have begun developing innovative solutions to help companies mitigate and adapt. Tools like green bonds, climate risk insurance, and carbon pricing mechanisms are becoming increasingly important. As we move through this book, we will explore how these financial instruments are reshaping corporate finance and enabling companies to transition toward a more sustainable future.

As companies begin to feel the weight of the climate crisis, financial markets are also evolving. Traditional models based on short-term profit maximization are increasingly being challenged by the need for longer-term, sustainable growth. This shift is not just driven by ethical considerations but also by changing investor demands for transparency around Environmental, Social, and Governance (ESG) factors. Later chapters will explore how ESG criteria are becoming an integral part of corporate valuation, risk management, and decision-making processes.

Finally, effective corporate governance is key to navigating these uncharted waters. Boards must bring expertise not only in finance but in environmental science, policy, and stakeholder engagement. Corporate finance in a climate-changed world requires holistic governance that balances shareholder interests with the urgent demands of the planet and society. We end with wondering about the great shift that is underway: Addressing the climate crisis requires not merely incremental change but a rethinking of market and governance systems.

Climate change is no longer an externality to be managed on the periphery of corporate strategy. It is an issue that requires businesses to rethink their entire value proposition. The transition to a low-carbon economy is not just about adjusting operations; it is about reimagining business models to create long-term resilience in the face of climate risk. This transition will be explored in detail in later chapters, where we will examine real-world examples of companies that have successfully navigated this shift and others that have struggled.

In this book, you will discover how corporate finance is being reshaped to meet the climate challenge. You will explore practical tools, case studies, and real-world insights into this evolving field—empowering you to contribute to a future where finance drives resilience and positive change.

1.1   Foundations of the Business Response to the Climate Crisis

In this introductory chapter, we provide an overview of key themes that will recur throughout the book, forming the foundation for understanding the business response to the climate crisis:

- **The interdependence of business, climate degradation, and policy**: The impact of climate change on business operations is inseparably linked to environmental damage and the policies addressing it. From international agreements to local regulations at the city, town, or provincial level, every effort contributes to containing global warming.
- **The escalating nature of climate risks**: As the planet gets hotter, climate risks are intensifying. At the same time, increasingly stringent climate policies are reshaping the way firms operate, make investments, and access financing.

- **The end of "business as usual"**: Once a reassuring phrase symbolizing continuity during crises, "business as usual" now highlights the dangers of inertia in the face of climate change. The status quo is no longer sustainable, and transformative action is required to adapt to a rapidly changing environment.
- **Transitionand physical risks as existential threats**: Climate change presents dual threats to businesses—transition risks from adapting to low-carbon policies and technologies, and physical risks from extreme weather and environmental degradation. Both challenge profitability and even the survival of firms, necessitating deep insights and strategic responses.
- **Opportunities within the crisis**: Alongside the risks, climate change offers abundant opportunities. Businesses and society as a whole can act as engines of innovation, driving solutions to contain global warming while unlocking new markets and value.
- **The dual tragedies shaping the response**: The *tragedy of the commons*—where collective resources are overexploited—and the *tragedy of the horizon*—where short-term incentives overshadow long-term risks—are central to the world's struggle to address the climate crisis. Business is no exception. As things stand, we are far off-track in preventing a climate catastrophe, underscoring the urgency for immediate and coordinated action.

To start things off, we present a compelling history lesson—the science of global warming is anything but new!

1.2   A Brief History Lesson

The physical risks of a warming planet are becoming increasingly evident, with new examples emerging almost daily. From devastating hurricanes battering coastlines to catastrophic floods submerging entire communities and deadly droughts crippling agriculture and water supplies, these events are wreaking havoc on societies worldwide. But what is causing all of this? Decades of scientific research have shown that the primary driver is the accumulation of greenhouse gases in the atmosphere, largely due to human activities such as burning fossil fuels, deforestation, and industrial processes. What is more striking is that we have known about these causes—and their potential consequences—for a long time.

As far back as 1956, the New York Times drew attention to the link between global warming and escalating carbon dioxide emissions when the paper published an article by Waldemar Kaempffert: "Warmer climate on the Earth may be due to more Carbon Dioxide in the Air" (Kaempffert, 1956). Mr. Kaempffert referenced the work of Dr. Gilbert Plass who had just published an article in American Scientist in which he linked carbon dioxide emissions and global warming (Plass, 1956). His work, at the time, challenged prevailing views on carbon dioxide and climate change. In fact, Dr. Plass pointed out that at the beginning of the twentieth century the link between carbon and climate appeared well received by the scientific community. However, fifty years later when Plass's paper was published, this view had fallen out of favor. Plass did not invent the linkage. His article built upon the work of John Tyndall who in 1859 published the results of experiments on the heat absorptive capacities of various atmospheric gasses, concluding that carbon dioxide would be linked to a warming climate (Tyndall, 1859).

Kaempffert concluded his piece in the New York Times with a prescient statement: "Even if our oil and coal reserves will be used up in one thousand years, seventeen times the present amount of carbon dioxide in the atmosphere must be reckoned with. The introduction of nuclear energy will not make much difference. Coal and oil are still plentiful and cheap in many parts of the world, and there is every reason to believe that both will be consumed by industry as long as it pays to do so" (Kaempffert, 1956).

Jumping forward 70 years, the earth is warmer, and emissions continue to rise. It would seem Kaempffert was right: it still pays to burn fossil fuels. Despite national and supranational regulation, it has been pretty much Business as Usual.

1.3 Global Warming

The planet is now 1.1 degrees warmer than it was in the late 1800s. Reliable record keeping beginning in 1880 shows that the earth warmed approximately 0.07 C every ten years until 1980. Since that time, the pace of warming has more than doubled. Rising temperatures have provoked violent storms across the globe, drought, heat waves, and wildfires and there is agreement that global warming can be traced directly to increased greenhouse gas (GHG) emissions. Figure 1.1 shows this temperature rise since 1850.

1.4 Increasing GHG Emissions

Figure 1.2 illustrates the unrelenting upward trajectory of emissions and tells a story of two distinct phases in the history of global carbon dioxide (CO_2) emissions, marked first by the industrial revolution and later by the rise of emerging markets.

In the early part of the timeline, the United Kingdom and other European nations were at the forefront of industrialization, powered largely by coal. As the nineteenth century progressed, the United States emerged as the dominant emitter, with its rapid industrial expansion and heavy reliance on fossil fuels driving a steep rise in CO_2 emissions. By the mid-twentieth century, the industrialized West had firmly established itself as the primary contributor to global emissions, a reflection of energy-intensive manufacturing, transportation, and urbanization.

However, the second half of the twentieth century tells a different story. As industrialized nations began to plateau or even reduce their emissions thanks to cleaner technologies and a shift toward service-based economies, emissions in emerging markets began to rise. China, in particular, saw its emissions skyrocket as it underwent rapid industrialization and urbanization, becoming the world's

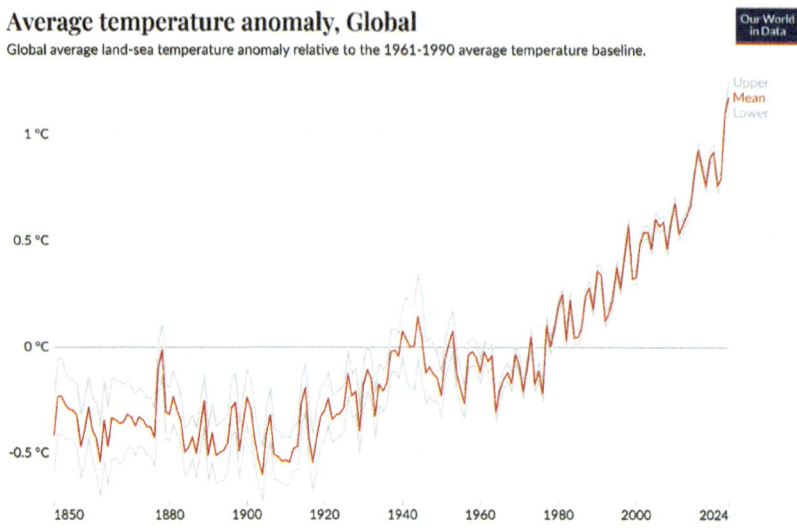

Fig. 1.1 Global rise in temperature

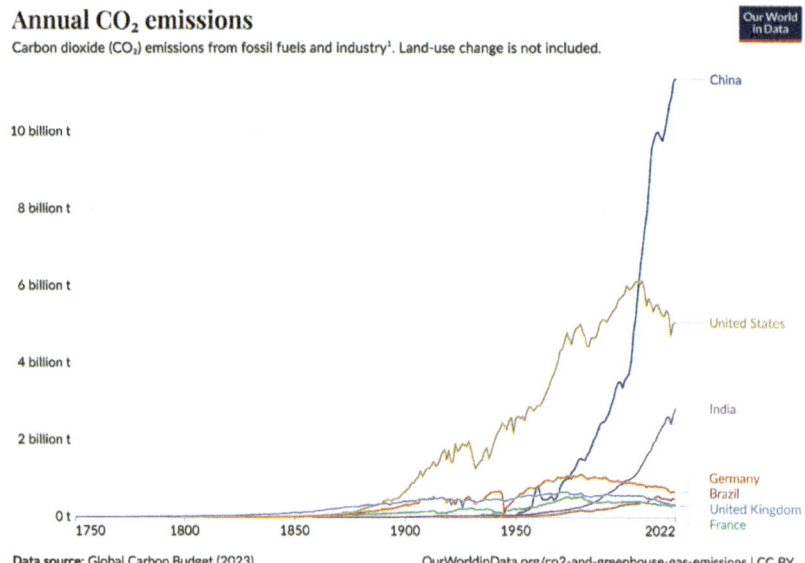

Fig. 1.2 Global GHG Emissions from 1750 to 2023

manufacturing hub. India followed a similar, though less steep trajectory, with its growing economy and energy demands driving consistent increases in emissions. Meanwhile, smaller but steady contributors such as Germany, Brazil, and France reflect varied timelines of industrialization and energy transitions.

The visualization in Fig. 1.2 underscores how the global burden of emissions has shifted over time. While the early industrializers bear a historical responsibility for much of the accumulated CO_2 in the atmosphere, the dramatic rise of emissions in emerging economies like China and India reveals the modern challenge of balancing economic development with climate action.

1.5   The Uneven Footprint of Global Warming

The Geography of Climate Change

As with emissions, global warming has not occurred evenly across the planet, and this uneven distribution of warming illustrates a stark and growing problem. Figure 1.3 illustrates the uneven warming pattern, with some regions experiencing far greater temperature increases than others.

Figure 1.3 shows that the Arctic and sub-Arctic regions have experienced the most intense warming, with recent evidence showing that since 1979, the Arctic has warmed nearly four times faster than the global average (Rantanen et al., 2022). This phenomenon, known as Arctic amplification, is driven by feedback

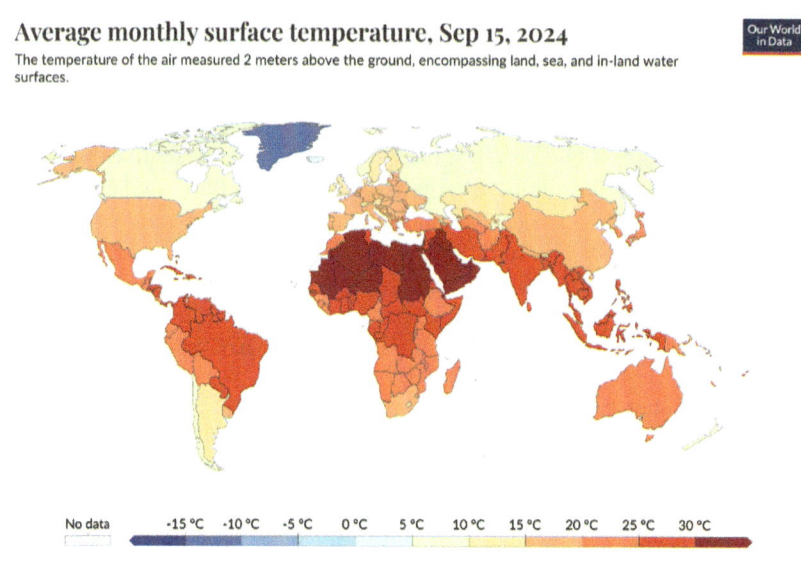

Fig. 1.3 Global surface temperatures in 2024

mechanisms like the loss of reflective ice and snow cover, which exposes darker surfaces that absorb more solar radiation, compounding the warming effect. The consequences are far-reaching: the accelerated melting of sea ice and glaciers is contributing to rising sea levels, altering ocean currents, and destabilizing global climate patterns. Permafrost thaw releases potent greenhouse gases such as methane and carbon dioxide, further amplifying climate change (IPCC, 2021).

Outside the polar regions, the impacts of warming are equally devastating, if less geographically concentrated. In Asia, for instance, the catastrophic 2022 floods in Pakistan submerged a third of the country, displacing millions and causing $30 billion in economic losses and flood damage (World Bank, 2022). The flooding, fueled by unprecedented monsoon rains and glacial melt, wreaked havoc on infrastructure, agriculture, and livelihoods, with global repercussions for food security. Similarly, in Europe, the summer of 2023 saw southern regions endure record-breaking heatwaves, with temperatures soaring above **45 °C** in parts of **Italy, Greece, and Spain**. These extreme conditions triggered wildfires, forcing mass evacuations and overwhelming public health systems. In **July 2023**, temperatures in Italy were expected to reach **48 °C** on the islands of **Sicily and Sardinia**, potentially the hottest temperatures ever recorded in Europe. In **Greece**, devastating wildfires continued into **August 2023**, exacerbated by extreme summer heat, leading to widespread evacuations and significant infrastructure damage (European Space Agency, 2023a; European Space Agency, 2023b). In North America, recurring heatwaves and prolonged droughts severely impacted agricultural productivity, water supplies, and public health, particularly in the western United States (Climate Change, 2022).

The Costs of Climate Change

The costs of climate change are enormous and on the rise. A recent study highlights that over the past two decades, extreme weather events such as hurricanes, floods, and heatwaves have inflicted approximately $2.8 trillion in damages globally. This equates to an average of $143 billion annually, or about $16.3 million per hour (World Economic Forum, 2023 October). The research attributes around 53% of these damages directly to human-induced climate change, with storms accounting for the majority, followed by heatwaves, floods, droughts, and wildfires. Additionally, the study estimates that climate change has contributed to nearly 61,000 deaths during the analyzed period.

While the Global South has historically contributed the least to global greenhouse gas emissions, it faces a disproportionate share of the consequences. Regions in the Global South are often more vulnerable to climate impacts due to their geographical exposure, reliance on climate-sensitive industries like agriculture, and limited financial and infrastructural resources for adaptation. At the same time, emerging economies such as China and India, which are part of the Global South, are now among the largest emitters due to rapid industrialization and population growth. This complexity highlights the uneven distribution of both historical responsibility and climate-related damages, with many low-emission nations still bearing the worst effects despite having contributed little to the problem.

1.6   The Tragedy of the Commons

The inherent unfairness of the climate crisis and the disparity between its costs and benefits were starkly outlined by Sir Nicholas Stern in his groundbreaking work, *The Economics of Climate Change: The Stern Review* (Stern, 2007). Stern argued that the crisis we face today is a direct consequence of the *Tragedy of the Commons*, where "the problem of climate change involves a fundamental failure of markets: those who damage others by emitting greenhouse gases generally do not pay" (Stern, 2007). Without regulations to limit individual greenhouse gas emissions, there is little incentive for emitters to reduce their contributions, leading to unrestricted growth in emissions. Stern famously called climate change the "greatest market failure the world has ever seen," highlighting how its far-reaching consequences touch every aspect of human life.

China exemplifies this modern-day Tragedy of the Commons. As the world's largest producer of greenhouse gas emissions, China emits nearly double the amount of the next-largest emitter, the United States, based on 2019 data (Ritchie & Roser, 2020).

Yet neither China nor the United States face the greatest vulnerability to the destructive effects of climate change. Instead, the countries most impacted—such as Afghanistan, Bangladesh, Chad, and Haiti—are part of the developing world, where resources to combat or adapt to climate catastrophes are sparse. While G20 nations are responsible for the majority of global emissions, these developing nations bear the brunt of the consequences.

Despite verbal commitments to reduce emissions, China's actions tell a different story. In 2022, new coal power capacity under development in China surged by 38%, even as such capacity fell by 20% in the rest of the world (South China Morning Post, 2023). This highlights the tension between economic development and environmental responsibility, a challenge not unique to China but emblematic of the difficulties in addressing the global tragedy of climate change. Without stronger collective action and enforceable commitments, the disparities in both responsibility and vulnerability will continue to widen, leaving the most vulnerable nations to suffer disproportionately.

1.7   Global Response to the Climate Crisis

United Nations Framework Convention on Climate Change (UNFCCC)

The global response to climate change began in a coordinated way with the establishment of the United Nations Framework Convention on Climate Change (UNFCCC) in 1992.[1] This foundational treaty brought together 199 member nations to stabilize atmospheric greenhouse gas concentrations and prevent dangerous disruptions to the climate system (UNFCCC, n.d.). The UNFCCC's annual Conferences of the Parties (COPs) became a platform for nations to negotiate targets, review progress, and design strategies. Building on this framework, the Kyoto Protocol emerged as the first major agreement to set binding emission reduction targets for industrialized nations.

Kyoto Protocol

Adopted in 1997, the Kyoto Protocol marked a turning point in global climate policy by legally binding developed countries to reduce their GHG emissions by an average of 5% below 1990 levels during its first commitment period (2008–2012) (United Nations, 1998). It introduced mechanisms like emissions trading and carbon credits, allowing countries to meet their targets through market-based solutions, including Certified Emission Reductions (CERs) from projects in developing nations. However, Kyoto's effectiveness was limited by its exclusion of major emitters like China and the United States and challenges with compliance and participation (Victor, 2001). Despite these limitations, Kyoto laid the groundwork for more inclusive and flexible agreements, such as the Paris Agreement.

The Paris Agreement

It seems like it took the world from 1859 (Tyndall's research) to 2015 (the Paris Agreement) to solidly grasp the unfolding climate crisis. The Paris Agreement, adopted in 2015, represented a significant evolution in global climate governance. Unlike Kyoto, it required participation from all nations, acknowledging that addressing climate change demands a collective effort. Its primary goal is to limit global warming to well below 2 °C above pre-industrial levels, with an aspirational target of 1.5 °C (UNFCCC, n.d.). The agreement introduced Nationally Determined Contributions (NDCs), allowing countries to set and periodically update

---

[1] The 1992 Rio Earth Summit led to the adoption of three key environmental treaties: the United Nations Framework Convention on Climate Change (UNFCCC), the Convention on Biological Diversity (CBD), and the United Nations Convention to Combat Desertification (UNCCD). These conventions are designed to be mutually reinforcing, addressing interconnected challenges related to climate change, biodiversity loss, and land degradation (United Nations, 1992).

their own climate targets, and mechanisms like Internationally Transferred Mitigation Outcomes (ITMOs) to promote global cooperation (Rajamani & Bodansky, 2019). Additionally, it emphasized both mitigation and adaptation, recognizing the need to reduce emissions while building resilience to unavoidable climate impacts.

Regional Agreements: Fit for 55

While global agreements have provided the framework for action, regional leadership has emerged as a critical driver of progress. The European Union (EU) has led the way with its ambitious *Fit for 55* package, unveiled in 2021. This initiative aims to reduce the EU's net emissions by at least 55% by 2030, with measures such as expanding the EU Emissions Trading System (EU ETS) to cover marine transport, introducing a Carbon Border Adjustment Mechanism (CBAM) to address emissions embedded in imports, and tightening energy efficiency standards for buildings and transportation (European Commission, 2021). The EU's comprehensive approach reflects a deep commitment to decarbonization, setting an example for other regions while addressing challenges like carbon leakage and promoting clean energy innovation.

Summing Up Climate Policy

The evolution of climate policy—from the UNFCCC's foundational framework to Kyoto's binding commitments, Paris's global inclusivity, and the EU's Fit for 55 package—illustrates the increasing urgency and complexity of tackling climate change. These agreements and initiatives highlight the necessity of coordinated action at every level, from supranational collaboration to regional implementation. However, they also present significant *transition risks* for firms, that must adapt to shifting regulations, market demands, carbon pricing pressures alongside *generational opportunities* associated with mitigation and adaptation efforts.

1.8   Capital Investment Required for Net Zero

Reducing GHG in the global economy requires a massive disruption in Business as Usual. The world's energy mix, which stands around 80% fossil fuel, must shift to about 20% fossil fuel. This shift will require significant investments in R&D and substantial capital expenditures to develop and implement new technologies. Nonetheless, adaptation and mitigation efforts to confront climate change present generational opportunities. As of 2022, Bloomberg analysis indicates investment opportunities totaling almost $200 trillion by 2050 in electric vehicles, low-carbon power, power grids, as well as major investment in carbon capture and storage, nuclear energy, and things we cannot even think of today (M&G Investments, 2024). For perspective, the GDP of the United States in 2022 was $20.89 trillion. By 2050 Bloomberg estimates that current climate policies in place will still demand annual investment of $4.1 trillion; versus $6.7 trillion if policy initiatives become more vigorous in order to achieve Net Zero. To interpret what these numbers mean: in 2022, Japan, the world's third largest economy, had GDP of $5.06 trillion.

Climate Change Progress and Business as Usual

Progress on containing global warming is not on track to achieve the desired goal of limiting temperature rise to 1.5 °C by 2050. As we will see in subsequent chapters, even at the highest level, Nationally Determined Contributions under the Paris Agreement, the strategies contained therein are incompatible with sustainable temperature rise. Broadly speaking, lack of progress can be pinned to "the tragedy of the commons," the "tragedy of the horizon," and the nature of the climate change problem itself—a determined economic and political environment that is unable or unwilling to abandon "Business as Usual."

Conclusion

Tragedy of the Horizon

In a famous speech delivered to Lloyd's of London in 2015 by Mark Carney, at the time Governor of the Bank of England, and Chair of the Financial Stability Board,[2] Carney outlined barriers to meaningful action on climate change. "Climate change is the Tragedy of the Horizon. We do not need an army of actuaries to tell us that the catastrophic impacts of climate change will be felt beyond the traditional horizons of most actors— imposing a cost on future generations that the current generation has no direct incentive to fix. That means beyond: the business cycle; the political cycle; and the horizon of technocratic authorities, like central banks, who are bound by their mandates. The horizon for monetary policy extends out to 2–3 years. For financial stability it is a bit longer, but typically only to the outer boundaries of the credit cycle—about a decade (Carney, 2015)."

Business as Usual

"Business as usual" historically served as a reassurance, conveying stability and continuity during periods of uncertainty or crisis. In 19th-century Britain, it reflected confidence in governance, industrial progress, and imperial control, assuring citizens and markets that daily operations would

---

[2] The Financial Stability Board (FSB) is an international body that monitors and makes recommendations about the global financial system. It was established in 2009 as a successor to the Financial Stability Forum (FSF) and includes all G20 major economies, FSF members, and the European Commission.

proceed uninterrupted despite political instability, economic disruptions, or global conflicts. However, in the context of climate change, the phrase has taken on a pejorative meaning. Rather than signifying resilience, "business as usual" has become synonymous with complacency and inertia in addressing the climate crisis. It embodies the dangers of maintaining the status quo in the face of escalating environmental risks, emphasizing the urgent need for transformative action to mitigate global warming and its impacts.

Climate Change Is a Wicked Problem[3]

" Climate change is an issue that presents great scientific and economic complexities, some very deep uncertainties, profound ethical issues, and even lack of agreement on what the problem is" (World Bank, 2014). This is how the World Bank characterizes climate change as a wicked problem.

To address climate change, countries must cooperate to reduce greenhouse gas emissions by implementing measures such as putting a price on carbon, incentivizing low-carbon industrial processes, shifting the energy mix from fossil fuels to green alternatives, and investing in technologies to remove carbon from the atmosphere. However, the benefits of climate change mitigation are not equally shared, as developing countries, who are often the most vulnerable to climate events, are also the least capable of implementing costly strategies. Therefore, international efforts must be aware of the need for a fair transition to a low-carbon economy.

Climate change is a "wicked problem" because it is a complex global issue that involves a wide range of stakeholders and is influenced by a multitude of social, economic, and environmental factors. This means that there is no clear solution that can be easily identified or agreed upon. The impacts of climate change are widespread, and the effects can be slow and incremental, making it difficult for people to perceive the urgency of the problem. There are also trade-offs and unintended consequences associated with different solutions, which can make it challenging to reach a consensus among stakeholders. Lastly, there is high uncertainty and disagreement among stakeholders regarding the magnitude, timing, and most effective strategies for addressing climate change.

As we move through the decade of the 2020s, we see tremendous effort to ensure growth, equity, and sustainability in a carbon-constrained world. Yet the complexities of this wicked problem imply many tentacles, often

---

[3] The term "wicked problem" was first coined by planning theorists Horst Rittel and Melvin Webber in the 1970s to describe complex social issues that are difficult to define and solve due to their many interconnected and unpredictable factors (Rittel & Webber, 1973).

times working against each other. As emissions rise and climate degradation continues, the urgency to address climate change becomes ever more pressing. Climate change is a market failure that can only be mitigated through regulation which affects how business will operate in a carbon-constrained world. Understanding corporate finance in climate change requires a deep understanding of the risks of inaction as well as the inherent risks associated with the actions necessary to adapt/mitigate climate risk. The science is clear—the threat is real. Climate change is a global emergency.

In the chapters that follow, we will show how business must transform and adapt to a decarbonized future. Our book is about this transformative change and how the practice of corporate finance must do a lot of things differently. "We cannot solve our problems with the same thinking we used when we created them (Attributable to Albert Einstein)."

Key Takeaways Chapter 1

1. Integration of Climate Change in Corporate Finance

Climate change is no longer an externality but a strategic issue for corporate finance. Companies must incorporate climate risks and opportunities into their financial decisions, including capital allocation, risk management, and governance.

2. Dual Nature of Climate Risks
   • **Physical risks**: Extreme weather events, rising sea levels, and other disruptions affect business operations and supply chains.
   • **Transition risks**: Regulatory changes, market shifts, and technological adaptation challenge traditional business models.
3. Opportunities Amid the Crisis

The shift to a low-carbon economy opens avenues for innovation and investment in renewable energy, sustainable infrastructure, and green technologies.

4. Economic and Ethical Complexities

Climate change represents both a scientific and market failure. Mitigation requires global cooperation, equitable transitions, and significant investments in low-carbon solutions.

5. Historical Perspective on Emissions

Industrialization led to rising GHG emissions, with early contributions from Western nations and recent surges in emerging economies like China and India. This underscores complexities in balancing development and climate action.

6. Global Responses to Climate Change

Key milestones include the Kyoto Protocol and the Paris Agreement. Regional initiatives, like the EU's "Fit for 55" package, illustrate leadership in decarbonization efforts.

7. Tragedies of Commons and Horizon
   * **Tragedy of the Commons**: Overuse of shared resources leads to climate degradation.
   * **Tragedy of the Horizon**: Short-term incentives conflict with long-term climate risks, as articulated by Mark Carney.
8. Uneven Impact of Climate Change

Developing countries, contributing minimally to emissions, face disproportionate climate impacts due to limited resources for adaptation.

9. Capital Investment Needs

Transitioning to a low-carbon economy requires trillions in global investments, emphasizing the importance of sustainable finance solutions.

10. The Wicked Problem of Climate Change

Climate change involves diverse stakeholders, trade-offs, and uncertainties, making it highly complex. Incremental, slow impacts complicate consensus and effective solutions.

Questions

1. What was the significance of the Paris Agreement, and what are its key goals?
2. Describe the "Tragedy of the Commons" in the context of climate change.
3. Explain "Arctic amplification" and its effects on climate change.
4. What are the primary physical impacts of climate change on the environment?
5. How does the Kyoto Protocol differ from the Paris Agreement in its approach to emissions reduction?
6. Discuss the economic costs of climate change, providing examples of recent estimates.
7. What are "transition risks" for businesses, and what are some examples?
8. How might climate change exacerbate global inequality?
9. What is the "Tragedy of the Horizon" as described by Mark Carney, and how does it relate to climate finance?
10. Why is climate change considered a "wicked problem"?

References

Carney, M. (2015, September 29). *Breaking the tragedy of the horizon—Climate change and financial stability*. Speech presented at Lloyd's of London. Bank of England. (February 24, 2025) from https://www.bankofengland.co.uk/-/media/boe/files/speech/2015/breaking-the-tragedy-of-the-horizon-climate-change-and-financial-stability.pdf

Climate Change (2022). Impacts Adaptation and Vulnerability Working Group II Contribution to the Sixth Assessment Report of the Intergovernmental Panel on Climate Change North America Cambridge University Press 1929–2042.

European Commission. (2021). Fit for 55: Delivering the EU's 2030 climate target on the way to climate neutrality. (February 24, 2025) from https://ec.europa.eu/commission/presscorner/detail/en/ip_21_3541

European Space Agency. (2023a, July 13). Europe braces for sweltering July. European space agency. (February 24, 2025) from https://www.esa.int/Applications/Observing_the_Earth/Copernicus/Sentinel-3/Europe_braces_for_sweltering_July

European Space Agency. (2023b, August 24). Wildfires continue to rage in Greece. European space agency. (February 24, 2025) from https://www.esa.int/ESA_Multimedia/Images/2023/08/Wildfires_continue_to_rage_in_Greece

IPCC (2021). Sixth assessment report: Climate Change 2021 – The Physical Science Basis. (February 24, 2025) from https://www.ipcc.ch/report/ar6/wg1/

M&G Investments. (2024, February). Investment opportunities on the path to net zero. M&G Investments. (February 24, 2025) from https://www.mandg.com/investments/professional-investor/en-gb/insights/mandg-insights/latest-insights/2024/02/investment-opportunities-on-the-path-to-net-zero

Plass, Gilbert N. *Carbon dioxide and the climate*. American Scientist, vol. 44, no. 3, 1956, pp. 302–16. JSTOR, http://www.jstor.org/stable/27826805. 5 Feb. 2023.

Rantanen, M., Karpechko, A. Y., Lipponen, A., Nordling, K., Hyvärinen, O., Ruosteenoja, K., & Vihma, T. (2022). *The Arctic has warmed nearly four times faster than the globe since 1979*. Communications earth & environment, *3*(1), 1–10. https://doi.org/10.1038/s43247-022-00498-3

Rajamani, L., & Bodansky, D. (2019). The Paris rulebook: Balancing international prescriptiveness with national discretion. *International and Comparative Law Quarterly, 68*(4), 1023–1040. https://doi.org/10.1017/S0020589319000320

Ritchie, H., & Roser, M. (2020). *CO_2 and greenhouse gas emissions*. Our world in data. https://ourworldindata.org/co2-and-other-greenhouse-gas-emissions

South China Morning Post. (2023, April 6). *China's aggressive expansion of coal power undermining global efforts to wean off dirty fuel: Study*. (February 24, 2025), from https://www.scmp.com/business/article/3216164/chinas-aggressive-expansion-coal-power-undermining-global-efforts-wean-itself-dirty-fuel-study

Stern, N. (2007). *The Economics of Climate Change: The Stern Review*. Cambridge University Press. https://doi.org/10.1017/CBO9780511817434

Tyndall, John. *Note on the Transmission of Radiant Heat through Gaseous Bodies*. Proceedings of the royal society of London, *10*(1859), 37–39. JSTOR, (5 Feb. 2023).http://www.jstor.org/stable/111604

United Nations. (1992). *Report of the United Nations Conference on Environment and Development* (Rio de Janeiro, 3–14 June 1992), Volume I: Resolutions adopted by the conference. United Nations. (February 24, 2025) from https://www.un.org/en/conferences/environment/rio1992

United Nations. (1998). *Kyoto Protocol to the United Nations Framework Convention on Climate Change*. United nations framework convention on climate change, (November 4, 2024).https://unfccc.int/resource/docs/convkp/kpeng.pdf

Victor, D. G. (2001). The Collapse of the Kyoto Protocol and the Struggle to Slow Global Warming. *Princeton University Press*. https://doi.org/10.1515/9780691187473

World Bank. (2014) *A Wicked Problem*: Controlling global climate change. (November 4, 2024) at https://www.worldbank.org/en/news/feature/2014/09/30/a-wicked-problem-controlling-global-climate-change

World Bank. (2022, October 28). Pakistan flood damages and economic losses over USD 30 billion, and reconstruction needs over USD 16 billion: New assessment. World Bank. (February 24, 2025) from https://www.worldbank.org/en/news/press-release/2022/10/28/pakistan-flood-damages-and-economic-losses-over-usd-30-billion-and-reconstruction-needs-over-usd-16-billion-new-assessme

World Economic Forum. (2023, October). Climate loss and damage costs $16 million per hour. World Economic Forum. (February 24, 2025) from https://www.weforum.org/stories/2023/10/climate-loss-and-damage-cost-16-million-per-hour

Climate Change: Key Concepts for the Corporate Sector

Introduction

This chapter offers essential insights into the science behind greenhouse gas emissions, tipping points, and the far-reaching climate risks that are reshaping the economic landscape.

Greenhouse gas emissions have surged since the Industrial Revolution, with the most dramatic increases occurring in the twentieth century as industrialization, population growth, and reliance on fossil fuels reshaped our atmosphere. This rapid rise in emissions is transforming the climate, with significant impacts on ecosystems, economies, and human well-being. Over the past 50 years, advances in climate science have broadened our understanding of climate change, highlighting the profound and lasting consequences of a warming planet, from disrupted weather patterns and biodiversity loss to economic instability and health risks. In the context of the climate crisis, understanding the sources and growth of emissions, their impacts, and the urgent need for mitigation is now essential for everyone, whether policymakers, advocates, or business leaders.

This chapter begins with an overview of the global rise in emissions, their primary sources, and the need for substantial emissions reduction. We also examine the concept of tipping points—thresholds beyond which changes in the climate become irreversible, leading to profound disruptions across natural and economic systems. For business, this chapter lays the groundwork for navigating the strategic risks and opportunities created by climate change.

Subsequent chapters build on this foundation, covering topics such as disclosure frameworks, emissions accounting, risk assessment, financial markets and risk pricing, investment strategies to adapt to a warming planet, and

© The Author(s), under exclusive license to Springer Nature Switzerland AG 2025
S. Dow and Y. Shi, *Corporate Finance Under Climate Crisis*,
https://doi.org/10.1007/978-3-031-83487-5_2

mitigation efforts to reduce greenhouse gas emissions. These scientific fundamentals are vital for businesses today, providing a basis to understand and respond to emerging risks and build resilience in an uncertain future.

2.1   Overview of Greenhouse Gas Emissions[1]

The amount of GHG emissions vary widely across countries. Industrialized countries emit the most GHG. The Group of Twenty (G20), comprises the world's major economies, representing all inhabited continents; 80% of world GDP; 75% of global trade; and 60% of the world's population. The G20 are also the world's largest emitters of GHG. Collectively, the top ten countries in the world contribute about 60% of total global emissions (World Resources Institute (WRI), 2023). On this list, only Iran is not part of the G20. Figure 2.1 shows GHG emissions for the top ten emitters in 2019.[2] China, the world's largest emitter of carbon dioxide, has about two and a half times the level of emissions compared to second rank United States. India's greenhouse gas (GHG) emissions began to take off in the 1990s, following a period of rapid economic growth and industrialization to the point that India is now the third largest emitter of carbon dioxide globally.

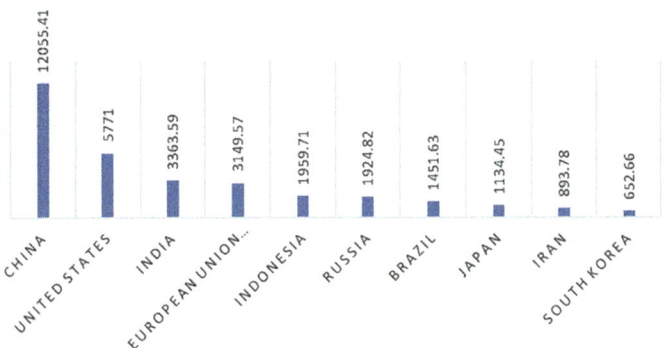

Fig. 2.1   GHG emission in 2019 for top 10 emitters. *Source* WRI (2023)

---

[1] The World Resources Institute provides excellent emissions data. Climate Watch has an open data commitment and provides information free of constraints and restrictions on use.

[2] Data are from World Resources Institute (WRI). WRI uses a three-year lag in reporting emissions to ensure comparability across countries.

2.2   Climate Science

Understanding climate change, its causes, and the potential impacts on different sectors and regions is critical to identifying a firm's exposure to climate risk. A good place to launch our learning is with an explanation of the Greenhouse Effect and radiative forcing.

2.3   The Greenhouse Effect

There is an agreement that Global warming can be traced directly to increased greenhouse gas emissions (GHG). Greenhouse gases act like a greenhouse, trapping heat in the Earth's atmosphere and warming the planet. The gases allow sunlight to pass through, but they prevent much of the resulting heat from escaping back into space, similar to the way a greenhouse traps heat to warm its interior. This process is called the greenhouse effect and the gases responsible for it are called greenhouse gases (mainly CO_2, methane, nitrous oxide, and F-gases). Greenhouse gases play a significant role in regulating the temperature of the Earth. Without any greenhouse gases, the average temperature on Earth would be much colder, making it difficult for life to exist. The problems we currently confront, however, relate to the concentration of greenhouse gases in the atmosphere that has increased significantly since pre-industrial times, due to human activities such as burning fossil fuels methane emissions, deforestation, and agriculture, causing the Earth's temperature to rise and leading to the current phenomenon of global warming.

Figure 2.2 shows the growth of GHG emissions from 1860 through 2022. We can clearly see the rapid rise of carbon emissions since the second half of the twentieth century, corresponding to an unprecedented pace of industrialization, particularly for Asian countries, including China, India, and South Korea, all of which made the list as Top Ten Emitters.

2.4   Radiative Forcing

To have a more complete understanding of climate risk and the impact on business, we have more work to do. We need a framework to understand how climate change could evolve and thus what potential physical and transition risks lie ahead. A lot of discussions of business risk in climate change compare scenarios that include business as usual (BAU) or "current policies," and then postulate various scenarios that differ in terms of the extent of warming. Many of these representations are straightforward to understand, as we are familiar with the concept of 1.5 °C and have an intuitive understanding of temperature from a thermometer. However, there is another more robust approach that considers current and future emissions paths and rather than using the thermometer, we measure climate change in terms of **Radiative Forcing**.

Annual greenhouse gas emissions by world region, 1850 to 2022

Greenhouse gas emissions[1] include carbon dioxide, methane and nitrous oxide from all sources, including land-use change. They are measured in tonnes of carbon dioxide-equivalents[2] over a 100-year timescale.

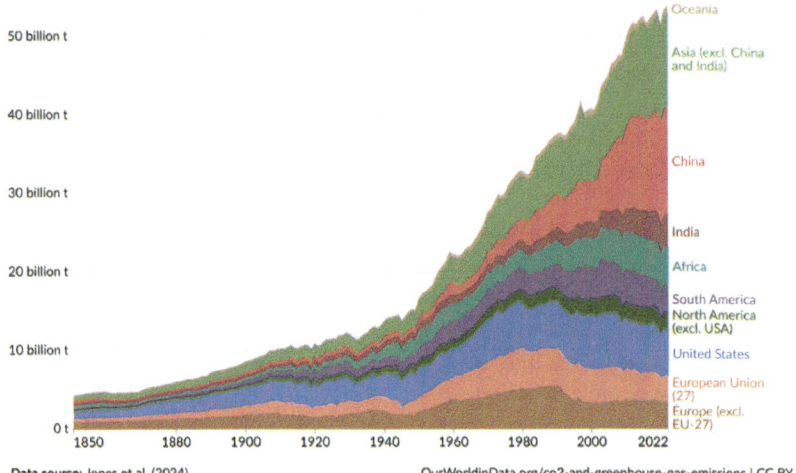

Data source: Jones et al. (2024)   OurWorldinData.org/co2-and-greenhouse-gas-emissions | CC BY

1. **Greenhouse gas emissions:** A greenhouse gas (GHG) is a gas that causes the atmosphere to warm by absorbing and emitting radiant energy. Greenhouse gases absorb radiation that is radiated by Earth, preventing this heat from escaping to space. Carbon dioxide (CO_2) is the most well-known greenhouse gas, but there are others including methane, nitrous oxide, and in fact, water vapor. Human-made emissions of greenhouse gases from fossil fuels, industry, and agriculture are the leading cause of global climate change. Greenhouse gas emissions measure the total amount of all greenhouse gases that are emitted. These are often quantified in carbon dioxide equivalents (CO_2eq) which take account of the amount of warming that each molecule of different gases creates.

2. **Carbon dioxide equivalents (CO_2eq):** Carbon dioxide is the most important greenhouse gas, but not the only one. To capture all greenhouse gas emissions, researchers express them in "carbon dioxide equivalents" (CO_2eq). This takes all greenhouse gases into account, not just CO_2. To express all greenhouse gases in carbon dioxide equivalents (CO_2eq), each one is weighted by its global warming potential (GWP) value. GWP measures the amount of warming a gas creates compared to CO_2. CO_2 is given a GWP value of one. If a gas had a GWP of 10 then one kilogram of that gas would generate ten times the warming effect as one kilogram of CO_2. Carbon dioxide equivalents are calculated for each gas by multiplying the mass of emissions of a specific greenhouse gas by its GWP factor. This warming can be stated over different timescales. To calculate CO_2eq over 100 years, we'd multiply each gas by its GWP over a 100-year timescale (GWP100). Total greenhouse gas emissions – measured in CO_2eq – are then calculated by summing each gas' CO_2eq value.

Fig. 2.2 Annual GHG emissions worldwide

Radiative Forcing (or Climate Forcing) is the change in the net, downward minus upward, radiative flux watts per square meter (W/m^2) at the top of the atmosphere (tropopause) due to a change in an external driver of climate change, such as, a change in the concentration of carbon dioxide (CO_2) (anthropogenic sources) or the output of the Sun (natural sources). A clear explanation (for the layperson) of Radiative Forcing is found at MIT's Climate Portal. "Radiative forcing is what happens when the amount of energy that enters the Earth's atmosphere is different from the amount of energy that leaves it. Energy travels in the form of radiation: solar radiation entering the atmosphere from the sun, and infrared radiation exiting as heat. If more radiation is entering Earth than leaving—as is happening today—then the atmosphere will warm up. This is called radiative forcing because the difference in energy can force changes in the Earth's climate. Before the industrial revolution, there was a balance between energy coming into the earth and energy leaving. By 1950, radiative forcing was 0.57 W/m^2 compared to 1750, and by 1980 it had jumped to 1.25 W/m^2, and in 2011, it reached 2.29 W/m^2. Anthropogenic climate forcings have been driving this increase since 1750.

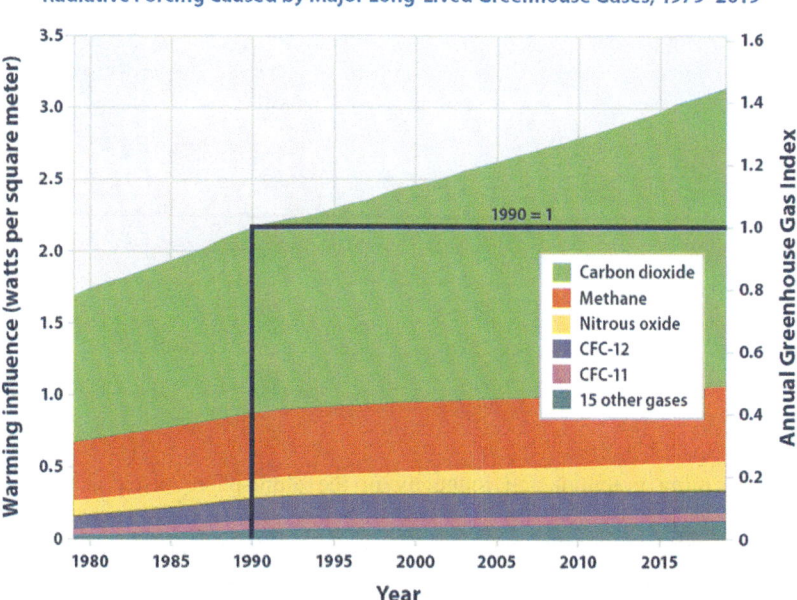

Data source: NOAA (National Oceanic and Atmospheric Administration). 2020. The NOAA Annual Greenhouse Gas Index. Accessed December 2020. www.esrl.noaa.gov/gmd/aggi.

For more information, visit U.S. EPA's "Climate Change Indicators in the United States" at www.epa.gov/climate-indicators.

Fig. 2.3 Growth in radiative forcing

Figure 2.3 shows the sources of radiative forcing due to fossil fuel emissions since 1979.

2.5   GHG Emissions

GHG emissions are not just carbon (CO_2) emissions. In fact, there are a number of various sources of GHG emissions, some of which are produced naturally. The main types of GHG are:

- Carbon Dioxide (CO_2)
- Methane (CO_4)
- Nitrous Oxide (N_2O)
- F-Gases: Chlorofluorocarbons (CFCs) and other trace gases (HFCs, CFCs, SF6).

Figure 2.4 shows a pie chart of the different GHGs identified in the atmosphere as a percentage of total GHG emissions in 2019.

Fig. 2.4 Global emissions by gas in 2019. *Source* WRI (2023)

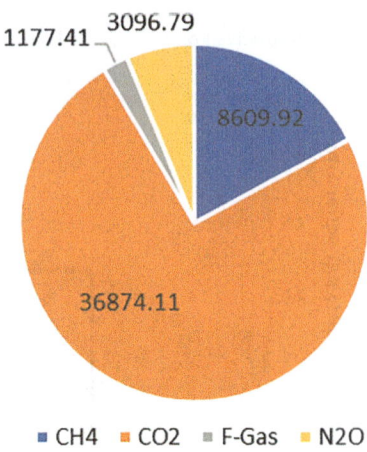

The warming potential and longevity in the atmosphere differ across these gases. Researchers use "Global Warming Potential (GWP)" as a metric to compare the warming effects of different greenhouse gases relative to carbon dioxide (CO_2) over a specific time horizon, typically one hundred years. This allows for a standardized comparison of gases like methane, fluorinated gases, and nitrous oxide in terms of "carbon dioxide equivalents" (CO_2e), accounting for both their warming potential and atmospheric lifespan. Simply curtailing carbon dioxide emissions, for example, does not reverse or halt global warming because carbon stays around for 100 years. Methane, on the other hand, hangs around for only about 10 years but has far more warming potential than carbon dioxide. Nitrous Oxide, like carbon dioxide, stays around for one hundred years once emitted but has a warming potential of about three hundred times that of carbon dioxide. When you see numbers for GHG emissions, they will have been corrected for warming potential and longevity.

2.6   Carbon Dioxide

Carbon emissions (CO_2) are emissions associated principally with the burning of fossil fuels: coal and oil for electricity and heating as well as powering internal combustion engines in transportation. Certain industrial processes, namely concrete and steel manufacturing, oil refining, fermentation attributable to alcohol and certain pharmaceutical processes, and deforestation also contribute to CO_2 emissions.

Forests and oceans are natural carbon sinks. They play a dual role in GHG emissions. Forests absorb and store carbon dioxide (CO_2) from the atmosphere through the process of photosynthesis, in which trees and other plants convert CO_2 into organic matter such as wood, leaves, and roots. This process locks away carbon in the form of plant biomass and soil organic matter, where it can be stored

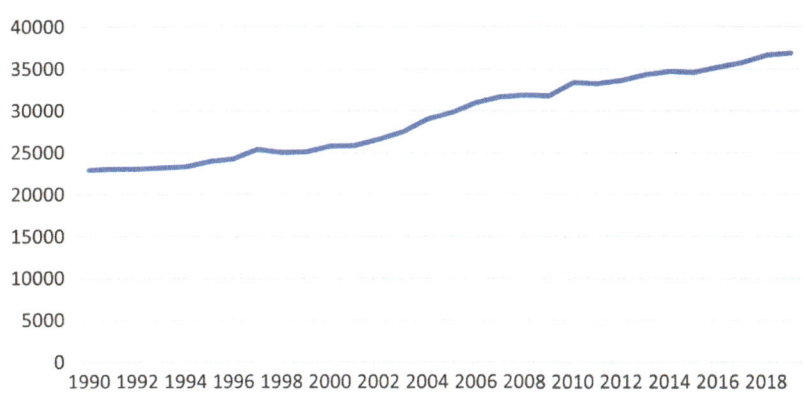

Fig. 2.5 CO_2 Emissions 1990–2019. *Source* WRI (2023)

for long periods of time. For example, a mature forest can contain up to several hundred tons of carbon per hectare. However, deforestation, land-use change, and forest degradation can release large amounts of carbon back into the atmosphere, contributing to the overall increase in atmospheric CO_2 concentrations. This is a principal concern associated with deforestation of the Amazon.

Oceans act as a sink for about one-third of the carbon dioxide that is emitted into the atmosphere from human activities, such as the burning of fossil fuels. When CO_2 enters the ocean, it reacts with seawater to form carbonic acid, that can then dissolve into bicarbonate ions and other forms of dissolved inorganic carbon. However, the ocean's capacity to absorb CO_2 also has consequences for marine life. The increasing acidification of the ocean due to the absorption of CO_2 is affecting the ability of some species to build and maintain their shells and skeletons, which can have far-reaching impacts on the food chain and the overall health of the ocean. Figure 2.5 shows the growth of carbon emissions from 1990 to 2019.

The growth in CO_2 emissions since 1990 as depicted in Fig 2.5 should be more than worrisome as they are the leading source of GHG emissions accounting for nearly ¾ of all GHG emissions. Adding to the warming threat is that CO_2 stays in the atmosphere for a very long time, perhaps hundreds or even thousands of years. NASA puts the longevity of carbon between 300 and 1000 years (NASA, n.d.), meaning that even when emissions are curtailed, the warming effect caused by these emissions are with us for a very long time. Natural carbon sinks (forests and oceans) cannot keep pace with the rate of CO_2 emissions and hence global warming accelerates. When comparing the carbon dioxide (CO_2) emissions of different fossil fuels (coal, oil, and natural gas), it is important to consider their carbon content and the efficiency of combustion. Here is a breakdown of the emissions associated with coal, oil, and natural gas:

- Coal: Coal is the most carbon-intensive fossil fuel. It has the highest carbon content among these three fuels and produces the most CO_2 emissions per unit of energy when burned. On average, coal combustion emits about 2.2–2.7 times more CO_2 than natural gas per unit of energy generated.
- Oil: Oil, including petroleum and its refined products, has slightly lower carbon content than coal. The CO_2 emissions from oil combustion are higher than those from natural gas but lower than coal emissions.
- Natural gas: Natural gas is the least carbon-intensive fossil fuel. It has the lowest carbon content and, therefore, emits fewer CO_2 emissions per unit of energy when compared to coal and oil. Natural gas combustion emits approximately 50% less CO_2 than coal for the same amount of energy produced.

The sources responsible for almost all carbon emissions along with their annual growth in emissions from 1750 to 2022 are shown in Fig. 2.6. These are probably very conservative estimates since the trajectory was interrupted by the COVID-19 pandemic.[3] Coal takes up the lion's share of fossil fuel emissions, but its growth has slowed a lot in recent years. Both gas and cement, however, display strong growth trends. Natural gas is primarily composed of methane, but when natural gas is extracted, processed, transported, and burned, it can release both methane and carbon dioxide into the atmosphere. The main reason for the high emissions from cement production is the chemical process involved in making cement, which requires large amounts of heat and energy. During cement production, limestone is heated to high temperatures to produce clinker, which is then ground into a fine powder and mixed with other materials to make cement. This process requires significant amounts of energy, most of which is generated from the combustion of fossil fuels.

2.7   Methane

Methane emissions arise from agricultural practices, fossil fuel production, landfills, wetlands, and rice production. This gas is produced during the digestive process of livestock and is also released from manure management systems. The extraction, processing, and transportation of fossil fuels releases methane, as does the decomposition of organic waste in landfills. However, methane also occurs naturally during the decomposition of organic matter in wetland areas and in rice paddy fields. The natural production of methane is amplified by changes in land use and management practices, such as the expansion of rice cultivation and drainage of wetlands. According to current estimates, approximately 60% of methane production is attributed to human activities.

---

[3] Flaring refers to the controlled burning of natural gas that is released during oil and gas exploration and production processes. When oil and gas are extracted from the ground, they often contain associated natural gas that is released as a byproduct. If this natural gas is not captured and transported to market, it may be burned off at the wellsite through a process called flaring.

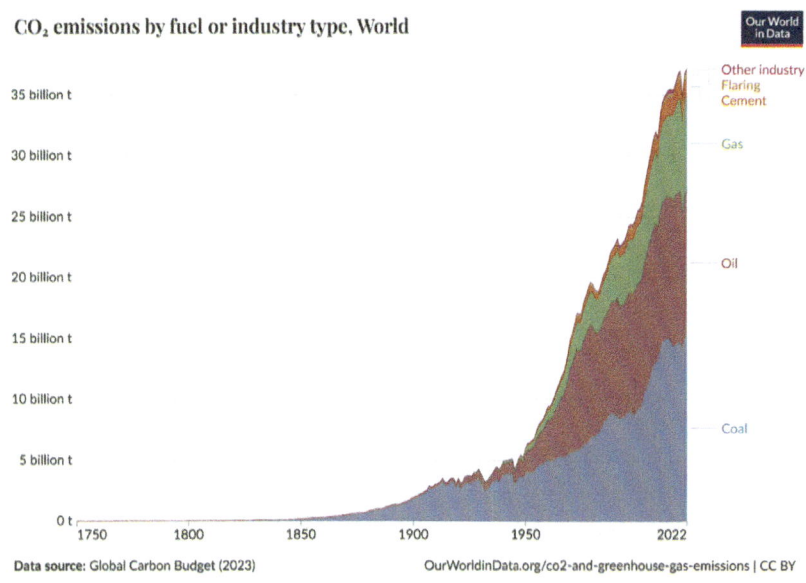

Fig. 2.6 Emissions by fuel type/sector

Methane has a much shorter atmospheric lifetime compared to carbon dioxide, only about 10 years. However, despite its relatively short atmospheric lifetime, methane is much more effective at trapping heat—about one hundred times more effective than CO_2. So, methane emissions are a significant contributor to global warming, but they are also far more challenging to compute when compared to CO_2 due primarily to their source. In fact, it was not until 1990 that methane was recognized as a contributor to GHG emissions and was included as part of the GHG inventory of emissions (Intergovernmental Panel on Climate Change (IPCC), 1990). The IEA attributes about 30% of current global warming to methane, while adding the disclaimer that methane estimates are highly uncertain (International Energy Agency (IEA), 2022). So, while methane's contribution to global warming seems well-understood, the scale of methane emissions is clouded in ambiguity. As more robust and homogenous methods for methane measurement mature, we can expect more stringent and targeted legislation to curtail these emissions. Figure 2.7 shows the growth in methane emissions since the 1990s and highlights its rapid rise. By 2021, methane emissions rose more than any other year on record, surpassing increases in 2020 that were record-breaking then (NOAA, 2021). Agriculture is the largest contributor, emitting over three billion tons annually, driven mainly by livestock digestion and rice production. Fugitive emissions, largely from fossil fuel extraction, follow closely, showing a steady rise with the growth in energy demand. Waste management practices contribute significantly as well, while emissions from other sectors like transport, buildings, and industry remain relatively

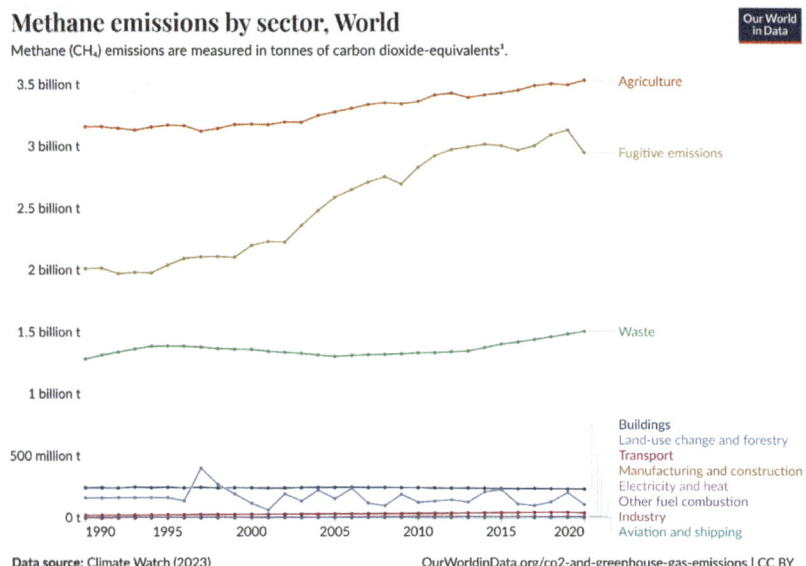

Fig. 2.7 Annual rise in methane emissions and their sources

low in comparison. Understanding these trends underscores the importance of targeted climate action to reduce methane emissions, especially in agriculture and energy sectors, to mitigate global warming.

While scientists may not have formed a consensus on why the rapid increase in methane emissions has occurred, this increase alone would seem to point to more stringent regulation of methane emissions in the future.

2.8   Natural Gas and Methane Emissions

Natural gas is often promoted as a lower carbon alternative to coal and oil. However, it is primarily composed of methane, a potent greenhouse gas with a short atmospheric lifespan but a significantly higher warming potential than carbon dioxide. Similarly, some biofuels intended to reduce greenhouse gas (GHG) emissions may indirectly contribute to methane emissions. For example, growing biofuel feedstocks such as corn, sugarcane, or soybeans often requires fertilizers, which can release nitrous oxide (N_2O)—a greenhouse gas with a Global Warming Potential nearly three hundred times that of carbon dioxide. Improper fertilizer application can also lead to methane emissions from soil. Additionally, land-use changes associated with biofuel production, such as converting pastures or forests into croplands, can displace carbon-storing ecosystems and increase methane release.

Methane is often considered less problematic because it dissipates within 10–12 years. However, it is responsible for approximately 30% of current global warming. If left unchecked, methane emissions could drive near-term temperature spikes, exacerbating climate instability. Recent scientific assessments indicate that natural methane emissions are rising at unprecedented rates, particularly from thawing permafrost, wetlands, and tropical regions. Observations show that methane release from permafrost degradation is accelerating faster than climate models previously projected, raising concerns about self-reinforcing feedback loops that could further amplify global warming (Intergovernmental Panel on Climate Change (IPCC), 2021). Atmospheric monitoring data confirm that global methane levels have surged to record highs in recent years, underscoring the risks posed by both natural and anthropogenic sources (NOAA, 2022).

In parallel, agricultural methane emissions are expected to rise as global incomes grow and dietary patterns shift toward greater meat consumption. While behavioral changes—such as reducing meat consumption—could mitigate some of this growth, the energy sector faces increasing scrutiny for methane emissions, not just carbon dioxide. Given methane's high short-term warming potential, even modest reductions could yield significant climate benefits. Failure to curb these emissions risks derailing the Paris Agreement's temperature targets and increasing the likelihood of dangerous overshoot scenarios.

Scientific uncertainty remains regarding the speed at which warming could trigger tipping points and feedback loops, as well as the feasibility of reversing certain effects even if net-zero emissions are eventually achieved. However, models suggest that exceeding temperature targets may lead to severe and potentially irreversible environmental and economic consequences. This reinforces the urgency of immediate methane mitigation efforts, particularly in the energy, agriculture, and land-use sectors.

2.9 Nitrous Oxide

Nitrous oxide (N_2O) is a powerful greenhouse gas that is about three hundred times more effective at trapping heat in the atmosphere than carbon dioxide, and much more effective at heat-trapping than methane. Nitrous oxide can stay in the atmosphere for about one hundred years. The gas is released into the atmosphere through various natural processes, such as the activity of microbes in soils and oceans and can also be reduced through chemical reactions in the stratosphere and troposphere. However, nitrous oxide is also released through human activity. Maybe you think of laughing gas when you hear nitrous oxide. You would not be wrong, but that is not the real source of the problem. Nitrogen-based fertilizers used in agriculture and manure management are the main source of anthropogenic Nitrous Oxide emissions. The role of nitrous oxide emissions has received far less attention than those of carbon and even methane. BBC ran a story in 2021 with the headline: "The World's Forgotten Greenhouse Gas" (Tiseo, 2021). In this news article they report the findings of a study published in Nature in 2020 (Tian et al., 2020): "…despite its important contribution to climate change, N_2O emissions have largely been ignored in climate policies and the gas continues to accumulate. A 2020 review of nitrous oxide sources and sinks found that emissions rose 30% in the last four decades and are exceeding all, but the highest potential emissions scenarios described by the Intergovernmental Panel on Climate Change. Agricultural soil—especially because of the globe's heavy use of synthetic nitrogen fertilizer—is the principal culprit." In other words, the growth of nitrous oxide emissions can derail the Paris Agreement's global warming target of 1.5 °C.

Nitrous oxide emissions have soared since 1940, as indicated in Fig. 2.8.

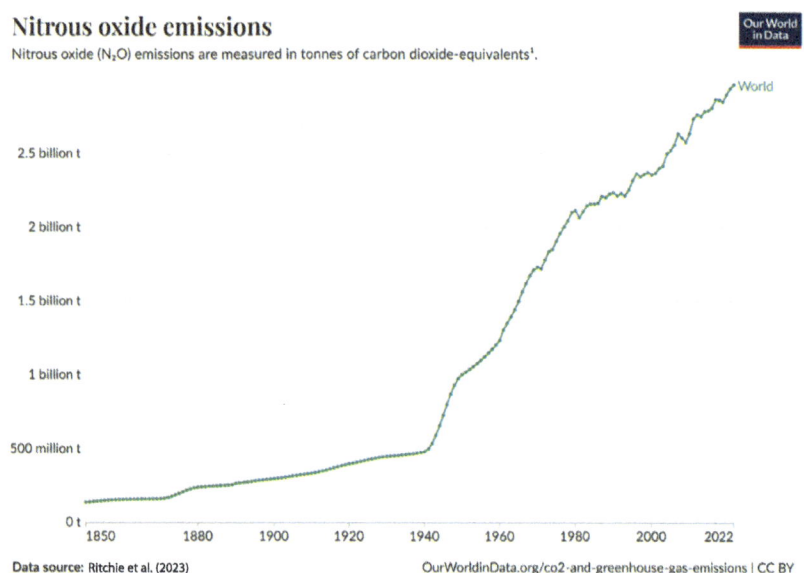

Nitrous oxide emissions

Nitrous oxide (N_2O) emissions are measured in tonnes of carbon dioxide-equivalents[1].

Data source: Ritchie et al. (2023)   OurWorldinData.org/co2-and-greenhouse-gas-emissions | CC BY

1. **Carbon dioxide equivalents (CO_2eq):** Carbon dioxide is the most important greenhouse gas, but not the only one. To capture all greenhouse gas emissions, researchers express them in "carbon dioxide equivalents" (CO_2eq). This takes all greenhouse gases into account, not just CO_2. To express all greenhouse gases in carbon dioxide equivalents (CO_2eq), each one is weighted by its global warming potential (GWP) value. GWP measures the amount of warming a gas creates compared to CO_2. CO_2 is given a GWP value of one. If a gas had a GWP of 10 then one kilogram of that gas would generate ten times the warming effect as one kilogram of CO_2. Carbon dioxide equivalents are calculated for each gas by multiplying the mass of emissions of a specific greenhouse gas by its GWP factor. This warming can be stated over different timescales. To calculate CO_2eq over 100 years, we'd multiply each gas by its GWP over a 100-year timescale (GWP100). Total greenhouse gas emissions – measured in CO_2eq – are then calculated by summing each gas' CO_2eq value.

Fig. 2.8 Nitrous Oxide Emissions since 1990

2.10   FluorinatedGases (F-gases)

Fluorinated gases (F-gases) are synthetic compounds not found in nature, produced exclusively through human activities. The primary categories of F-gases include hydrofluorocarbons (HFCs), perfluorocarbons (PFCs), and sulfur hexafluoride (SF_6). These gases are extensively utilized in industrial applications such as refrigeration, air conditioning, insulation, and as solvents in the electronics industry (European Environment Agency, 2021).

F-gases are potent greenhouse gases with Global Warming Potentials (GWPs) significantly higher than that of carbon dioxide (CO_2). For example, SF_6 has a GWP 23,500 times greater than CO_2 over a 100-year period. Additionally, F-gases can remain in the atmosphere for extended durations; PFCs can persist for thousands of years, thereby contributing to long-term warming.

In recent years, emissions of F-gases have shown varying trends across different regions.

In the European Union (EU), F-gas emissions nearly doubled between 1990 and 2014. However, since the implementation of the 2014 F-gas Regulation, emissions have been consistently decreasing each year from 2015 onwards. This regulation introduced measures such as a quota system to phase down the use of HFCs, leading to a significant reduction in emissions.

In other parts of the world, though, F-gas emissions have been on the rise. For example, in the United States, emissions of fluorinated greenhouse gases have increased due to their use as substitutes for ozone-depleting substances. The U.S. Environmental Protection Agency (EPA) reported that in 2021, fluorinated gases accounted for 3% of total US greenhouse gas emissions (U.S. Environmental Protection Agency, 2023).

2.11   Recap: Abundance of GHG

So far, we have seen increases in the main components of GHG in about the past 30 years, despite the Kyoto Protocol and the Paris Agreement. It is not just about the rise in annual emissions, however. In looking at the different gases we have seen that they differ in both warming potential and longevity. Even if we could hit a pause button on all GHG emissions, these gases would still be with us causing global warming. One way of visualizing this is to examine the amount (abundance) of GHGs recorded in 2020 as compared to pre-industrial levels (1750). Methane abundance is 261% greater than pre-industrial levels; carbon sits at 149% greater; and nitrous oxide is 123% more abundant today (World Meteorological Organization, 2021).

2.12   The Climate Tipping Point: A Journey to the Edge

Imagine walking on a thin layer of ice across a vast frozen lake. Every step seems stable, but beneath lies the constant threat of a crack—one that could send shock-waves through the entire surface, with unpredictable and irreversible consequences. This metaphor captures the essence of climate tipping points. As we have seen, greenhouse gases have been accumulating in the atmosphere, pushing the Earth's climate ever closer to thresholds that, once crossed, could alter our world in ways we may not be able to undo. This journey to the edge of a tipping point is not just about rising temperatures or melting ice; it is about reaching a "point of no return" where climate phenomena take on a life of their own, setting in motion changes that cannot simply be reversed by reducing emissions.

2.13   Tipping Elements

"Tipping elements" refer to critical parts of the Earth's climate system—such as large ice sheets, ocean circulation patterns, and essential ecosystems like the Amazon rainforest—that can undergo abrupt and often irreversible changes if they reach certain thresholds, or "tipping points." When these points are crossed, they can lead to significant regional or global changes, potentially destabilizing climate patterns. For example, the melting of massive ice sheets could drastically raise sea levels, while the collapse of major forests may severely impact global carbon storage. In the context of the climate, this means that once certain thresholds are crossed, the resulting changes—like the melting of polar ice or the collapse of rainforests—will continue autonomously, driven by feedback loops that make these changes self-sustaining. The effects of crossing a tipping point could be catastrophic, impacting global weather patterns, sea levels, ecosystems, and human societies in ways that would be impossible to fully reverse.

Figure 2.9 shows tipping elements against the backdrop of a map of the world. The arrows in Fig. 2.9 illustrate potential interactions between these tipping elements, indicating that a change in one element, like the Greenland Ice Sheet, could trigger changes in others, creating a "cascading" effect.

2.14   Consequences of Surpassing Tipping Points

Crossing climate tipping points can have far-reaching and often irreversible consequences on natural systems and human societies. For instance, the collapse of the Greenland and West Antarctic Ice Sheets would contribute significantly to sea level rise, potentially displacing millions of people in coastal regions. Similarly, a disruption in the Atlantic Meridional Overturning Circulation (AMOC) could lead to drastic shifts in weather patterns, affecting rainfall and agriculture across regions such as Europe and the tropics. Ecosystem collapses, like the dieback of the Amazon rainforest, would not only reduce biodiversity but also release large

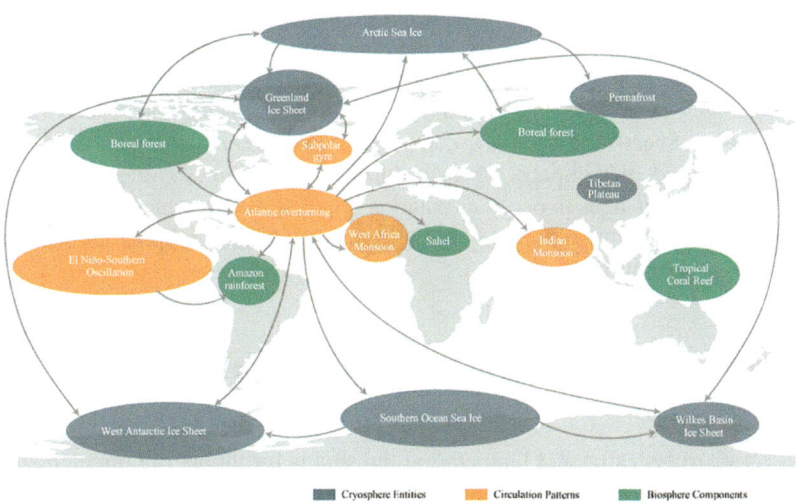

Fig. 2.9 Tipping elements in the climate system. *Source* OECD (2022)

amounts of carbon, exacerbating global warming. Figure 2.10 illustrates the potential for cascading effects, where one tipping point could set off a chain reaction that destabilizes other systems, underscoring the high stakes involved.

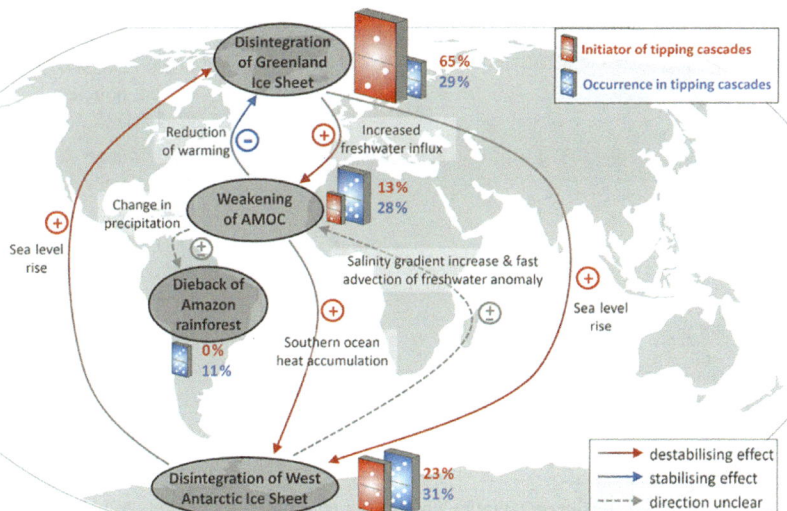

Fig. 2.10 Interactions between climate tipping elements and their roles in tipping cascades. *Source* OECD (2022)

While scientific understanding of tipping points is still evolving, the gravity of their potential impact demands attention. It is also important to understand that while some tipping points, such as those in ice sheets, may not result in abrupt changes, even slow shifts can have significant short-term impacts. While the Greenland Ice Sheet may not thaw completely for 10,000 years once the tipping point is triggered, the impacts of even partial melting are significant including rising sea levels, changes in ocean circulation, feedback loops which can accelerate further melting; and even extending to the necessity of new shipping routes which could impact geopolitical tensions!

Temperature Thresholds and Uncertainty Ranges of Tipping Points

Table 2.1 provides estimated temperature thresholds for global tipping elements.

Table 2.1 Temperature thresholds and uncertainty ranges of global tipping points

Tipping Point	Central threshold (°C)[4]	Uncertainty range[5] (°C)	Risk level
Greenland ice sheet collapse	1.5	0.8—3	Red—risk within paris agreement range (1.5–2 °C)
West Antarctic ice sheet collapse	1.5	1—3	Red—risk within paris agreement range (1.5–2 °C)
Amazon rainforest dieback	3.5	2—6	Dark orange—risk with current policies (2–4 °C)
Boreal permafrost collapse	4	3 - 6	Dark orange—risk with current policies (2–4 °C)
AMOC collapse	4	1.4—8	Dark orange—risk with current policies (2–4 °C)
East Antarctic ice sheet collapse	7.5	5—10	Light orange—risk above 4 °C

Source Adapted from OECD (2022)

---

[4] The central estimate is the temperature at which a tipping point is most likely to be crossed. It is based on current scientific models and represents a best estimate for when a critical change might occur.

[5] This range reflects variability in scientific estimates, acknowledging that each tipping point might be crossed at temperatures somewhat lower or higher than the central estimate. The range accounts for the complex and sometimes unpredictable responses of climate systems to temperature increases.

- Red Indicator: Tipping point risks are within the 1.5–2 °C range, meaning even limited warming could initiate irreversible change. The Greenland and West Antarctic Ice Sheets are at risk of collapse within this threshold, posing a significant threat to sea levels.
- Dark Orange Indicator: Tipping point risks are within the 2–4 °C range. This range is linked with current policy trajectories. If the Amazon rainforest is not preserved, a 3.5 °C rise could trigger significant carbon release.
- Light Orange Indicator: Tipping points with higher thresholds, above 4 °C, are not projected with current policies. The East Antarctic Ice Sheet has a high threshold of 7.5 °C, representing risks under extreme warming.

2.15 Triggered Tipping Points and the Paris Agreement

One of the main conclusions of research on tipping points is that the Paris Agreement goal of limiting temperature rise to well below 2 °C and striving for 1.5 °C is too weak an ambition. Studies are showing that even limiting temperature rise to 1 °C, which we have already surpassed, puts us at risk of triggering some tipping points. It is possible that the Greenland Ice Sheet and the West Antarctic Ice Sheet may already have crossed critical thresholds. Evidence that the Greenland Ice Sheet thaw is underway was provided by a group of researchers at Ohio State University in 2020 (King, et al., 2020. Equally troubling is evidence published in Nature Communications in July 2023 that warns of a potential imminent collapse of the Atlantic meridional overturning circulation (AMOC), despite earlier studies projecting the AMOC to collapse 50 years from now, although a range of 15–300 years was considered possible (Ditlevsen & Ditlevsen, 2023). Today, however, was not a reference point.

Is Overshooting Temperature Targets Forgivable?

In another recent research piece published in Nature Climate Change in December 2022, the authors determine that even temporarily overshooting the climate targets of 1.5–2 °C Celsius (Paris Agreement) could increase the tipping risk of several Earth system elements by more than 70% (Wunderling et al., 2022). The authors of this study recommend that global warming should not exceed 1 °C. Another study indicates that overshooting carries a hefty price tag: Although the investment required to substantially curtail emissions in the near term are high, avoiding overshoot could add 2% to global GDP by the end of the century. In brief, overshooting seems unforgiveable. Where do we stand now? In this Nature Climate Change research, the authors place current global warming at 1.2 °C. Their analysis indicates that current climate change policies (the Paris Agreement) place the Earth on a path to 2–3 C of global warming by the end of this century.

2.16   Methane Bombs and the Pace of Climate Change

A "methane bomb" refers to a sudden, large-scale release of methane gas that could dramatically speed up global warming. One example is the release of methane from Arctic permafrost as it thaws. Permafrost holds vast amounts of organic material that, when warmed, decomposes, and emits methane. Rapid thawing could trigger tipping points and feedback loops, accelerating warming unexpectedly.

In August 2021, *Smithsonian Magazine* warned of a "ticking methane bomb" in Siberia, where permafrost is thawing 70 years ahead of projections (Smithsonian Magazine, 2021). *The Guardian* reported in 2023 that more than 1000 "super-emitter" sites, mainly oil and gas facilities, leak enough methane to equal millions of cars' emissions. The article also identified fifty-five fossil fuel sites that could release methane equivalent to 30 years of U.S. greenhouse gas emissions (Guardian, 2023).

2.17   Implications for Policymakers and Business

The conclusion that the Earth is on a path to 2–3.6 °C global warming by the end of the century despite the Paris Agreement's policies has significant implications for both future policy and business. From a policy perspective, this conclusion underscores the urgency of increasing ambition and implementing more aggressive climate change policies to limit global warming to well below 2 °C, as called for in the Paris Agreement. Indeed, what you have just read indicates that the earth may not be able to tolerate a temperature rise of greater than 1.0 °C, and we are too late for this. The consequences of weak ambition and faster temperature rise will undoubtedly lead to establishment of more ambitious emission reduction targets, acceleration of clean energy technologies, increases in the "price of carbon" via taxation and through carbon markets, and implementation of policy to promote climate-resilient infrastructure and adaptation measures.

Of course, such shifts in climate policy will force businesses to take a strategic approach to climate change risk management, incorporating climate risks into their decision-making processes and developing strategies to reduce emissions, adapt to physical risks, and transition to a low-carbon economy. This will involve a broad range of new mandates from investing in renewable energy and energy efficiency, to increasing R&D and CAPX to develop and implement green technologies such as CCUS, developing climate-resilient supply chains, and disclosing climate-related risks and opportunities to stakeholders.

Conclusion

As this chapter demonstrates, the science of climate change underscores a pressing reality: accelerating greenhouse gas emissions and the risks of crossing critical tipping points bring existential challenges to businesses and society. The path to net-zero emissions is not merely an environmental ambition; it is a strategic imperative that finance professionals and corporate leaders must understand to navigate the coming era. Each greenhouse gas, from long-lived CO_2 to potent but short-lived CH_4, plays a unique role in shaping future climate scenarios. Meanwhile, tipping elements—like polar ice sheets and rainforests—warn us of the irreversible shifts that could transform ecosystems and economies alike if we fail to act.

For the corporate sector, adapting to climate risks means reevaluating traditional financial strategies, embracing new resilience frameworks, and understanding that long-term stability now depends on climate-conscious decision-making. This chapter lays the foundation for those steps, equipping finance professionals to grasp both the profound risks and the transformative potential in steering business toward sustainability. In the chapters ahead, we will delve deeper into risk assessment, mitigation strategies, and the financial tools that can align business objectives with climate resilience.

Key Takeaways for This Chapter

1. **Greenhouse Gas Emissions**

The G20 countries are responsible for approximately 60% of global emissions, with CO_2 from fossil fuels being the largest contributor. Emissions have surged dramatically since the Industrial Revolution, with the pace accelerating post–1950.

2. **Greenhouse Effect**

Greenhouse gases trap heat in Earth's atmosphere, which is essential for life but is now intensified due to human activities such as fossil fuel combustion and deforestation. This intensification is driving global warming, with severe implications for ecosystems and economies.

3. Radiative Forcing

This concept measures the imbalance in Earth's energy system caused by greenhouse gases, aiding in quantifying their warming impacts. Radiative forcing has increased significantly since the Industrial Revolution, driven by anthropogenic emissions.

4. GHG Longevity and Warming Potential

CO_2 persists in the atmosphere for centuries, CH_4 for a decade but with higher immediate warming potential, and N_2O for a century with the highest warming potential among the main greenhouse gases. These differences influence climate strategies and prioritization of emission reductions.

5. Climate Tipping Points

Critical thresholds in the Earth's climate system, such as ice sheet collapses or rainforest diebacks, could lead to irreversible changes and cascading effects on ecosystems and economies. Some tipping points, like the Greenland Ice Sheet, may already be close to being triggered.

6. Methane and Methane Bombs

Methane emissions, especially from agriculture and thawing permafrost, represent a potent but short-lived driver of warming. Sudden, large-scale methane releases ("methane bombs") pose a significant risk to meeting climate goals.

7. The Role of Carbon Sinks

Forests and oceans absorb CO_2, but their capacity is being outpaced by emissions. Ocean acidification and deforestation are undermining these natural carbon sinks.

8. Policy and Business Implications

Climate tipping points and rapid warming demand urgent action to enhance climate policies and enforce stricter emission reductions. Businesses must integrate climate risks into decision-making, focusing on mitigation, adaptation, and sustainable investments.

9. The Cost of Overshooting Climate Targets

Exceeding temperature goals could lead to dramatic risks and costs, both environmental and economic. Avoiding overshoot requires significant near-term investments but offers long-term financial and climate benefits.

10. Strategic Importance for Businesses

Understanding the scientific foundations of climate change equips businesses to anticipate risks and leverage opportunities for sustainability and resilience.

These key takeaways provide a basis for addressing the challenges posed by climate change while navigating the risks and opportunities it creates for the corporate sector.

Questions

1. Describe the Greenhouse Effect.—What role does it play in Earth's climate system, and how has human activity intensified it?
2. What is radiative forcing, and how does it relate to global warming?—Provide an example of how changes in greenhouse gas concentrations affect radiative forcing.
3. Identify and briefly explain the primary greenhouse gases contributing to climate change.
4. How do their warming potentials and atmospheric lifetimes differ?
5. Explain the concept of "carbon dioxide equivalent" (CO_2e)
6. Why is it useful when discussing greenhouse gas emissions?
7. What are "tipping points" in the context of climate science?
8. Provide an example of a tipping element and describe the potential consequences of crossing its tipping point.
9. Why is the Group of Twenty (G20) significant in the context of global greenhouse gas emissions?
10. How might their collective influence shape climate policies?
11. How does methane's shorter atmospheric lifetime but higher warming potential impact climate policy?
12. What might be the implications for businesses, particularly in the agricultural and energy sectors?
13. Describe the risks associated with "methane bombs" and their potential impact on climate goals.
14. How might the corporate sector incorporate these risks into business and investment strategies?
15. Discuss the strategic importance of understanding tipping points for businesses.
16. How could tipping points affect market stability, supply chains, and investment risks?
17. Explain the difference in climate risk posed by long-lived greenhouse gases like CO_2 versus short-lived gases like methane.
18. How should businesses prioritize emission reductions based on these differences?
19. 11. Evaluate the potential business implications of the conclusion that climate change tipping points could be nearer than previously estimated.
20. What strategies could the corporate sector implement to mitigate the financial risks of reaching these tipping points?
21. The chapter suggests that a temperature rise above 1 °C could still trigger some tipping points. How might this finding affect future climate policies and business practices?
22. Consider the "methane bomb" scenario as described in the chapter. If permafrost thaw accelerates methane release, what are the potential cascading risks to global markets and economies?

23. 14. Based on your understanding of greenhouse gases' longevity and warming potential, what might a balanced approach to emissions reduction look like for a global company?
24. What short-term and long-term measures could be implemented?
25. Discuss the potential financial impacts of failing to address radiative forcing and tipping points on a corporation's long-term strategy and stability.
26. How could businesses proactively manage these risks?

References

Ditlevsen, P., Ditlevsen, S (2023). *Warning of a forthcoming collapse of the Atlantic meridional overturning circulation.* Nat Commun 14, 4254. (November 4, 2024) from https://doi.org/10.1038/s41467-023-39810-w

European Environment Agency. (2021). *Fluorinated greenhouse gases 2021.* (November 4, 2024) from https://www.eea.europa.eu/publications/fluorinated-greenhouse-gases-2021

Guardian. (2023, March 6). *Revealed: 1000 super-emitting methane leaks risk triggering climate tipping points.* (November 4, 2024) from https://www.theguardian.com/environment/2023/mar/06/methane-leaks-risk-triggering-climate-tipping-points

International Energy Agency (IEA). (2022). *Global methane tracker 2022.* Paris, France: IEA. (November 4, 2024) from https://www.iea.org/reports/global-methane-tracker-2022

Intergovernmental Panel on Climate Change (IPCC). (1990). *Climate change: The IPCC scientific assessment.* In J. T. Houghton, G. J. Jenkins, & J.J. Ephraums (Eds.), Cambridge University Press.

Intergovernmental Panel on Climate Change (IPCC). (2021). *Climate change 2021: the physical science basis.* Contribution of working group i to the sixth assessment report of the intergovernmental panel on climate change. Cambridge University Press.

King, M. D., Howat, I. M., Candela, S. G., et al. (2020). Author Correction: dynamic ice loss from the greenland ice Sheet driven by sustained glacier retreat. *Commun Earth Environ, 1,* 14. https://doi.org/10.1038/s43247-020-00019-0

NASA. (n.d.). *The atmosphere: Getting a handle on carbon dioxide.* (November 29, 2024) from https://science.nasa.gov/earth/climate-change/greenhouse-gases/the-atmosphere-getting-a-handle-on-carbon-dioxide/

National Oceanic and Atmospheric Administration (NOAA). (2022). *Increase in atmospheric methane set another record during 2021.* (November 4, 2024) from https://www.noaa.gov/news-release/increase-in-atmospheric-methane-set-another-record-during-2021.

OECD. (2022). *Climate tipping points: Insights for effective policy action.* OECD Publishing. (November 4, 2024) from https://doi.org/10.1787/abc5a69e-en

Ritchie, H., Rosado, P., & Roser, M. (2023). *Annual nitrous oxide emissions.* In CO_2 and greenhouse gas emissions. Data adapted from Jones et al. OurWorldInData.org. (November 29, 2024) from https://ourworldindata.org/grapher/nitrous-oxide-emissions

Smithsonian Magazine: "Permafrost Thaw in Siberia Creates a Ticking 'Methane Bomb' of Greenhouse Gases, Scientists Warn," August 5, 2021.

Tian, H., Xu, R., Canadell, J. G., et al. (2020). A comprehensive quantification of global nitrous oxide sources and sinks. *Nature, 586,* 248–256. https://doi.org/10.1038/s41586-020-2780-0

Tiseo, A. (2021, June 3). *Nitrous oxide: The world's forgotten greenhouse gas.* BBC Future. (November 29, 2024) from https://www.bbc.com/future/article/20210603-nitrous-oxide-the-worlds-forgotten-greenhouse-gas

U.S. Environmental Protection Agency. (2023). *Inventory of U.S. Greenhouse gas emissions and sinks: 1990–2021*. (November 4, 2024) from https://www.epa.gov/system/files/documents/2023-04/Data-Highlights-1990-2021.pdf

Wunderling, N., Winkelmann, R., Rockström, J., Loriani, S., Armstrong McKay, D.I., Ritchie, P.D.L., Sakschewski, B., Donges, J.F. (2022). *Global warming overshoots increase risks of climate tipping cascades in a network model*. Nature climate change. See the discussion at Stockholm Resilience Centre.

World Meteorological Organization. (2021). *Greenhouse gas bulletin: The state of greenhouse gases in the atmosphere based on global observations through 2020*. (November 4, 2024) from https://library.wmo.int/index.php?lvl=notice_display&id=21880#.YbKxQpPMJD8

World Resources Institute (WRI). (2023). Climate Watch: Country greenhouse gas emissions data and methodology. https://www.wri.org/research/climate-watch-country-greenhouse-gas-emissions-data-andmethodology

Climate Change and the Challenges for Central Banks: Implications for Monetary Policy and Financial Stability

Introduction

Monetary policy, long regarded as the guardian of economic stability, is emerging as an essential pillar in the global response to the climate crisis. Much like the media's role as the "fifth estate" in shaping public discourse, central banks are increasingly recognized as pivotal actors within the climate crisis ecosystem. This ecosystem is shaped by four primary forces:

(1) climate science, which provides the foundation for understanding climate risks;
(2) environmental and public policy, which set the regulatory and societal framework for action;
(3) corporate strategy, which operationalizes adaptation and mitigation efforts;
(4) Monetary policy which adds a distinct and vital dimension, influencing the allocation of resources and the stability of financial systems in a world reshaped by climate change.

As climate risks—ranging from physical damage to the devaluation of carbon-intensive assets—permeate financial systems, central banks have expanded their traditional focus on price stability and financial stability to include climate considerations. This evolution positions monetary policy as both a stabilizer of financial systems and a catalyst for the global transition to a low-carbon economy.

In this chapter, we explore how monetary policy integrates into the broader climate response, examining the tools central banks are deploying, the systemic risks they aim to mitigate, and the challenges they face in maintaining independence while addressing these emerging threats. The "climate Minsky moment," a sudden market-wide repricing of climate-related risks,

S. Dow and Y. Shi, *Corporate Finance Under Climate Crisis*,
https://doi.org/10.1007/978-3-031-83487-5_3

illustrates the urgency of this integration. As central banks reshape their policies to address one of the most significant economic challenges of our time, businesses must adapt to these new financial realities.

3.1   Climate Change as a Systemic Economic Risk

Climate change has increasingly come to be recognized not just as an environmental challenge but as a profound and systemic economic risk. As economist Nicholas Stern (2007) famously stated, climate change is "the greatest market failure the world has ever seen," a statement that underscores the deep interlinkages between climate dynamics and economic stability. Unlike more conventional financial risks, climate risks are pervasive, spanning physical, economic, and transitional dimensions that collectively challenge the resilience of the financial system. While a full exploration of physical and transition risks follows in a dedicated chapter, this section focuses on understanding climate change as a systemic economic threat and introduces the concept of a potential "climate Minsky moment" to illustrate how climate risks may trigger widespread financial instability if left inadequately addressed.

Systemic Economic Impacts of Climate Change

Climate change affects economic activity through disruptions in infrastructure, shifts in agricultural yields, and increased adaptation costs. Extreme weather events like floods and wildfires reduce productivity and raise maintenance costs, while the transition to a low-carbon economy risks devaluing carbon-intensive assets. These "stranded assets" can destabilize banks and investors if losses cascade through financial markets, underscoring the need for proactive management of transition risks (Caldecott et al., 2015; IMF, 2020).

The Minsky Moment?

The notion of a "Minsky moment," named after economist Hyman Minsky, refers to a sudden and severe market collapse following an extended period of unsustainable investment and risk accumulation. Applied to climate change, a "climate Minsky moment" could arise if markets suddenly reprice climate risks, triggering a rapid devaluation of assets tied to carbon-intensive activities (Carney, 2015). This revaluation could result from growing recognition of climate risks, new policy mandates, or changes in societal and investor sentiment, prompting a sell-off of high-emission assets.

The 2008 financial crisis offers instructive parallels. Leading up to the crisis, housing assets were significantly overvalued, and the associated risks were underappreciated. Similarly, the current financial system tends to underprice climate-related risks, creating the potential for a comparable systemic collapse as climate realities become inescapable. In a scenario where regulatory actions tighten abruptly or climate impacts become more severe, markets could respond with sudden risk adjustments, leading to significant losses across multiple sectors. As highlighted by the Bank for International Settlements, climate change represents a "green swan" event—an unpredictable but potentially devastating risk that has "far-reaching and irreversible consequences for humanity" and the economy (Bolton et al., 2020).

Climate Risk and Financial Stability

Central banks and financial regulators are increasingly concerned with the destabilizing potential of climate-related financial risks. The Network for Greening the Financial System (NGFS), a coalition of central banks and financial supervisors, emphasizes that climate risks are a source of systemic financial instability requiring immediate attention and global cooperation (Network for Greening the Financial System (NGFS), 2019). Climate stress testing, green financial disclosures, and the integration of climate risk assessments in financial stability frameworks are some of the tools central banks are adopting to monitor and mitigate climate risks.

However, unlike traditional financial risks, climate risks are characterized by long-term, nonlinear effects that are challenging to incorporate within existing risk models. Climate risks often unfold over decades, with impacts that are difficult to predict or quantify within conventional risk assessment frameworks (Vermeulen et al., 2018). Consequently, central banks face the task of rethinking their approaches to financial stability, incorporating climate-related variables that account for long-term changes and sector-specific vulnerabilities. This shift calls for enhanced data collection, sophisticated modeling, and new forms of collaboration with governments and international organizations.

As these systemic risks grow, central banks are adapting their mandates and tools to address the profound economic challenges posed by climate change. However, this shift introduces complex policy implications and challenges that merit closer examination.

Policy Implications and the Role of Central Banks

To address climate-related economic risks, central banks are gradually expanding their mandates to integrate climate considerations into their monetary and financial policies. Historically, central banks have focused primarily on price stability and financial stability without venturing into environmental policy areas. However, as climate risks increasingly affect the financial system, central banks are adapting to this new reality. Many central banks are now conducting climate stress tests, requiring disclosures of climate exposures, and encouraging investment in green finance (IMF, 2020).

Nevertheless, these efforts raise questions about the appropriate boundaries of central bank mandates. Critics argue that by expanding their role to include climate objectives, central banks risk overstepping their traditional remit, which could compromise their independence and lead to "mission creep" (Campiglio et al., 2018). While central banks can support climate adaptation by managing climate risks within financial stability frameworks, effective climate policy ultimately requires comprehensive fiscal actions, such as carbon pricing, green subsidies, and direct regulation, which fall within the purview of government authorities.

The Need for Global Coordination

Given the global nature of climate risks, international coordination among central banks and financial institutions is essential to address the systemic economic impacts of climate change effectively. The NGFS and similar initiatives provide platforms for central banks to share best practices, develop standardized approaches to climate risk assessment, and harmonize policy measures across borders (NGFS, 2019). By working together, central banks can create a more resilient financial system capable of withstanding climate risks while ensuring a level playing field across different jurisdictions.

However, achieving global policy alignment remains challenging due to varying national economic interests, energy dependencies, and political stances on climate action. Countries that rely heavily on fossil fuels, for example, may be less inclined to adopt stringent climate policies, creating potential competitive imbalances and regulatory arbitrage. To mitigate these risks, central banks must adopt flexible frameworks that allow for regional adaptation while maintaining alignment with international climate goals.

As central banks navigate these complexities, it is crucial to understand the various channels through which climate risks impact key economic variables such as inflation, output, and employment. The next section examines these direct and indirect transmission channels, detailing how climate risks permeate the economic landscape and influence the stability of financial systems.

3.2   Channels of Transmission

Climate change affects key economic variables such as inflation, output, and employment through multiple transmission channels. These impacts manifest both directly and indirectly, posing challenges to central banks in maintaining price stability and financial stability. Understanding these transmission channels is essential for central banks and policymakers as they evaluate the broad economic consequences of climate change and develop tools to mitigate its impact.

Physical Risks and Economic Disruption

Physical risks from climate change—such as extreme weather events, rising sea levels, and temperature fluctuations—directly disrupt production processes and supply chains. These events lead to increased costs, particularly in sectors like agriculture, energy, and construction, which are highly sensitive to weather conditions (IMF, 2020). For example, severe droughts reduce agricultural yields, driving up food prices and creating inflationary pressures that ripple through the economy. Similarly, extreme weather events can damage infrastructure and halt industrial activities, leading to reduced output and increased repair costs (Batini et al., 2021).

In addition to affecting physical infrastructure, climate-related disruptions create volatility in commodity prices, particularly for food and energy, which are essential components of consumer price indices. This volatility complicates central banks' ability to maintain price stability, as inflation becomes more challenging to predict and manage in an environment of frequent climate shocks (NGFS, 2019). Rising energy demand due to temperature extremes, such as increased cooling during heatwaves or heating during cold snaps, further strains energy supplies and raises prices, adding to inflationary pressures (IMF, 2020).

Transition Risks and Structural Shifts in Employment

Transition risks reflect the economic shifts driven by policies like carbon pricing and subsidies for renewable energy. These changes reduce reliance on fossil fuels, impacting employment in carbon-intensive sectors while creating opportunities in renewables and green technologies. The resulting labor market disparities and skill mismatches challenge central banks to balance inflation trends with long-term employment stability (McKibbin et al., 2017).

Inflation Dynamics and Price Instability

The impact of climate change on inflation is complex, as climate-related disruptions and the transition to a low-carbon economy influence prices both directly and indirectly. Climate-induced price volatility can destabilize inflation expectations, especially when food and energy prices become unpredictable due to extreme weather or regulatory shifts (Bolton et al., 2020). For example, a sudden increase in food prices due to a drought can lead to higher inflation, which may require central banks to adjust interest rates even if the core inflationary trend remains stable.

In response to volatile inflation, central banks face the challenge of distinguishing between temporary, climate-induced price increases and longer-term inflationary trends. Traditional monetary policy tools, such as adjusting interest rates, may be less effective in managing inflation driven by climate risks, as these risks are often exogenous and difficult to control through standard policy measures

(Dafermos et al., 2018). Furthermore, as central banks attempt to stabilize infla-tion, they must also consider the broader economic impacts of their policies on employment and investment in green sectors.

Financial Stability and Asset Revaluation

Climate change also poses risks to financial stability through the revaluation of assets exposed to climate risks. For example, properties and infrastructure located in areas vulnerable to flooding or wildfires may experience declines in value, affecting mortgage-backed securities and other financial instruments linked to these assets (Caldecott et al., 2015). Transition risks also impact asset values, as industries reliant on fossil fuels may face asset stranding if climate policies accel-erate. This can lead to significant losses for investors, insurers, and banks with high exposure to carbon-intensive sectors.

The revaluation of climate-exposed assets creates systemic risks, particularly if financial institutions are unprepared for sudden changes in asset prices. As markets adjust to climate realities, a sharp correction could destabilize the financial system, similar to the systemic shock experienced during the 2008 financial crisis (Bolton et al., 2020). For central banks, managing these risks requires developing stress testing and risk assessment frameworks that incorporate climate scenarios, allow-ing them to assess potential impacts on financial stability and adjust regulatory requirements accordingly (NGFS, 2019).

As central banks adapt to the transmission channels through which climate risks impact economic stability, they face the challenge of integrating these risks into their policy frameworks. In the following section, we explore how central banks are incorporating climate risk into their mandates, assessing the tools and strategies available to mitigate these emerging risks while balancing traditional objectives.

3.3   Climate Risk and Financial Stability

Central banks and financial regulators are increasingly concerned with the desta-bilizing potential of climate-related financial risks. The Network for Greening the Financial System emphasizes that climate risks are a source of systemic financial instability requiring immediate attention and global cooperation (NGFS, 2019). To address these challenges, central banks are implementing a range of tools, including stress testing, enhanced climate-related disclosures, and adjustments to financial stability frameworks. However, climate risks differ from conventional financial risks due to their extended time horizons and complex, nonlinear effects, making them particularly difficult to quantify and integrate into standard risk models (Vermeulen et al., 2018). Consequently, central banks face the task of rethinking their approaches to financial stability, incorporating climate-related variables that account for long-term changes and sector-specific vulnerabilities.

This shift calls for enhanced data collection, sophisticated modeling, and new forms of collaboration with governments and international organizations.

As these systemic risks grow, central banks are adapting their mandates and tools to address the profound economic challenges posed by climate change. However, this shift introduces complex policy implications and challenges that merit closer examination.

3.4 Policy Implications, Critiques, and Challenges of Climate-Responsive Monetary Policy

As climate-related risks become increasingly evident, central banks are evolving their roles to address these challenges. The integration of climate considerations into monetary policy is reshaping traditional frameworks, bringing both opportunities and significant challenges.

Expanding Central Bank Mandates

The systemic risks posed by climate change—such as stranded assets and market disruptions—compel central banks to go beyond their traditional mandates of price and financial stability (Carney, 2015). Tools like climate stress testing and green quantitative easing (QE) are central to managing these risks. For example, the European Central Bank (ECB) and the Bank of England (BoE) have conducted climate stress tests to evaluate the resilience of financial institutions under various climate scenarios (Bank of England, 2021; ECB, 2021).

However, this expansion raises concerns about "mandate overreach." Critics argue that addressing climate change should fall under fiscal policy, not monetary policy. By engaging in climate-related interventions, central banks risk compromising their independence, inviting political pressures, and straying from their core objectives (Campiglio et al., 2018; Dafermos et al., 2018).

Tools and Their Limitations

Central banks employ various tools to address climate risks, including:

- **Climate Stress Testing**: These tests simulate the financial impacts of climate scenarios but face limitations in accurately modeling long-term, nonlinear risks, given the inherent uncertainties surrounding climate trajectories (Vermeulen et al., 2018).
- **Green QE**: By purchasing green bonds, central banks lower financing costs for sustainable projects. However, this tool risks distorting asset prices and creating bubbles in green markets, especially given the limited supply of green bonds (Andersson et al., 2016).

- **Climate-Adjusted Collateral Frameworks**: By favoring green assets over carbon-intensive ones, central banks incentivize low-carbon investments. Yet, the lack of standardized definitions for green assets raises the risk of greenwashing, undermining policy credibility (Monnin, 2018).

While these tools demonstrate innovation, they cannot directly address the structural causes of climate change, which depend on fiscal measures such as carbon pricing, subsidies, and regulation (Campiglio, 2016).

Balancing Trade-Offs

Central banks must navigate significant trade-offs between climate objectives and their traditional mandates:

- **Market Distortions:** Preferential treatment for green assets could divert capital from sectors reliant on carbon-intensive processes, potentially slowing broader economic growth (Bolton et al., 2020).
- **Inflationary Pressures:** Climate policies like green QE could drive up asset prices, complicating inflation management (Dafermos et al., 2018).

Balancing these priorities requires central banks to carefully calibrate their interventions to support climate goals without undermining financial stability or economic growth.

Addressing Data Challenges

Reliable data are essential for effective climate-responsive monetary policy. However, the lack of standardized metrics for climate risks and green assets complicates central banks' efforts. Without rigorous definitions, greenwashing—the mislabeling of assets as sustainable—can erode the credibility and impact of monetary interventions (Monnin, 2018). Collaboration with international organizations, such as the Network for Greening the Financial System (NGFS), is critical to developing robust data standards and frameworks for assessing climate risks (NGFS, 2019).

3.5   Monetary Policy Tools and Climate Change

Building on these broader policy implications, this section explores the specific monetary tools central banks are employing to address climate risks and their evolving roles. One of the challenges central banks face in addressing climate change is that traditional monetary policy tools are designed to operate within business cycles, which are short-term, cyclical fluctuations in economic activity.

Climate change, however, is a long-term structural risk that does not align with typical business cycle patterns. Instead of periodic fluctuations, climate change imposes persistent and often nonlinear impacts on economic stability, which complicates the central bank's usual approach of managing inflation and output stability (IMF, 2020). This section explores how central banks are adapting their tools, both conventional and unconventional, to address the unique challenges posed by climate change.

Conventional Tools: Interest Rates and Reserve Requirements

Interest rates remain one of the primary tools central banks use to influence economic activity. By raising or lowering rates, central banks impact borrowing costs, consumer spending, and business investment. Some economists have proposed using interest rates to encourage green investment, either by lowering rates on green projects or by imposing "green" reserve requirements that encourage banks to prioritize climate-friendly investments (Campiglio, 2016). This approach, known as "green targeting," would incentivize financial institutions to shift capital toward low-carbon activities, potentially accelerating the transition to a sustainable economy.

However, the effectiveness of using interest rates for climate purposes is debatable. Adjusting interest rates to influence climate-related investments could conflict with the core objective of price stability, as it might create inflationary pressures or distort investment across other sectors. Moreover, climate impacts are often unpredictable and long-term, meaning that adjusting interest rates to respond to climate risks would require a forward-looking approach that goes beyond typical business cycle considerations (Campiglio, 2016).

Reserve requirements, another conventional tool, involve setting the amount of capital that banks must hold against their loans. By adjusting these requirements, central banks can influence the availability of credit in the economy. In a climate context, central banks could implement differentiated reserve requirements, lowering them for green assets and raising them for carbon-intensive assets. This approach has the potential to reduce financial sector exposure to climate risks while also promoting sustainable investments (Bolton et al., 2020). However, as with interest rates, the challenge remains in aligning this tool with central banks' traditional objectives without creating unintended economic distortions.

Unconventional Tools: Green Quantitative Easing and Climate-Adjusted Collateral Frameworks

In recent years, central banks have employed unconventional monetary policy tools such as quantitative easing (QE) to support economic stability. QE involves central banks purchasing financial assets, typically government bonds, in order to inject liquidity into the economy and lower borrowing costs. To address climate change, some central banks are exploring "green QE," which involves buying green bonds—debt issued specifically to fund environmentally beneficial projects (Andersson et al., 2016). Green QE could increase demand for green bonds, lowering financing costs for sustainable projects and signaling central bank support for the low-carbon transition.

However, green QE introduces complexities that warrant careful consideration. By focusing on green assets, central banks risk distorting asset prices and creating financial imbalances. Furthermore, the relatively small market for green bonds may constrain the scope of green QE's economic impact. Nonetheless, when paired with complementary policies, green QE could contribute significantly to the transition toward a sustainable economy.

Another emerging approach involves climate-adjusted collateral frameworks. Central banks typically require commercial banks to pledge collateral in exchange for loans, and by adjusting these frameworks to favor green assets, central banks can incentivize environmentally sustainable investments. For example, central banks could apply "haircuts" (reductions in collateral value) to high-carbon assets while accepting green assets at full value. This would make green investments more attractive for banks, indirectly encouraging a shift toward lower carbon portfolios (NGFS, 2019). While this tool may promote green investments, it also requires reliable definitions and standards for green assets to prevent greenwashing and maintain the credibility of central bank policies (Monnin, 2018).

Forward Guidance and Climate Communication

Forward guidance is a communication strategy used by central banks to signal their future policy intentions, shaping expectations in financial markets and the broader economy. In the context of climate change, forward guidance could be used to signal the central bank's commitment to addressing climate risks, potentially encouraging private sector actors to align with low-carbon objectives. For example, central banks could announce that they will consider climate risks in their policy decisions or prioritize green investments in their asset purchase programs, encouraging businesses and investors to transition to sustainable practices (IMF, 2020).

While forward guidance offers a means of promoting climate stability, it also carries risks. If central banks overstate their ability to manage climate risks, they may undermine their credibility, especially given the uncertainties surrounding climate impacts. Furthermore, effective climate-related forward guidance requires clear communication strategies to avoid misinterpretation in financial markets. As such, central banks must carefully balance their messaging to maintain public trust while advancing climate goals (Bolton et al., 2020).

Balancing Climate Goals with Monetary Objectives

A recurring challenge in integrating climate considerations into monetary policy is the potential conflict with central banks' traditional objectives of price stability and financial stability. As central banks seek to address climate risks, they must navigate the trade-offs that arise from simultaneously managing cyclical economic fluctuations and long-term structural risks. Unlike inflationary pressures or recessions, climate risks are less predictable and require a proactive approach that may, at times, conflict with short-term economic objectives.

Furthermore, as central banks introduce climate-responsive tools, they must avoid creating excessive dependencies on green sectors, which could result in new forms of market distortion. For instance, if green assets receive preferential treatment without clear standards, there is a risk of greenwashing, which could weaken the central bank's policy efficacy and erode trust (Dafermos et al., 2018). The balancing act between advancing climate goals and maintaining economic stability is therefore central to the evolution of climate-responsive monetary policy.

As central banks continue to adapt their monetary tools to address climate risks, they must also consider the broader implications for financial stability. In the next section, we examine how central banks are approaching financial stability in a climate context, including their role in promoting sustainable finance, managing the systemic risks posed by stranded assets, and preparing for potential financial shocks related to climate risks.

3.6   Financial Stability and the Role of Central Banks in Climate Mitigation

Climate change introduces risks that extend beyond the traditional responsibility of central banks, threatening the stability of financial systems in ways that demand new responses and strategies. As previously discussed, climate risks represent systemic threats that could destabilize the financial sector through abrupt asset revaluations, economic disruption, and increased volatility. In response, central banks are evolving from passive observers of climate-related financial stability to proactive agents, leveraging tools such as climate stress testing, enhanced regulatory frameworks, and green finance initiatives to mitigate these risks.

Climate Stress Testing and Risk Assessment

Climate stress testing has emerged as a foundational tool for central banks aiming to assess the resilience of financial institutions to climate risks. Unlike traditional stress tests, which evaluate resilience to economic downturns, climate stress tests simulate a range of climate-related scenarios, including both physical and transition risks. For example, the Bank of England's (BoE) 2021 climate stress test evaluated the potential impacts of early, delayed, or no action on climate change, assessing the effects of these scenarios on banks and insurers (Bank of England, 2021).

By analyzing how different climate scenarios might impact credit portfolios, asset values, and insurance liabilities, climate stress tests help central banks identify vulnerabilities in the financial system. These assessments inform regulatory measures, such as capital requirements for institutions with high climate exposure, which aim to enhance financial resilience against climate risks (Bolton et al., 2020). However, climate stress testing presents challenges, particularly in accurately modeling long-term, nonlinear climate impacts and incorporating uncertainties around policy and technology pathways (Vermeulen et al., 2018).

Green Finance and Incentivizing Sustainable Investment

Central banks are also leveraging their influence to support the development of green finance, which channels capital into projects and industries that contribute to environmental sustainability. By encouraging financial institutions to allocate resources toward green investments, central banks help mitigate transition risks and support the low-carbon economy. This often involves developing frameworks that define and standardize "green" assets, allowing investors and financial institutions to more easily identify sustainable investments (NGFS, 2019).

In addition to setting standards, central banks are integrating green finance incentives into their operational frameworks. For instance, the European Central Bank (ECB) has introduced favorable collateral treatment for green bonds, encouraging banks to hold more environmentally sustainable assets (ECB, 2021). By providing incentives, central banks can promote sustainable financing within the private sector, gradually shifting capital flows toward projects that align with climate goals. However, the effectiveness of these measures relies on the accurate and consistent classification of green assets, a challenging task given the risks of greenwashing and inconsistent standards across jurisdictions.

Prudential Regulation and Managing Financial Stability Risks

Beyond encouraging sustainable finance, central banks are adapting their prudential regulation frameworks to account for climate-related financial risks. Traditional prudential measures—such as capital adequacy ratios, liquidity requirements, and risk-weighted asset calculations—are being updated to include climate considerations. For instance, some central banks are requiring financial institutions to hold additional capital against high-emission assets, which are more vulnerable to transition risks as regulatory policies tighten (Dafermos et al., 2018).

Climate-adjusted prudential regulation helps to reduce the exposure of banks and insurers to climate risks, supporting financial stability by ensuring that these institutions have adequate capital buffers to absorb climate-related shocks. These regulations also create incentives for financial institutions to diversify their port-folios away from carbon-intensive sectors, promoting long-term resilience within the financial system. However, implementing climate-adjusted prudential measures remains a complex task, as central banks must balance the need for stability with the risk of over-regulation, which could hinder credit availability and economic growth in the short term (NGFS, 2019).

Addressing Potential Financial Shocks from Climate Risks

A crucial aspect of central banks' evolving role in climate mitigation is preparing for potential financial shocks linked to climate risks. While systemic risks were discussed in earlier sections, it is important to note that central banks must develop robust mechanisms to manage the specific financial shocks that could arise from abrupt climate-related events, such as rapid revaluations in the housing market due to flooding risks or spikes in energy prices due to extreme weather events. By enhancing risk management frameworks and coordinating with other regulatory bodies, central banks can better prepare for these shocks, minimizing potential ripple effects across the financial system (IMF, 2020).

Furthermore, central banks are increasingly engaging in international collabo-ration to build a globally resilient financial system. The Network for Greening the Financial System (NGFS) plays a pivotal role in these efforts, providing a platform for central banks to share best practices, harmonize regulatory approaches, and develop climate risk assessment standards. Global coordination is essential, as cli-mate risks are transnational and require consistent frameworks to avoid competitive imbalances and regulatory arbitrage (Bolton et al., 2020).

As central banks continue to implement measures to promote financial stability and mitigate climate risks, they also face the need for international alignment to ensure effective policy harmonization. In the following section, we explore the importance of international coordination in managing climate risks, focusing on the efforts of global institutions and the challenges of achieving consistent standards across diverse economic and regulatory environments.

3.7   International Coordination and Policy Harmonization

The global nature of climate change requires coordinated international efforts to manage its economic and financial impacts effectively. While individual central banks can make progress within their own jurisdictions, climate-related finan-cial risks do not respect national boundaries. Financial systems across countries are interconnected, meaning that climate risks in one region can have ripple effects elsewhere. Consequently, central banks, financial regulators, and interna-tional organizations are increasingly collaborating to harmonize climate-related policies, share best practices, and create standardized frameworks. This section explores the importance of international coordination in addressing climate risks and discusses current initiatives aimed at fostering policy alignment.

The Need for Harmonized Approaches to Climate Risk

As climate risks threaten global financial stability, inconsistencies in regulatory standards across countries pose significant challenges to effective risk management. Differing definitions, metrics, and regulatory frameworks for green finance create uncertainty for investors and hinder the efficient flow of capital to sustainable projects. Without harmonized policies, countries with more stringent climate regulations may face competitive disadvantages, as businesses and financial institutions gravitate toward regions with weaker standards. This "regulatory arbitrage" risks undercutting global climate goals and weakening the financial sector's resilience to climate risks (Campiglio et al., 2018).

Addressing these inconsistencies requires coordinated efforts to harmonize approaches at the global level. Harmonized policies facilitate more accurate climate risk assessments and enable cross-border investments to align with climate objectives. Standardized frameworks and consistent metrics for climate disclosures allow investors to compare assets across countries, enhancing transparency and reducing the likelihood of greenwashing. For example, the adoption of globally recognized environmental, social, and governance (ESG) standards helps central banks and investors alike to assess the climate impact of investments and promote sustainable finance (OECD, 2021).

The Role of the Network for Greening the Financial System (NGFS)

The Network for Greening the Financial System (NGFS) has been instrumental in promoting international coordination on climate risks. Formed in 2017, the NGFS is a coalition of central banks and supervisory authorities dedicated to addressing climate risks in the financial sector. The NGFS provides a forum for central banks to share best practices, conduct joint research, and develop guidelines for managing climate risks. One of its key contributions has been the development of climate scenarios, which enable central banks to perform climate stress testing with standardized assumptions and variables (NGFS, 2019).

The NGFS has also published a series of recommendations for integrating climate risk into financial stability assessments, prudential regulation, and monetary policy. These recommendations include encouraging climate-related disclosures, supporting green finance, and promoting the adoption of common frameworks for climate risk management (Bolton et al., 2020). Through its work, the NGFS has become a central force in aligning central banks' climate efforts and fostering a unified response to climate-related financial risks.

Basel Committee and Cross-Border Regulatory Collaboration

Beyond the NGFS, other international organizations, such as the Basel Committee on Banking Supervision, are addressing climate risks in regulatory frameworks. The Basel Committee is examining how climate risks affect banking stability and is exploring ways to incorporate these risks into its regulatory standards, particularly regarding capital adequacy and risk-weighted assets (Basel Committee, 2020). By updating its guidelines to reflect climate considerations, the Basel Committee can encourage banks worldwide to account for climate risks in their risk management practices, helping to mitigate systemic vulnerabilities.

Cross-border regulatory collaboration also extends to efforts like the Financial Stability Board (FSB),[1] which established the Task Force on Climate-related Financial Disclosures (TCFD) to promote consistent climate-related reporting across jurisdictions. The TCFD provides recommendations for companies and financial institutions to disclose their climate exposures, enabling central banks and regulators to better assess systemic risks and inform policy decisions. Standardized disclosures also empower investors to make informed choices, driving capital toward projects that support climate resilience and sustainability (TCFD, 2017).

Challenges in Achieving Policy Harmonization

Despite the progress in international coordination, achieving full policy harmonization remains challenging due to diverse national economic interests, energy dependencies, and political landscapes. For instance, countries that rely heavily on fossil fuels may be reluctant to adopt strict climate regulations, fearing adverse economic impacts on key industries. In contrast, nations with established renewable energy sectors may advocate for ambitious climate policies that promote green finance and sustainability. These differences can lead to fragmented approaches, creating barriers to cohesive action and complicating the implementation of standardized climate regulations (Monnin, 2018).

Additionally, climate risks are complex and uncertain, making it difficult to design universal standards that apply to all countries and sectors. Developing economies, in particular, may require flexible regulatory frameworks that balance climate goals with economic growth and poverty reduction. This need for adaptability underscores the importance of developing climate policies that account for regional and sectoral nuances, allowing countries to align with global standards while addressing their unique challenges (IMF, 2020).

---

[1] The Financial Stability Board (FSB) is an international organization that monitors and advises on global financial stability. It was established in 2009 to strengthen regulatory and supervisory frameworks following the global financial crisis. The FSB created the Task Force on Climate-Related Financial Disclosures (TCFD) in 2015 to improve corporate climate risk reporting.

Moving Toward Global Financial Resilience

Achieving effective international coordination on climate risks is crucial for building a resilient global financial system. Collaborative efforts through organizations like the NGFS, Basel Committee, and TCFD enhance transparency, reduce regulatory fragmentation, and promote sustainable investment across borders. However, sustained political will and commitment from national governments and central banks are essential to drive meaningful progress. As the impacts of climate change intensify, international collaboration will play an increasingly central role in managing the global financial risks posed by climate change.

As central banks and regulators work toward harmonizing climate-related policies, they must also address the critiques and limitations of climate-responsive monetary policy. In the next section, we examine the potential challenges of integrating climate considerations into central banking, including concerns about mandate overreach, market distortions, and the long-term efficacy of climate-focused monetary policy.

3.8   Opportunities for Integrating Financial Stability Tools with Climate Initiatives

As central banks and financial regulators expand their focus to address climate-related risks, they are also exploring ways to integrate financial stability tools with broader climate initiatives. This integration is pivotal to managing systemic risks associated with climate change while supporting the transition to a low-carbon economy. By aligning financial stability policies with climate objectives, central banks can foster resilience within the financial system and promote sustainable investments. This section explores some of the emerging opportunities for central banks, including the use of climate-aligned financial stability frameworks, risk mitigation strategies, and collaboration with the corporate sector.

Climate-Aligned Financial Stability Frameworks

One of the promising areas for integrating climate initiatives into financial stability policy lies in adapting central banks' regulatory frameworks to consider climate risks. This shift has involved enhancing traditional financial stability assessments to incorporate physical and transition risks related to climate change. For example, the Network for Greening the Financial System (NGFS) recommends that central banks account for climate-related risks in their macroprudential frameworks, which guide financial institutions on the risks they must manage to maintain systemic stability (NGFS, 2019).

A climate-aligned financial stability framework would adjust capital requirements based on the exposure of assets to climate risks, incentivizing banks and insurers to shift their portfolios toward low-carbon sectors. By incorporating these risks, central banks aim to prevent "carbon bubbles"—asset bubbles associated with carbon-intensive industries—that could burst as climate policies intensify, potentially destabilizing the financial system (Bolton et al., 2020). In this way, climate-aligned frameworks not only bolster resilience but also encourage the reallocation of capital toward sustainable industries.

Risk Mitigation Through Diversified Portfolios and Capital Buffers

Central banks are also exploring risk mitigation strategies that encourage financial institutions to diversify their portfolios and maintain capital buffers against climate shocks. Diversification of assets can reduce banks' exposure to high-risk, carbon-intensive industries, thereby mitigating the potential impact of climate risks on financial stability. Capital buffers, which require banks to hold additional capital to absorb losses, can be adjusted to reflect climate risk exposure, ensuring that financial institutions remain resilient in the face of climate-induced disruptions (Vermeulen et al., 2018).

Another approach involves climate "stress buffers," which require higher capital reserves for institutions with significant climate exposures. This form of prudential regulation, aimed at supporting financial stability, encourages banks to limit their investments in high-carbon industries and increase their resilience to climate-related risks. For instance, the European Central Bank (ECB) has implemented policies that incentivize financial institutions to improve their climate resilience, signaling the growing importance of climate factors in regulatory assessments (ECB, 2021).

Collaboration with the Corporate Sector and Sustainable Finance

The transition to a low-carbon economy requires significant investment in sustainable projects, and central banks are working to facilitate this transition by collaborating with the corporate sector. Central banks have introduced policies that promote green finance, encouraging corporate issuers to pursue sustainable initiatives. By supporting green bond markets, central banks enable corporations to access capital for projects that reduce greenhouse gas emissions and improve environmental sustainability (NGFS, 2019).

Additionally, forward guidance on climate-related financial stability can signal central banks' commitment to climate goals, influencing corporate behavior. By providing clear guidance on how climate factors will be integrated into regulatory policies, central banks create an environment in which corporations are incentivized to align their strategies with sustainable financial goals. This collaboration can accelerate corporate investments in renewable energy, green technologies, and sustainable infrastructure, contributing to a more resilient financial system (Campiglio et al., 2018).

Technology and Data for Climate Risk Analysis

The integration of financial stability tools with climate initiatives has underscored the importance of reliable data and technology to accurately assess climate risks. Central banks are increasingly investing in technologies that enable more precise measurement and forecasting of climate-related risks. Advances in climate modeling, artificial intelligence, and data analytics are essential for identifying sectoral exposures and assessing the potential impact of extreme weather events on financial stability (Campiglio et al., 2018). Data standardization is also crucial for enabling consistent climate disclosures across sectors and regions. As central banks leverage financial stability tools to address climate risks, the implications for the broader financial ecosystem, including the corporate sector, are significant. In the final section, we reflect on the transformative changes in the relationship between climate change and monetary policy, examining what these developments mean for corporate finance and the role of businesses in a climate-responsive financial system.

Conclusion

As the economic and financial implications of climate change grow more urgent, central banks are redefining their roles in the global economy. No longer confined to traditional mandates of price stability and financial stability, they are stepping into uncharted territory by addressing the long-term systemic risks posed by climate change. This evolution reflects a recognition that climate risk is not a distant challenge but an immediate and material threat to the global financial system.

By integrating climate risks into monetary policy and financial stability frameworks, central banks are not only mitigating potential crises but also shaping the transition to a low-carbon economy. Tools like climate stress testing, green quantitative easing, and climate-adjusted collateral frameworks represent a fundamental shift in how central banks approach risk, regulation, and investment. These measures signal a broader alignment between monetary policy and sustainability goals, reinforcing the interconnectedness of financial stability and climate resilience. However, this expanded role comes with challenges. Central banks must balance their climate initiatives with their core mandates, navigate uncertainties around climate impacts, and avoid overstepping their boundaries into fiscal policy domains. They also face the complex task of managing global coordination, ensuring that policies are both harmonized across jurisdictions and adaptable to regional differences.

For corporations, the message is clear: the financial system is increasingly aligned with climate objectives. Firms that adapt quickly—investing in sustainable technologies, transitioning to greener operations, and embracing climate risk disclosures—stand to benefit from improved access to capital and market opportunities. Conversely, those that resist these changes risk being left behind, facing higher financing costs, regulatory pressures, and diminished competitiveness. Central banks' focus on climate resilience places new demands on businesses to align their strategies with emerging sustainability standards. Among these demands, one critical area stands out: corporate climate disclosures. As central banks incorporate climate risks into their frameworks, the need for reliable, transparent, and standardized climate data is becoming paramount. Disclosures provide the foundation for assessing systemic risks, guiding sustainable investment, and ensuring accountability across the financial system.

The next chapter explores the emerging landscape of disclosure frameworks. From the evolution of global standards to the role of initiatives like the TCFD and SBTi, it examines how businesses can navigate the complexities of climate reporting and meet the expectations of central banks, investors, and other stakeholders. In doing so, it highlights the growing interplay between climate disclosures and the financial system, underscoring their importance in a world increasingly shaped by climate-aware decision-making.

Key Takeaways for This Chapter

1. Climate Change as a Systemic Risk

Climate change poses profound and systemic risks to the global economy and financial stability, with impacts spanning physical damages, transition risks, and systemic market disruptions.

2. Evolving Role of Central Banks

Central banks are expanding beyond their traditional mandates of price and financial stability to address climate risks, incorporating tools like stress testing, green quantitative easing (QE), and climate-adjusted collateral frameworks.

3. Mandate Challenges

While central banks are vital in managing climate-related financial risks, concerns about "mandate overreach" highlight the importance of aligning their actions with fiscal policies like carbon pricing and subsidies.

4. Critical Tools for Climate Risk Management for Central Banks
   - Stress Testing: Simulating physical and transition risk scenarios to assess financial institution resilience.
   - Green QE: Encouraging green investments by purchasing green bonds, with caution to avoid market distortions.
   - Prudential Regulation: Adjusting capital requirements to reflect climate risk exposures and promote financial system resilience.

5. Balancing Trade-Offs

Central banks must carefully manage trade-offs between climate objectives and their traditional roles, ensuring inflation stability and avoiding resource misallocations.

6. Importance of Data and Collaboration

Reliable, standardized climate data and international coordination are crucial for effective climate risk management. Initiatives like the NGFS and TCFD play critical roles in aligning global efforts.

7. Corporate Implications

Firms aligning with sustainable practices and enhancing climate risk disclosures will benefit from better access to capital, while those that delay risk falling behind in an increasingly climate-conscious financial system.

8. The Path Forward

As climate impacts intensify, central banks and corporations must innovate and collaborate to create a resilient, sustainable financial ecosystem, balancing long-term climate goals with immediate economic stability.

Questions

1. What are the systemic risks posed by climate change to the global financial system?
2. How have central banks adapted their mandates to address climate-related risks?
3. What are the critiques of central banks' involvement in climate policy?
4. What is green quantitative easing (QE), and what are its potential risks?
5. Why is reliable data essential for effective climate-responsive monetary policy?
6. How does international coordination support climate-related financial policies?
7. What are the potential trade-offs central banks face when integrating climate considerations into monetary policy?
8. How do climate stress tests differ from traditional stress tests conducted by central banks?

References

Andersson, M., Bolton, P., & Samama, F. (2016). Hedging climate risk. *Financial Analysts Journal,* *72*(3), 13–32.

Bank of England. (2021). *Climate-related financial disclosure report.* Retrieved from November 4, 2024 from https://www.bankofengland.co.uk/prudential-regulation/publication/2021/june/climate-related-financial-disclosure-2020-21

Basel Committee on Banking Supervision. (2020). *Climate-related financial risks: A survey on current initiatives.* Bank for International Settlements. Retrieved November 4, 2024, from https://www.bis.org/bcbs/publ/d502.pdf

Batini, N., Di Serio, M., Fragetta, M., Melina, G., & Waldron, A. (2021). *Building back better: How big are green spending multipliers?* (IMF Working Papers). Retrieved November 4, 2024, from https://www.imf.org/en/Publications/WP/Issues/2021/03/19/Building-Back-Better-How-Big-Are-Green-Spending-Multipliers-50264

Bolton, P., Despres, M., Pereira da Silva, L. A., Samama, F., & Svartzman, R. (2020). *The Green Swan: Central banking and financial stability in the age of climate change.* Bank for International Settlements. Retrieved November 4, 2024, retrieved from https://www.bis.org/publ/othp31.pdf

Caldecott, B., Dericks, G., & Mitchell, J. (2015). *Stranded assets and subcritical coal: The risk to companies and investors.* University of Oxford, Smith School of Enterprise and the Environment. Retrieved November 4, 2024, from https://www.smithschool.ox.ac.uk/sites/default/files/2022-04/Stranded-Assets-and-Subcritical-Coal

Campiglio, E. (2016). Beyond carbon pricing: The role of banking and monetary policy in financing the transition to a low-carbon economy. *Ecological Economics, 121,* 220–230.

Campiglio, E., Dafermos, Y., Monnin, P., Ryan-Collins, J., & Schotten, G. (2018). Climate change challenges for central banks and financial regulators. *Nature Climate Change, 8,* 462–468.

Carney, M. (2015). *Breaking the tragedy of the horizon—Climate change and financial stability.* Bank of England Speech. Retrieved November 4, 2024, from https://www.bankofengland.co.uk/speech/2015/breaking-the-tragedy-of-the-horizon-climate-change-and-financial-stability

Dafermos, Y., Nikolaidi, M., & Galanis, G. (2018). Climate change, financial stability and monetary policy. *Ecological Economics, 152,* 219–234.

European Central Bank (ECB). (2021). *ECB strategy on climate.* Retrieved November 4, 2024, from https://www.ecb.europa.eu/press/pr/date/2021/html/ecb.pr210708_1~f104919225.en.html

International Monetary Fund. (2020). *World economic outlook: A long and difficult ascent.* International Monetary Fund. Retrieved November 4, 2024, from https://www.imf.org/en/Publications/WEO/Issues/2020/09/30/world-economic-outlook-october-2020

McKibbin, W. J., Morris, A. C., Panton, A. J., & Wilcoxen, P. J. (2017). *Climate change and monetary policy: Dealing with disruption* (Brookings Climate and Economics Discussion Papers). Retrieved November 4, 2024, from https://www.brookings.edu/wp-content/uploads/2017/12/es_20171201_climatechangeandmonetarypolicy.pdf

Monnin, P. (2018). *Central banks and the transition to a low-carbon economy. Council on economic policies.* Retrieved November 4, 2024, from https://www.cepweb.org/central-banks-and-the-transition-to-a-low-carbon-economy/

Network for Greening the Financial System (NGFS). (2019). *A call for action: Climate change as a source of financial risk.* Retrieved November 4, 2024 from https://www.ngfs.net/sites/default/files/medias/documents/ngfs_first_comprehensive_report_-_17042019_0.pdf

OECD. (2021). *Policy guidance on market practices to strengthen ESG investing and finance a climate transition.* OECD Publishing. Retrieved from https://www.oecd.org/en/publications/policy-guidance-on-market-practices-to-strengthen-esg-investing-and-finance-a-climate-transition_2c5b535c-en.html

Stern, N. (2007). *The economics of climate change: The Stern review.* Cambridge University Press.

Task Force on Climate-related Financial Disclosures (TCFD). (2017). *Recommendations of the task force on climate-related financial disclosures*. Financial Stability Board. Retrieved November 4, 2024, from https://www.fsb-tcfd.org

Vermeulen, R., Schets, E., Lohuis, M., Kölbl, B., Jansen, D.-J., & Heeringa, W. (2018). An energy transition risk stress test for the financial system of the Netherlands. *De Nederlandsche Bank Occasional Studies, 16*(7).

Non-Financial Disclosures: Standards, Strategies, and Future Outlook

4

Introduction

This chapter explores the evolving non-financial disclosures (NFDs) landscape over the three momentous past decades. It investigates the drivers behind firms' adoption of NFDs and the current challenges, followed by a venture into future outlooks. This future-focused section consists of speculative potential scenarios, the impact of advanced technology, and advice on monitoring the integration of sustainability and financial accounting standards. Finally, the chapter recommends strategies companies are employing to manage the uncertainties in the evolution of NFDs, providing a comprehensive perspective on this crucial aspect of contemporary corporate management.

4.1 Three Decades of Evolving Non-Financial Disclosure Standards

Non-financial disclosures (NFDs) have evolved significantly, transitioning from initiatives by smaller entities to major global forces advocating for comprehensive sustainability practices. This section traces their development against the backdrop of global sustainability challenges, societal shifts, and evolving regulatory landscapes. More than a historical account, it reveals a narrative of continuous adaptation and innovation, emphasizing the persistent drive for corporate accountability in a rapidly changing world.

Fig. 4.1 First Fairtrade
Mark, 1988, and the Fairtrade
International Trademark
Today

The Early Development

Fairtrade Standards

A major effort to enhance global social justice and environmental sustainability through market forces, Fairtrade standards, their origin, evolution, and adoption represented an early, significant development in crossing the Global North/South divide.

The concept of Fairtrade began in the late 1940s and early 1950s. It emerged from various church groups and non-governmental organizations in Europe and North America that started selling crafts made by impoverished communities in the Global South. The first formal Fairtrade shop, called MUDAC in the Netherlands, was opened in 1959. This initiative was followed by Oxfam UK starting to sell crafts made by Chinese refugees in 1964, marking a significant step in the Fairtrade movement. During the 1960s and 1970s, the concept of "alternative trade" (later known as Fairtrade) was developed. It aimed to provide better trading conditions, improve sustainability, and offer higher prices to exporters in developing countries.

In the 1980s, the first Fairtrade labeling initiatives began. Stichting Max Havelaar, the world's first Fairtrade certification mark and later known as Fairtrade Nederland, became official on November 15, 1988.[1] The need for a global coordinating body led to the establishment of Fairtrade Labeling Organizations International (FLO, now Fairtrade International) in 1997. This organization helped standardize the Fairtrade criteria and certify products globally.[2] Figure 4.1 shows the original and current Fairtrade labeling.

Initially focused on coffee, the range of Fairtrade products expanded to include tea, chocolate, bananas, sugar, and many others, encompassing a wide array of agricultural and artisan goods.

The adoption of Fairtrade standards has been significantly driven by increasing consumer awareness and demand for ethically sourced products (Andofer & Liebe, 2012). Many mainstream brands and retailers have adopted Fairtrade products or created their own ethical sourcing programs (Pelsmacker et al., 2003). As a result,

---

[1] Source: https://en.wikipedia.org/wiki/Stichting_Max_Havelaar
[2] Source: https://en.wikipedia.org/wiki/Fairtrade_International

it has provided small-scale producers in developing countries with more equitable trade terms, such as fairer prices and better working conditions (Reed, 2009). Fairtrade has become a globally recognized standard, with certified products being sold in dozens of countries.

The adoption and evolution of Fairtrade standards reflect a growing global consciousness about the interconnectedness of trade, social justice, and environmental sustainability.[3] It symbolizes an ongoing effort to make global commerce more equitable and sustainable. Despite its successes, Fairtrade has faced critiques regarding its effectiveness, impact on poverty reduction, and market limitations. These critiques have prompted ongoing refinement and adaptation of Fairtrade standards (Hira & Ferrie, 2006). Moreover, pressures for more radical approaches remain to resolve the inherent tensions between the movement's historical vision to transform globalization and the established markets and global economy (Raynolds et al., 2007).

Global Reporting Initiative (GRI)

Crafted as code of environmental conduct for companies by Ceres, a US-based nonprofit organization, GRI was launched as the global standard for corporate sustainability by the United Nations Environment Program in 1997. GRI is now widely used globally, with thousands of organizations across various sectors adopting the GRI standards for sustainability reporting.

Beginning in 1989, shortly after the Exxon Valdez oil spill, a group of forward-looking investors and environmentalists led by Joan Bavaria, a pioneering socially responsible investor, formed Ceres. They were at the forefront of a transformative movement in business at the time and claimed to have established "climate risk" and more recently "water risk" as mainstream concepts for global investments.[4]

Initial adopters of GRI were larger corporations with high public visibility and those experiencing environmental scrutiny. Public outcry over environmental issues, notably the Exxon Valdez oil spill, was a major driver. The GRI standards began with a focus on environmental conduct and expanded to include social, economic, and governance issues. Starting with the first version of GRI Guideline (G1) in 2000, it has evolved through several iterations (G2 in 2002, G3 in 2006, and G4 in 2013) to provide a comprehensive framework for sustainability reporting. In 2016, GRI transitioned from providing guidelines to setting the first global standards for sustainability reporting.[5] To do that, the Global Sustainability Standards Board (GSSB) was established as an independent operating entity under the auspices of GRI, which has members who represent a range of expertise and multi-stakeholder perspectives on sustainability reporting.

Halkos and Nomikos (2021) trend analysis of the mean changes in GRI adoption for the period 1999–2017, across all continents, in all sectors, and in both large

---

[3] Source: Fairtrade International, https://www.fairtrade.net/standard

[4] Source: https://www.ceres.org/about-us

[5] Global Reporting Initiative - Mission and history, https://www.globalreporting.org/about-gri/mission-history/

multinational and small medium-sized enterprises, shows that Europe has passed from a full-grown to a downturn stage. So have Oceania and North America, to a similar but less pronounced degree. Asia, on the other hand, is in a spreading-out stage with a steady expansion. Latin America, the Caribbean, and Africa have reached a full-grown stage. The authors argue that this picture of GRI diffusion (spreading out, full grown, and downturn) may help decision-makers to recognize companies' understanding of their sustainability strategies and activities.

In 2022, GRI and the IFRS Foundation (see sections on CDP, SASB, IR, and TCFD below) announced collaboration agreement under which their respective standard-setting boards, the International Sustainability Standards Board (ISSB) and the GSSB, will seek to coordinate their work programs and standard-setting activities.[6]

Social Accountability International's SA8000

The focus of SAI is human rights at work. The global non-governmental organization was founded in 1997 with the vision of decent work everywhere through advancing an understanding that a socially responsible workplace builds sustainable business.[7] Using its multi-industry SA8000 Standard, along with other capacity-building programs, SAI works with a diverse group of stakeholders, including brands, suppliers, governments, trade unions, non-profits, and academia to empower workers and managers at all levels of businesses and supply chains.

Early adopters of SAI standards were primarily suppliers operating in developing economies, such as India. Requests by socially conscious customers, driven by the need to avoid the loss of business due to non-compliance with social standards, were a major reason for these suppliers to adopt SA8000. In 2023, SAI is undertaking a full revision of our SA8000 Standard for Decent Work, which is also used as a benchmark by many other standards and in company codes of conduct.

Analyzing the worldwide diffusion of the Social Accountability 8000 standard certification across sixty-six activity sectors in sixty-five countries from 1999 to 2011, Llach et al. (2015) found that the diffusion followed a similar pattern as other similar standards with a near-future growing trend in developing countries. Leading the pack were India, China, Romania, and Brazil. A separate study noted that the effectiveness of SA8000 had been influenced by the symbolic nature of adoptions and the challenges in creating a positive business case for its implementation, although the initial focus was on ensuring ethical working conditions (Koster et al., 2019).

United Nations Global Compact (UNGC)

UNGC is a voluntary initiative based on the commitments of CEOs of participating businesses in answering a call from the UN Secretary-General to align company strategies and operations with ten universal principles in the areas of human rights,

---

[6] Source: IFRS website, https://www.ifrs.org/about-us/who-we-are/#history
[7] Source: Social Accountability International, https://www.sa-intl.org/

labor, environment, and anti-corruption. Launched in 2000, UNGC has increased its membership to over 15,000 businesses and close to 4000 non-business entities from more than 160 countries. With broad-based support, UNGC remains a formidable global normative authority and reference point for action and leadership within the growing global corporate sustainability movement.

Early adopters were global corporations seeking to align with international norms on human rights, labor, environment, and anti-corruption, which evolved as a commitment to sustainable and responsible business practices, in line with the principles set out by the United Nations. The Ten Principles served as a blueprint for businesses to contribute toward the achievement of the 2030 Agenda for Sustainable Development and the Paris Agreement, which grew to include more comprehensive reporting and local networks for collaborative action.

For companies committed to sustainability, reporting to stakeholders in a transparent and public manner is fundamental. Non-financial information informs the decisions of mainstream investors, consumers, local communities, and civil society organizations. Since quality information is essential to decision-making, stakeholders' need for transparent and timely disclosures is the driving force that pushes companies to manifest their commitment to sustainability through non-financial disclosures.

A key component of this under UNGC is the annual Communication on Progress (COP). Once only a voluntary activity, UNGC now requires participating companies to produce an annual COP that details their work in embedding the Ten Principles into strategies and operations, as well as efforts in supporting societal priorities.[8] The COP is an expression of a company's commitment to sustainability and is visible to stakeholders through the company's profile page. It is in some cases used to meet government requirements. Participating companies that fail to report or to meet the criteria over time may be removed from the UNGC initiative. To date, over 47,000 COPs have been posted.

Underneath UNGC's core strategies of normative principles, networks for learning and cooperation, and communication and transparency mandates are its networked governance model (see Figure). It distributes governance functions among several entities to engage participants and stakeholders at the local and global levels, ensuring not only the role and participation of the stakeholders, but also the matters of greatest importance to the stakeholders. For example, the establishment of the UNGC Government Group is to formalize the role of governments in the initiative. The 2017 Governance Review consulted a total of three hundred stakeholders, including local networks and governments to deliver on the Compact's mandate to "mobilize a global movement of sustainable companies and stakeholder."

Runhaar and Lafferty (2009) studied the early UNGC participants, numbering only a quarter of today's total (Runhaar & Lafferty, 2009). They found the role of the initiative in shaping, implementing, and reporting about the companies' CSR

---

[8] Source: The UNGC reporting framework, https://unglobalcompact.org/participation/report/

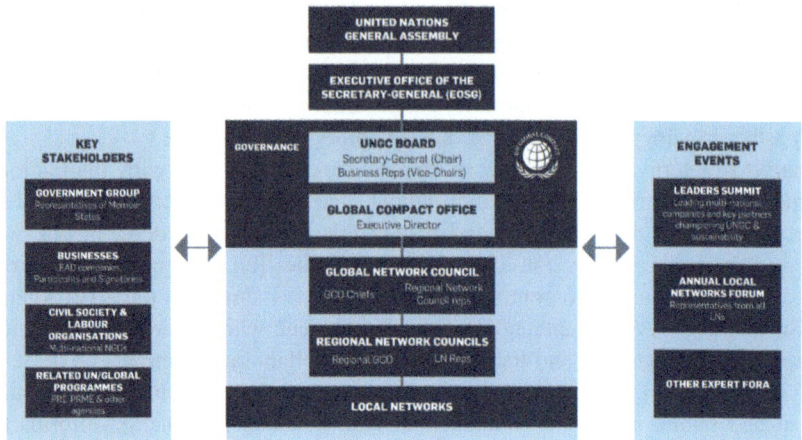

Fig. 4.2 UNGC Governance Framework (*Source UN Global Compact website,* https://unglobalc ompact.org/about/governance)

strategies was at most modest. In other words, companies did more than what the initiative required for some reason. There were two important reasons, speculated by the authors. One was that many of the CSR issues were industry specific and addressed in specific, often non-UNGC networks. The second reason was that the UNGC principles were perceived as minimum requirements, thus reporting on efforts made to meet these requirements was hardly a reputation booster (Fig. 4.2).

Carbon Disclosure Project

Established as the "**Carbon Disclosure Project**"[9] in 2000 to ask companies to disclose their climate impact, the not-for-profit charity shortened its name to CDP in 2013 to broaden the scope of environmental disclosure. CDP now runs the global disclosure system for investors, companies, cities, states, and regions to manage their environmental impacts, including GHG emissions, deforestation, and water security, and has become a global standard of environmental reporting with the richest and most comprehensive dataset on corporate and city action. In 2023, 12,455 companies disclosed climate change data, 837 companies disclosed forestry data, and 3251 companies disclosed water security data through CDP.

Like the history of other disclosure initiatives, early adopters were companies in sectors with significant environmental impacts, particularly those facing regulatory or public pressure regarding carbon emissions. But a confluence of factors has driven the inflection point around 2017 in the number of companies making climate disclosures (Fig. 4.3).

---

[9] CDP, London, UK, https://www.cdp.net/

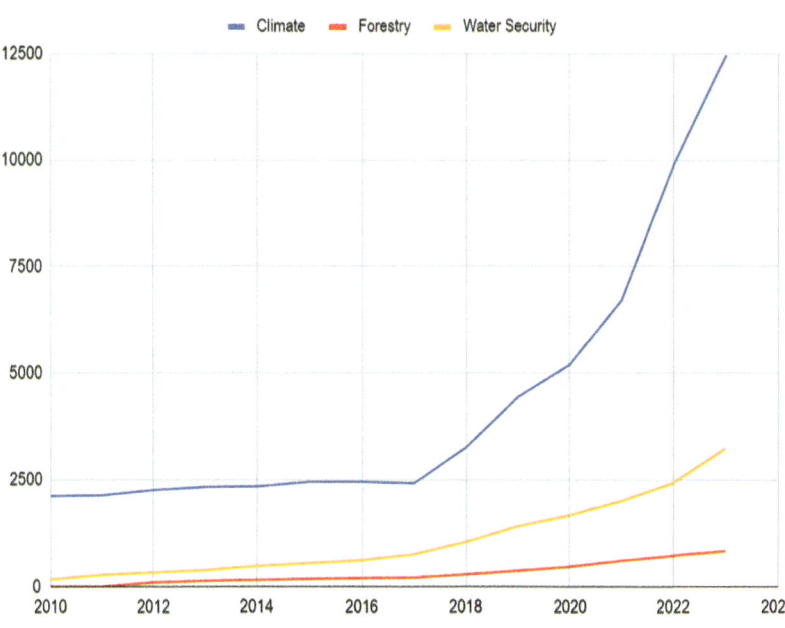

Fig. 4.3 Increasing Corporate Disclosures of Environmental Data (2010–2023) (*Source Submitted CDP responses from 2010 to 2023, compiled on 21 December 2023,* https://www.cdp.net/en/res ponses)

The Paris Agreement and Sustainable Development Goals

The adoption of the Paris Agreement in 2015 and the United Nations' 2030 Agenda for Sustainable Development marked significant global commitments to climate action. These initiatives emphasized the reduction of greenhouse gas emissions, the mitigation of climate change effects, and the enhancement of resilience to climate change impacts. Following these, the European Commission published its Action Plan "Financing Sustainable Growth" in 2018, setting out a comprehensive strategy on sustainable finance. This increased focus on sustainable development and climate change at a global policy level influenced corporate behavior significantly, leading to a surge in climate-related disclosures. The period post-2017 also saw an increase in regulatory actions and investor demand for consistent, comparable, and accurate financial disclosures addressing climate-related risks. This demand from investors and the evolving regulatory landscape in regions like the EU and UK prompted more companies to adopt climate-related disclosures to meet these new expectations and requirements.[10]

The Task Force on Climate-Related Financial Disclosures (TCFD)
The TCFD recommendations in 2017 called for more robust disclosures of climate-related risks. The awareness of climate-related financial risks prompted efforts to integrate knowledge of these risks into financial decision-making and disclosures. The increasing sophistication of climate analytics, which helps assess the financial implications of climate risks, also played a role in enhancing the quality and quantity of disclosures (Fiedler et al., 2021).

Financial Sector Pressures
In particular, the pressures on the financial sector contributed significantly. There was a growing understanding that disclosures about climate risks are not just environmental concerns but are integral to financial stability and risk management. This understanding led to an increased push for robust climate disclosures, aligning with the broader trend of integrating environmental considerations into the core business and financial strategies (Borghei, 2021).

These factors collectively contributed to the noticeable increase in climate disclosures around 2017, reflecting a broader shift in the corporate response to climate change, influenced by global policy developments, financial sector pressures, and evolving regulatory and investor expectations

CDP is not to be confused with CDSB, the Carbon Disclosure Standards Board. Incepted in 2007, CDSB is an international consortium of business and environmental NGOs, committed to advancing and aligning the global mainstream corporate reporting model to equate natural capital with financial capital. Ever since its inception, CDSB and CDP have worked together to provide a complete, dependable, and verified system for climate disclosure. CDP has been providing its global secretariat, leading the strategy delivery, and managing the day-to-day work program on behalf of the consortium of business and environment NGOs that make up CDSB. CDP companies and investors can disclose climate change and elements of environmental information through the CDP platform, which provides the structure for data collection and the content for reporting. Through its reporting framework, CDSB provides the guidance to communicate that content in mainstream reports, which helps companies inform their investors and stakeholders, while providing regulators with a comprehensive set of information. CDP is also represented on CDSB's Board and its Technical Working Group. On January 31, 2022, CDSB was consolidated into the IFRS Foundation to support the work of the newly established International Sustainability Standards Board (ISSB).

---

[10] Investors and Regulators Turning up the Heat on Climate-Change Disclosures, Harvard Law School Forum on Corporate Governance, Oct 4, 2021, https://corpgov.law.harvard.edu/2021/10/page/8/

The Equator Principles (EP)[11]

Initiated in 2003, EP is intended to serve as a common baseline and risk management framework for financial institutions to identify, assess, and manage environmental and social risks when financing projects. Recognizing large infrastructure and industrial projects can have adverse impacts on people and on the environment; EP applies to five financial products: project finance advisory services, project finance, project-related corporate loans, bridge loans, project-related refinance, and project-related acquisition finance.

As of September 1, 2023, 140 financial institutions in thirty-nine countries are members of the EP Association. 34 of the 39 countries are *Designated Countries* that are deemed to have robust environmental and social governance, legislation systems and institutional capacity designed to protect their people and the natural environment, although the EP Association makes no independent assessment of each country's performance in these areas. As a proxy for such an assessment, the EP Association requires that a country must be both a member of the Organization for Economic Co-operation and Development (OECD) and appear on the World Bank High Income Country list[12] to qualify as a Designated Country. These data sets are reviewed quarterly by the EP Association Secretariat to ensure that any change in status is reflected in the Designated Countries list.

The principles have been revised several times to address evolving best practices and new regulatory requirements and the latest edition (EP4) came into effect for all EP financial institutions on October 1, 2020. The Equator Principles, voluntarily adopted by banks for project risk management, primarily serve banks' strategic interests with environmental benefits being a potential secondary outcome. The real impact of these principles on the environment remains unclear due to the lack of detailed disclosures and a standardized performance evaluation framework (Malve & Chen, 2010).

Principles for Responsible Investment (PRI)

In 2005, UN Secretary-General Kofi Annan initiated the development of the PRI, engaging a 20-person investor group from twelve countries, supported by seventy experts from various sectors. As of 2021, the PRI had 3,826 signatories, managing $121.3 trillion in assets.[13]

The PRI does not mandate specific disclosure requirements. Instead, it provides a framework of six principles: emphasizing the integration of ESG issues into investment analysis and decision-making, being active owners, seeking appropriate ESG disclosures from entities invested in, promoting the principles in the investment industry, collaborating to enhance effectiveness in applying the principles, and reporting on activities and progress toward implementing the principles. While signatories are encouraged to report on their implementation of these principles,

---

[11] https://equator-principles.com/about-the-equator-principles/
[12] Source: World Bank, https://data.worldbank.org/country/XD
[13] https://www.unpri.org/

the PRI focuses on promoting best practices rather than enforcing strict disclosure requirements. Signatories are expected to work toward enhancing their responsible investment strategies and report their progress, but the specifics of disclosures are guided by the individual approaches of the institutions.

The PRI has two UN partners, UN Global Compact and the UN Environment Program Finance Initiative (UNEP FI), which hold a seat each on the PRI Board and provide additional avenues for signatories to learn, collaborate, and act toward responsible investment.

ISO 26000: Guidance on Social Responsibility

The Geneva, Switzerland-based International Organization for Standardization, published ISO 26000 in 2010 following five years of negotiations between many different stakeholders across the world. Representatives from government, NGOs, industry, consumer groups, and labor organizations around the world engaged in its development, which means it represents an international consensus.

ISO 26000 is widely referenced globally by companies across various industries that need to translate principles into effective actions and share best practices relating to social responsibility.[14] While unlike other well-known ISO standards, ISO 26000:2010 provides guidance rather than requirements. The guidance helps businesses clarify what social responsibility is and how it should be operationalized, instead of what to follow to gain an ISO certification. The ISO name brand may lend confidence and trust as well.

Sustainability Accounting Standards Board (SASB)

Founded in 2011, SASB merged with the International Integrated Reporting Council (IIRC, see below) to form the Value Reporting Foundation (VRF) in June 2021. On August 1, 2022, the IFRS Foundation officially consolidated VRF,[15] exactly six months after the consolidation of CDSB by IFRS (see the CDP section above).

SASB was founded as a nonprofit organization to help businesses and investors deal with the growing complexity in the landscape of corporate sustainability disclosures through developing a common language about the financial impacts of sustainability. Sustainability issues are global business issues that affect the entity's cash flows, access to finance, and cost of capital (Rogers & Herz, 2013). For example, data security, a social issue, is important to companies in the service and software industries. Water management, an environmental issue, is essential to a food and beverage producer. Managing conflicts of interest among stakeholders, a governance issue, is critical for an investment bank. Effectively managing these issues means enhanced long-term financial performance.

For decades, however, it has been challenging for businesses and investors to connect meaningfully sustainability issues with their mandatory financial reporting. Although financial accounting standards have provided a common language

---

[14] Source: ISO website, https://www.iso.org/iso-26000-social-responsibility.html
[15] Source: https://www.sasb.org/

for companies and investors to talk about financial performance, they were developed in a world where tangible assets comprised most of the market valuation of companies and the material effects of their intangibles were only meagerly accounted for. By focusing on materiality of sustainability issues, SASB standards have become more industry specific, offering tailored guidance to businesses and investors across different sectors (Hales, 2018).

The recent merger with the IIRC indicates a move toward integrated financial and non-financial reporting. Together IIRC and SASB (VRF) offer a comprehensive suite of resources, including the Integrated Thinking Principles, Integrated Reporting Framework (IR), and SASB Standards, designed to help businesses and investors develop a shared understanding of enterprise value.

Integrated Reporting Framework (IR)

International Integrated Reporting Council (IIRC), now part of the Value Reporting Foundation after merging with SASB, released IR in 2013.

From a vague concept of a new form of "multi-capitalism," attempting to address the misallocation of resources and capital that had led to the financial crisis, came the formation of IIRC in 2010. Today, the concept of integrated financial and non-financial reporting has been embedded by over 2500 companies in more than seventy countries. Over forty stock exchanges refer to it in their guidance. The concepts within IR have been woven into the fabric of corporate governance reform in countries across the world.[16]

Task Force on Climate-Related Financial Disclosures (TCFD)

A task force formed in 2015 by the Financial Stability Board (FSB) TCFD introduced recommendations for climate-related financial disclosure to enhance market transparency and informed capital allocation in 2017.

FSB is an international body that monitors and makes recommendations about the global financial system. Its initial mandate (1999) as a forum for finance ministers and central bank governors of G7 was broadened to include the heads of state and government of G20 during the "London Summit" on April 2, 2009, shortly after the onset of the financial crisis.

The TCFD recommendations focused on four key areas: governance, strategy, risk management, and metrics and targets, comprising eleven detailed disclosures. After releasing the recommendations, the TCFD promoted their adoption, offering guidance, monitoring practices aligned with these recommendations, and preparing annual status reports until 2023.

On October 12, 2023, concurrent with the release of the 2023 Status Report, the FSB deemed that the TCFD had fulfilled its purview and was thus disbanded. The FSB has asked the IFRS Foundation to take over the monitoring of the progress of companies' climate-related disclosures.[17]

---

[16] https://www.integratedreporting.org/10-years/10-years-summary/
[17] TCFD website, https://www.fsb-tcfd.org/about/#history

Taskforce on Nature-Related Financial Disclosures (TNFD)

While still in the initial stages of development, the TNFD is gaining attention and support from financial institutions and businesses focused on integrating nature-related risks and opportunities into their decision-making processes.

The TNFD aims to provide a framework for organizations to report and act on evolving nature-related risks, aligning with the increasing global emphasis on biodiversity and ecosystem services in financial decision-making. Launched in June 2021, the final TNFD Recommendations were released in September 2023. Over 1000 organizations joined the TNFD Forum in March 2023.

The taskforce has forty non-contractual, informal members, who are senior executives from financial institutions, corporates, and market service providers with a market capitalization of over US$2.3 trillion, over US$20.6 trillion in assets under management and offices in over 180 countries. It is led by a Stewardship Council with members from sixteen organizations like Global Canopy, United Nations Development Program (UNDP), United Nations Environment Program Finance Initiative (UNEP FI), and World Wide Fund for Nature (WWF).

The 23-member TNFD Secretariat provides operational support in communication and coordination among taskforce members, collaboration with the Stewardship Council, knowledge partners, and consultation groups, and outreach to potential additional funding partners. A U.K.-based private firm hosts the Secretariat.[18] Building on current market practice for climate transition planning, the taskforce recently published a discussion paper for open comments ending early 2025 to guide corporates and financial institutions that develop and disclose a transition plan in line with TNFD recommended disclosures.

4.2   From Voluntary to Mandatory Disclosure

For over forty years, the sustainability disclosure world has been buzzing with the dynamism of ecosystems and the thrum of corporate engines. Non-financial disclosure standards like Fairtrade, UNGC, GRI, and others have emerged as the vanguards of a new era. They began as focused beacons, illuminating specific sustainability challenges, but have since blossomed into comprehensive frameworks encapsulating the entire spectrum of ESG concerns. Similarly, the initial concentration on corporate social responsibility has transitioned into a structured recognition of sustainability challenges as systemic risks, crucial to the long-term value of firms (Bartolacci et al., 2022).

This evolution, driven by a collage of individuals and organizations from titans like the Financial Stability Board, UN Secretary-General, or then Prince of Wales to spirited pioneers like Ceres, its co-founder Joan Bavaria, or the Carbon Disclosure Project (now CDP), echoes a deepened realization of the multifaceted living web of global sustainability.

---

[18] Source: TNFD website, www.tnfd.global

Historically, these standards were the voluntary banners under which companies could parade their corporate responsibility and transparency. Yet, this narrative is gradually shifting too, a shift that has been punctuated by seismic events like catastrophic oil spills, roaring forest fires, unprecedented heat waves, or destructive financial meltdowns. What was once voluntary is morphing into the mandatory, as epitomized by the European Union's Non-Financial Reporting Directive (NFRD), promulgated in 2014, and its successor, the EU Corporate Sustainability Reporting Directive (CSRD), heralding a transformative epoch in corporate disclosure, where transparency and accountability are not just applauded or expected but required by law for all large companies and almost all companies listed in regulated markets.[19]

The other aspect of this evolution is the converging standards. In 2022, IFRS Foundation consolidated the Climate Disclosure Standards Board (CDSB) and the Value Reporting Foundation (VRF). CDSB Framework formed the basis for the TCFD recommendations while VRF maintained the Integrated Reporting (IR) Framework and SASB Standards. IFRS's standards body, ISSB, assumed these frameworks and standards and encouraged their continued use. Now under the same banner of the IFRS Foundation and in parallel with its widely adopted financial accounting standards developed by the International Accounting Standards Board (IASB), IFRS Sustainability has made IFRS S1 and IFRS S2 effective on January 1, 2024. IFRS S1 requires a business entity to disclose information about its sustainability-related risks and opportunities to inform decision-makers for the same purposes as the financial reports. IFRS S2 serves the same objective but focuses on climate-related risks and opportunities.

Disclosure: Why Do Firms Do It?

Non-financial disclosure standards signify a shift in corporate reporting, extending beyond traditional financial metrics to encompass environmental, social, and governance (ESG) aspects. These standards are pivotal in the contemporary business landscape for a multitude of interactive reasons:

Transparency and Accountability: Non-financial disclosures provide transparency about a company's impact on society, the environment, and its governance practices. In turn, this transparency holds companies accountable for their actions beyond just financial performance.

Investor Decision-Making: Investors increasingly consider ESG factors as part of their investment decisions. Non-financial disclosures provide the necessary information to assess risks and opportunities related to environmental and social issues, which can significantly impact long-term profitability and sustainability.

---

[19] Source: European Commission, Corporate Sustainability Reporting, https://finance.ec.europa. eu/capital-markets-union-and-financial-markets/company-reporting-and-auditing/company-report ing/corporate-sustainability-reporting_en

Risk Management: It helps in identifying and managing risks that are not immediately apparent in financial statements, such as environmental risks (e.g., climate change impact, resource scarcity), social risks (e.g., labor practices, community impact), and governance risks (e.g., board diversity, executive compensation).

Regulatory Compliance: Increasingly, governments and regulatory bodies are mandating non-financial disclosures to ensure companies are accountable for their role in societal and environmental issues. This compliance is becoming a fundamental part of business operations.

Market Trends and Consumer Demand: As consumers become more environmentally and socially conscious, they demand more transparency and net positive impact from businesses. Non-financial disclosures help meet these demands and align business practices with consumer values.

Corporate Reputation and Trust: Companies that actively disclose non-financial information tend to build greater trust with their stakeholders, including customers, employees, and the public. This enhanced reputation can lead to a competitive advantage.

Societal Impact and Long-term Firm Value: Non-financial disclosures encourage companies to actively consider their impact on society and the environment, leading to not only more sustainable business practices, but also new strategic directions. This shift is critical in addressing global challenges like climate change, social inequality, and ethical governance. It also enables companies to be better positioned to thrive in the long term by being adaptive to a changing world where sustainability is key to business success.

Non-financial disclosure standards transcend mere compliance, illustrating a commitment to sustainable, ethical, and responsible business practices, now a hallmark of global economic value.

Building on the understanding of non-financial disclosure standards and their multifaceted importance in the modern business environment, Table 4.1 presents a typology of the strategic objectives behind these disclosures. This table categorizes these objectives based on whether they are externally oriented or internally driven and distinguishes between proactive and reactive approaches. Such a categorization aids in comprehending how different aims of non-financial disclosures are prioritized and balanced, reflecting a company's strategy and stakeholder expectations. It serves as a framework for understanding the diverse and dynamic motivations that drive organizations to go beyond traditional financial reporting, aligning their operations with the broader imperatives of sustainability, accountability, and societal impact.

The Externally Oriented, Proactive strategies focus on shaping market and societal perceptions or behaviors. They are about setting trends, contributing positively to society, and enhancing the company's public image. Their Reactive twins are about conforming to external expectations or regulations, ensuring transparency and accountability to stakeholders outside the organization.

The Internally Driven, Proactive strategies concentrate on internal strategic decisions and forward-looking actions that build confidence among investors

Table 4.1   Non-Financial Disclosure: A Typology of Strategic Objectives

	Externally Oriented	Internally Driven
Proactive	● Enhancing Brand Image Influencing Market Trends ● Driving Sustainable Practices ● Contributing to Social Welfare	● Building Investor Confidence Supporting Sustainable Growth ● Future-Proofing the Business
Reactive	● Meeting Legal Requirements ● Standardizing Reporting Practices ● Enhancing Corporate Transparency ● Promoting Accountability	● Informing Investment Decisions ● Risk Identification and Management ● Building Stakeholder Trust

and ensure the long-term sustainability of the business. Their Reactive counterparts involve meeting internal compliance needs like risk management, investor reporting, and building trust within the organization.

Table 4.1 helps in understanding the multifaceted nature of non-financial disclosure objectives, highlighting how different aims can be prioritized and balanced based on a company's strategy and stakeholder expectations (Vallone, 2022).

Differential Adoptions

The adoption of non-financial disclosure standards varies significantly across industries, largely due to the unique environmental, social, and governance (ESG) challenges each sector faces. Industries like oil and gas, mining, and heavy manufacturing often experience intense scrutiny over their environmental impact. These sectors tend to focus on environmental disclosures, such as emissions data, waste management, and energy consumption. Occupational health and safety also become a key focus due to the high-risk nature of their operations.

In contrast, the financial services sector emphasizes governance and risk management in its non-financial disclosures, reflecting the importance of trust and stability. Issues like ethical lending, investment in sustainable projects, and transparency in financial practices are commonly highlighted. The technology and telecommunications sectors prioritize data privacy, cybersecurity, and innovation. Given the rapid pace of technological change, disclosures related to research and development, as well as the ethical implications of emerging technologies, are also crucial.

Consumer goods and retail companies frequently focus on supply chain management, particularly labor practices and sustainable sourcing. They also emphasize product responsibility, including safety and the environmental impact of their products. The healthcare and pharmaceuticals sectors typically disclose information about patient safety, ethical marketing, and equitable access to medicines. Environmental impact, particularly concerning pharmaceuticals and medical devices, and responsible clinical trials are also important areas of focus.

In the utilities and energy sectors, the transition to renewable energy and energy efficiency measures are becoming central to sustainability reporting. Similarly, the real estate and construction industries are increasingly disclosing green building practices, energy efficiency in buildings, and the impact of their operations on local communities.

These variations in non-financial disclosures reflect the unique ESG issues most material to each industry. The materiality of these issues determines the type of information disclosed and the standards adopted, ensuring relevance to stakeholders, including investors, customers, and regulators.

To define their distinct ESG focus areas, industries must first understand the specific risks and opportunities they face. Environmental concerns such as resource scarcity, emissions, and waste vary widely across sectors. Similarly, social risks—ranging from labor conditions to community impacts—differ in significance depending on the industry. Governance considerations also take different forms, whether it is financial transparency in the financial sector or supply chain ethics in the apparel industry. By identifying these sector-specific risks, industries can prioritize the ESG factors most critical to their long-term sustainability.

Beyond risk identification, aligning ESG strategies with relevant standards and frameworks is essential. Global guidelines such as the Global Reporting Initiative (GRI), the Sustainability Accounting Standards Board (SASB), and the Task Force on Climate-related Financial Disclosures (TCFD) help industries structure their reporting. Industry coalitions—such as the Responsible Business Alliance for electronics or the Climate Disclosure Standards Board for climate-related issues—provide further guidance on the most material ESG issues. By consulting these frameworks, businesses can focus their reporting on the metrics that matter most to their sector.

Understanding stakeholder expectations is also crucial in determining ESG priorities. Different stakeholders, including investors, consumers, and employees, emphasize varying ESG issues depending on the industry. For example, investors in the energy sector may prioritize climate change mitigation strategies, while retail consumers may focus on fair labor practices and supply chain sustainability. Engaging with stakeholders and understanding their needs enable industries to refine their ESG priorities and address the concerns that matter most to their audience.

Moreover, industries must consider the evolving regulatory landscape for ESG reporting. Many sectors are subject to specific environmental regulations that shape how they report and manage ESG risks. For instance, financial services must comply with governance and transparency regulations, while industries like energy and manufacturing face mandates for carbon emissions reductions or environmental reporting. By aligning their ESG reporting with these regulations, companies can ensure compliance and demonstrate leadership in addressing critical sustainability issues.

Ultimately, industry-specific ESG reporting is designed to drive long-term value creation. Companies that understand and address their unique ESG challenges are better positioned to manage risks, tap into emerging markets, and build stronger stakeholder relationships. ESG reporting should focus not only on compliance but also on fostering sustainable growth through responsible and ethical practices. By adopting tailored ESG metrics, companies can align their operations with global sustainability goals while ensuring competitiveness and resilience in an ever-evolving market.

Below is a structured approach for industries to create ESG reporting frameworks that reflect their unique challenges and opportunities, while contributing meaningfully to global sustainability:

- **Identify relevant environmental, social, and governance risks** specific to the industry.
- **Consult industry standards and frameworks** to define focus areas.
- **Align ESG priorities with stakeholder interests** and expectations.
- **Ensure compliance with relevant regulatory frameworks** impacting ESG disclosures.
- **Focus on long-term value creation**, integrating ESG into broader corporate strategy.
- **Develop industry-specific key performance indicators (KPIs)** to measure relevant ESG metrics.
- **Communicate transparently and regularly** about ESG priorities and performance.

4.3 The Evolving Disclosure Environment

Market Indices and Rankings Based on Non-Financial Disclosures

An important part of the evolving non-financial disclosure ecosystem is the emergence of stock market indices and company rankings that are developed referencing the often-imperfect information. Below is a list of the major sustainability-related indices.

- *Fortune 100 Best Companies to Work For,*[20] Fortune Media IP Ltd, New York, New York, United States, launched in 1993, and referenced by employees, customers, competitors, and investors globally to benchmark firm competitiveness.
- *Dow Jones Sustainability Indices (DJSI),*[21] by S&P Dow Jones Indices, New York, New York, United States, launched in 1999, and used by investors globally to benchmark sustainability performance.

---

[20] https://fortune.com/ranking/best-companies/
[21] https://www.spglobal.com

- *FTSE4Good Index Series*,[22] by Financial Times Stock Exchange (FTSE), Russell, London, United Kingdom, created in 2001, and utilized by a wide range of investors for creating and assessing responsible investment funds.
- *Corporate Knights Global 100 Most Sustainable Corporations*,[23] by Corporate Knights Inc., Toronto, Canada, first published in 2005, and influencing corporations worldwide in sustainability practices.
- *MSCI ESG Leaders Indexes*,[24] MSCI Inc., New York, United States, introduced in 2007, and used globally for benchmarking ESG performance and developing investment products.
- *Bloomberg Gender-Equality Index (GEI)*,[25] Bloomberg L.P., New York, United States, launched in 2016, and used by investors to gauge company performance and commitment to gender equality.

Each of these rankings and indices plays a crucial role in shaping corporate behavior and investment strategies toward sustainability, social responsibility, and ethical practices. The influence of these rankings and indices varies, reflecting their relevance and impact across different industries, sectors, and geographical regions.

Corporate Responsibility and Accountability

Non-financial disclosures, such as those guided by the Global Reporting Initiative (GRI), compel companies to report on a wide range of sustainability issues. For instance, Unilever, through its comprehensive sustainability reporting, has shown improvements in various areas like reducing environmental footprints and enhancing social welfare programs. This kind of reporting promotes greater corporate responsibility and can lead to substantial societal benefits in terms of environmental protection and social well-being.

Studies and Reports. Numerous studies have analyzed the impact of sustainability reporting on corporate behavior. For instance, research published in journals like "Corporate Social Responsibility and Environmental Management" often shows a positive correlation between sustainability reporting and improved environmental and social performance in companies.

Aggregated Data Analysis. Meta-analyses of corporate sustainability reports can reveal trends in corporate behavior changes post-adoption of reporting standards.

---

[22] https://www.ftserussell.com/
[23] https://www.corporateknights.com/
[24] https://www.msci.com/
[25] https://www.bloomberg.com/gei

Changing Investor Behavior

The integration of ESG factors into investment decisions has become increasingly prevalent, as seen in the growth of sustainable investment funds. BlackRock, one of the world's largest asset managers, has significantly increased its ESG-focused fund offerings, influencing companies to improve their sustainability practices to attract and retain investment. This shift in investment strategies can lead to widespread adoption of sustainable practices across industries, contributing to societal sustainability goals.

Research by financial institutions and market analysts, such as reports by Morningstar or the Global Sustainable Investment Alliance, provides aggregated data on the growth of ESG investing and its influence on corporate behavior.

Investor Surveys. Surveys of institutional investors often reveal the growing importance of ESG factors in investment decision-making.

Consumer Choices

With rising consumer awareness, companies are increasingly being judged by their sustainability practices. Patagonia, an outdoor clothing company, has leveraged its commitment to sustainability as a key selling point, influencing consumer choices and encouraging other companies to adopt similar practices. This consumer-driven demand for sustainable products and services contributes to a shift toward more sustainable societal consumption patterns.

Market Research. Studies by market research firms can provide insights into consumer behavior changes in response to corporate sustainability practices.

Consumer Surveys. Surveys and polls conducted by organizations like Nielsen have demonstrated a growing consumer preference for sustainable brands.

Policy and Regulatory Frameworks

The insights gained from non-financial disclosures can inform public policy and regulatory frameworks. The European Union's Non-Financial Reporting Directive (NFRD) and its successor CSRD are examples where regulatory frameworks have been influenced by the growing emphasis on sustainability reporting. This directive has prompted companies operating in the EU to disclose information on how they manage social and environmental challenges, leading to broader societal impacts in terms of enhanced corporate sustainability.

Policy Analysis. Studies by legal and policy researchers often assess the impact of non-financial reporting requirements on corporate practices and their broader societal implications.

Comparative Regulatory Studies. Comparing different regulatory environments and their impact on corporate sustainability practices can provide substantial evidence of the influence of non-financial disclosures on policy.

Transparency and Dialogue

Non-financial disclosures can increase public awareness and dialogue about key sustainability issues (Ioannou & Serafeim, 2017). For example, disclosures related to carbon emissions under frameworks like the Task Force on Climate-related Financial Disclosures (TCFD) have heightened public and corporate awareness of climate change risks and mitigation strategies, contributing to a broader societal dialogue on climate action.

Public Awareness Studies. Research assessing the impact of corporate disclosures on public awareness and societal dialogue around sustainability issues.

Media Analysis. Studies analyzing media coverage and public discourse following major corporate sustainability disclosures can provide evidence of increased public engagement with these issues.

While these examples illustrate the role of non-financial disclosure standards in facilitating sustainability transitions, the evidence can be indirect and varies between regions and industries. The effectiveness of these standards in driving tangible change is influenced by factors such as the rigor of the standards, enforcement mechanisms, and corporate governance culture.

It is important to observe and explain varied impacts of non-financial disclosures across industries due to diverse material issues, competition, and industry-specific standards. For example, in the energy sector, environmental disclosures, particularly around emissions and renewable energy, are crucial, as seen in BP's sustainability reports. In contrast, the financial sector, like JPMorgan Chase, focuses more on governance and ethical practices. In industries like consumer goods, companies like Unilever use robust non-financial disclosures as a competitive edge, emphasizing their commitment to sustainability. The TCFD framework is widely adopted in industries with significant climate-related risks.

Variations across geographies are also an important consideration when examining the impacts. The European Union, with its strong regulatory framework for sustainability reporting, contrasts with regions like Southeast Asia, where such regulations are less stringent. In Scandinavian countries, social disclosures, especially related to workforce practices and community engagement, are often emphasized due to strong societal emphasis on social equality. Developed countries typically have more advanced non-financial disclosure practices than developing countries, partly due to more robust regulatory frameworks and higher stakeholder expectations. Also, markets with a high concentration of ESG-focused investors, such as certain European countries, see companies engaging in more detailed ESG reporting.

These real-world examples underscore the importance of non-financial disclosures in promoting sustainability at the societal level, contributing to a more sustainable and responsible global economy. While they provide a more robust and comprehensive view of the impact of non-financial disclosure standards, it is important to acknowledge that measuring such impacts can be complex and is often influenced by various external factors. Consequently, the evidence may not always be straightforward or uniform across different contexts. Additionally, the rapidly evolving nature of both non-financial disclosure practices and sustainability challenges means that ongoing research and updated analyses are crucial for a current understanding of these dynamics.

Impact and Impact Management

The adoption of non-financial disclosure standards has multifaceted impacts, varying across industries and geographies, reflecting the diverse nature of stakeholder expectations, regulatory environments, and material ESG (environmental, social, and governance) issues pertinent to each context. The other side of the non-financial disclosures is that they have played a significant role in advancing sustainability transitions at the societal level.

The substantiation of the impact of non-financial disclosure standards on sustainability transitions at the societal level primarily comes from a combination of academic research, industry analyses, and case studies. While individual examples provide illustrative insights, more comprehensive evidence is available through systematic studies and aggregated data analyses. Below is an overview of the types of evidence supporting each claim.

4.4 Challenges to Non-Financial Disclosures

The propagation and adoption of non-financial disclosure standards have faced several critiques, both in general terms and in specific industries or geographies. These critiques often center around the effectiveness, relevance, and implementation of these standards.

Top of the list is lack of standardization and comparability. A significant criticism is the lack of standardization across various non-financial disclosure frameworks, leading to difficulties in comparing data across companies. For example, most companies use general standards like CSDB, but methodologies can differ. Some report emissions based on operational control; others use equity shares. Additionally, how companies account for indirect emissions from purchased electricity or supply chains varies, leading to discrepancies in reported emissions. In the social sphere, companies face challenges in standardizing human rights impact assessments. Different firms might adopt varied approaches to measuring and reporting on labor practices, community impacts, or supply chain ethics. These diverse methodologies make it difficult for stakeholders to compare social performance across companies (Aluchna et al., 2023). In the financial sector, the absence of standardized ESG metrics can lead to inconsistent reporting, affecting the ability of investors to accurately assess the ESG performance of banks and financial institutions.

There is skepticism about whether the reported non-financial data is material and relevant to stakeholders' needs. Critics argue that some companies may engage in "selective reporting," focusing on positive sustainability aspects while neglecting more material issues. In emerging economies, where regulatory frameworks for non-financial disclosure are less developed, the materiality and relevance of reported data can be particularly questionable, with reports often seen as more of a marketing tool than a reflection of actual sustainability performance.

What is even worse, thus a major concern, is greenwashing, where companies exaggerate or falsely represent their sustainability practices. Non-financial disclosures that are not rigorously audited or verified can contribute to greenwashing, undermining the credibility of sustainability reporting. The energy sector, particularly fossil fuel companies, has faced accusations of greenwashing through their sustainability reports. This issue is highlighted by studies, which found a significant disconnect between the climate change rhetoric of major oil companies and their actual behavior. The research revealed that while these companies have increased their discussion of climate issues in their annual reports, there has been little action toward reducing fossil fuel production or shifting to clean energy (Boelders, 2020; Szadziewska & Kujawski, 2022). In fact, fossil fuel production remained relatively constant, and most oil majors increased their production from 2015 until 2020.

Furthermore, an investigation by the House Oversight & Reform Committee revealed that Big Oil companies have been engaged in greenwashing, claiming to embrace clean energy while continuing to invest heavily in fossil fuels. The internal communications of executives from major oil companies like BP, Chevron, Exxon, and Shell showed a resistance to limiting fossil fuel energy production, contradicting their public statements on promoting emissions cuts and clean energy.[26] These findings provide solid evidence for the greenwashing accusations against the fossil fuel industry, showing a stark contrast between their public commitments to sustainability and their actual investment and production activities. This contradiction is emblematic of the challenges in holding companies accountable for their contributions to climate change and the importance of transparent and honest environmental reporting.

For well-intended companies, the implementation of non-financial disclosure standards can still be challenging, especially for smaller companies with limited resources. The cost and complexity of reporting according to these standards can be prohibitive, leading to inconsistent quality and depth of reporting. In developing countries, where resources and expertise in ESG reporting are often limited, companies may struggle to implement non-financial disclosure standards effectively, resulting in reports that lack depth and rigor.

In jurisdictions where non-financial reporting is voluntary or where enforcement is weak, the impact of these disclosures on corporate behavior and sustainability outcomes can be limited. In industries like textiles and apparel, where supply chain

---

[26] Source: U.S. House of Representatives press release, December 9, 2022. https://oversightdem ocrats.house.gov/news/press-releases/oversight-committee-releases-new-documents-showing-big-oil-s-greenwashing

practices are critical, the lack of stringent regulations and enforcement mechanisms in some countries can lead to non-financial disclosures that do not fully capture the complexities and challenges of supply chain management.

A birth defect, critics argue that non-financial and financial reporting are often treated as separate entities, with limited integration between the two. This separation can undermine the holistic understanding of a company's overall performance and impact. In sectors like real estate, for example, where environmental sustainability is increasingly important, the lack of integration between financial and sustainability reporting can obscure the full impact of environmental factors on financial performance.

The challenges in implementing non-financial disclosure standards, such as lack of standardization, materiality concerns, greenwashing hazards, and limited resources, particularly for smaller companies and those in developing countries, can exacerbate each other, making effective implementation exceedingly difficult. For instance, the absence of standardized metrics not only leads to incomparable data but also fuels greenwashing, as companies can selectively report positive aspects while ignoring crucial issues. This situation is further aggravated in environments with limited regulatory frameworks and resources, where the complexity of reporting can lead to superficial disclosures lacking in depth and rigor. In such a setting, the gap between well-intentioned reporting and meaningful sustainability performance widens, as companies struggle to navigate the confusing web of non-financial disclosures amid resource constraints and varying regional standards. The result is a mosaic of reports with varied quality and depth, undermining the overall goal of transparent and accountable sustainability reporting.

4.5 Standards Proliferation: Are They Converging or Diverging

The proliferation of non-financial disclosure standards and accompanying indices is a response to the growing awareness and need for corporate transparency in environmental, social, and governance (ESG) aspects. This trend reflects the diverse interests and needs of various stakeholders, including investors, regulators, consumers, and the broader community.

Different stakeholders demand specific information. For instance, investors focus on risks and opportunities that impact financial performance, while regulators may be concerned with broader societal and environmental impacts. Different regions and industries face unique challenges and priorities, leading to the development of standards and indices that cater to these specificities. As understanding of ESG issues evolves, new standards emerge to address these complexities, like climate change (TCFD) or biodiversity (TNFD). Companies and financial institutions use various ESG standards or benchmarking different rankings to differentiate themselves in the market, showcasing their commitment to sustainability and social responsibility.

Nevertheless, current trends suggest a move toward convergence among non-financial disclosure standards. There have been several announcements indicating a convergence in sustainability reporting standards, financial and non-financial reporting standards, and market and regulatory drivers. A flurry of consolidation moves involving SASB, IIRC, CDP, CDSB, and IFRS during the past two years is a clarion illustration of the growing momentum for the convergence of non-financial frameworks, standards, and standards bodies, potentially leading to their inclusion in financial statements.

Academics and practitioners are advocating for the convergence and harmonization of non-financial disclosures, with evidence of de facto harmonization in the application of major standards like GRI, IR, and SASB. The development of high-quality sustainability reporting standards, based on principles like legitimacy, independence, and transparency, is fostering consistent, comprehensive, and comparable information.

Moreover, regulatory pushes are likely to amplify and accelerate standards convergence. The European Union is leading the pack, with the Corporate Sustainability Reporting Directive (CSRD) to enhance the scope and detail of non-financial disclosures by companies operating within the EU. The UK announced mandatory TCFD-aligned disclosures for large companies and financial institutions by 2025.

In the divisive political environment, the US Securities and Exchange Commission (SEC) has proposed rules to enhance and standardize climate-related disclosures for investors, focusing on the materiality of climate-related risks and impacts. There have been significant movements in Asia, including China, toward mandatory non-financial disclosure, particularly focusing on environmental, social, and governance (ESG) aspects.

China has been making strides in mandating ESG disclosures. Since January 2022, the Shanghai Stock Exchange and the Shenzhen Stock Exchange listing rules have required issuers to publish a CSR or ESG report and disclose the standard they use to make that disclosure and make timely disclosures around significant environmental and social incidents. Additionally, China's Five-Year Plans have increasingly emphasized sustainability and ESG issues, signaling a stronger regulatory push for ESG disclosures in the corporate sector. The Hong Kong Stock Exchange (HKEX) has been advancing its ESG reporting requirements and mandatory ESG disclosures are expected to come into effect from 2024.[27]

In March 2023, new rules designed by Japan's Financial Services Agency (FSA) were made effective, representing Japan's first stage of mandatory sustainability disclosure rules. These rules mandated the creation of a new section for sustainability-related information in the annual securities report. Under the rules, all listed companies in Japan (approximately 4000 including foreign companies

---

[27] Gaia Property Management website, https://gaiapm.com.hk/get-ready-for-mandatory-esg-rep
orting-in-hong-kong/#:~:text=Mandatory%20ESG%20disclosures%20are%20expected,from%
202024%20in%20Hong%20Kong

listed in Tokyo Stock Exchange) are required to disclose sustainability-related information using the TCFD pillars, i.e., Strategy, Metrics and Targets, Governance and Risk Management. The Singapore Exchange (SGX) has mandated since 2022 that all issuers include climate-related reporting in their sustainability reports, which is based on the recommendations of the Task Force on Climate-related Financial Disclosures (TCFD). Initially, companies refusing to disclose their social and environmental impact and performance are required to give an explanation, an enforcement mechanism known as "comply or explain." But for the financial industry all issues are required to comply in 2023, same for the agriculture, food, forest products industry, and energy industry. The materials and buildings industry and the transportation industry will be required to disclose in 2024.

Emerging technologies like AI, blockchain, and big data analytics are poised to revolutionize non-financial reporting. AI can aid in analyzing large datasets to identify trends and insights, blockchain can offer enhanced transparency and traceability in supply chains, and big data can enable real-time monitoring and reporting. These technologies could lead to more accurate, transparent, and timely disclosures and enable companies to enhance compliance and transparency in their public financial and sustainability disclosures.

Nevertheless, the convergence of non-financial disclosure standards, particularly in ESG reporting, is confronted with several challenges, as highlighted in recent research and industry insights. There is an ongoing debate about the challenges in establishing a unified set of ESG disclosure standards, encompassing both conceptual and practical aspects (Bose, 2020; Pizzi & Nuccio, 2023; Stolowy & Paugam, 2023).

The ESG agenda involves a wide range of stakeholders with varying interests, including regulators, businesses, investors, and the public. This diversity can complicate efforts to achieve a consensus on standards, resulting in the still rapidly evolving field of ESG reporting, with initiatives like the International Sustainability Standards Board (ISSB) and the European Financial Reporting Advisory Group (EFRAG) actively developing new standards. This dynamic environment can make it difficult to establish and maintain alignment.

Achieving comparability of ESG disclosures among different companies and aligning standards globally are critical yet challenging. Different regulatory requirements in regions like the United States and EU add complexity, particularly for companies with a global presence. These companies face the challenge of reporting under multiple standards, which can be complex and resource intensive. This is particularly difficult for businesses with operations in multiple jurisdictions or those with dual listings.

These challenges suggest that while there is a strong movement toward convergence in non-financial disclosure standards, achieving this goal requires navigating a complex landscape of stakeholder interests, regulatory environments, and practical reporting considerations.

The convergence of non-financial disclosure standards will simplify the reporting process for companies, provide more clarity for investors and other stakeholders, and enhance the overall quality and comparability of ESG data. However, this convergence will need to balance the diverse needs of various stakeholders and the specificities of different sectors and regions. While the proliferation of non-financial disclosure standards has been a response to the varied and evolving demands of the ESG landscape, the future is leaning toward a more harmonized and integrated approach to non-financial disclosure, facilitating better comparability and decision-making based on ESG factors (LaTorre et al., 2018; Zaid & Issa, 2023).

4.6   Outlook and Management Strategy

The intensifying environmental and social crises are catalyzing a paradigm shift in public and regulatory expectations, compelling a move toward more robust, action-oriented sustainability reporting. This shift is being propelled by technological innovations that enhance the precision and comprehensiveness of reporting, and a merging of financial and non-financial metrics in business accountability. In this rapidly evolving landscape, companies are compelled to adapt their strategies not just for compliance, but to effectively navigate the risks and opportunities presented by a new era of corporate sustainability and transparency.

Non-Financial Disclosure Future Outlook

In this section, we offer several future scenarios regarding non-financial disclosures, a dialectic perspective on the effect of advanced technology on the NFD future, and advice on monitoring the merging accounting standards, followed by recommendations for firms to manage the dynamic landscape. First, here are two contrasting scenarios for a 10-year future and two more for a 25-year future.

10-Year Future
Scenario 1: *Accelerated Adoption and Integration.* In this scenario, non-financial disclosure standards become more integrated with financial reporting, driven by strong regulatory mandates, and increasing investor demand for ESG data. Technologies like AI and blockchain improve data accuracy and transparency. By 2033, most companies globally are using a unified reporting standard that seamlessly integrates financial and non-financial data. Investors routinely use ESG data, derived from these reports, to make investment decisions, and sustainability performance is closely tied to a company's market value.

   Scenario 2: *Fragmented and Inconsistent Progress.* Here, the progress in non-financial disclosures is uneven, with significant disparities across regions and industries. Lack of standardization persists, and sustainability reporting is viewed skeptically due to greenwashing concerns. By 2033, some industries and regions have advanced in non-financial reporting, while others lag, leading to a patchwork of practices. Investors and consumers find it challenging to decipher and trust the sustainability reports due to inconsistent standards and lack of verification.

25-Year Future

Scenario 1: *Holistic and Dynamic Reporting Ecosystem*. Non-financial reporting standards have evolved to become dynamic and real time, with continuous data monitoring and reporting. Sustainability is deeply embedded in corporate strategies, and reporting standards adapt rapidly to emerging global challenges. By 2048, companies are using real-time dashboards for sustainability reporting, providing stakeholders with ongoing insights into their performance. Reporting standards have evolved to address new sustainability challenges, such as those related to climate change adaptation and circular economy practices.

Scenario 2: *Stagnation and Compliance-Centric Approach*. In this scenario, non-financial reporting standards stagnate, focusing more on compliance than on driving real change. Reporting is seen as a bureaucratic exercise, with minimal innovation or adaptation to new sustainability challenges. By 2048, non-financial reporting remains a box-ticking exercise for many companies, with little integration into strategic decision-making. Reports are produced annually, focusing on compliance with outdated standards that do not reflect contemporary sustainability challenges.

In each of these scenarios, the role of technology, regulatory environments, corporate culture, and stakeholder engagement is crucial in shaping the future of non-financial disclosure standards. These scenarios illustrate the potential divergent paths that the evolution of non-financial reporting could take, highlighting the importance of proactive and strategic approaches to ensure that these standards effectively contribute to global sustainability goals.

Technology as a Double-Edged Sword

Artificial intelligence (AI) and machine learning can analyze vast amounts of sustainability data to identify trends, risks, and opportunities, providing deeper insights into a company's ESG performance. A manufacturing company uses AI to analyze data from its global supply chain, identifying key areas where its environmental impact could be reduced. This leads to more informed sustainability reporting and targeted improvement strategies.

Blockchain technology can be used to create transparent and immutable records of a company's supply chain activities or its environmental impact data. To illustrate, a fashion retailer employs blockchain to track the sourcing of materials, ensuring ethical and sustainable practices. This information is then incorporated into its sustainability reports, offering stakeholders verifiable proof of the company's claims.

Big data analytics for real-time reporting can be leveraged to allow companies to monitor and report their sustainability performance in real time, rather than in annual reports. For example, a multinational corporation uses big data tools to continuously monitor and report its carbon emissions and energy usage across all operations, providing stakeholders with up-to-date information.

However, over-reliance on complex technologies could make sustainability reporting inaccessible or difficult to understand for some stakeholders, leading to a disconnect between the reported data and its practical interpretation. Imagine a small investor struggling to interpret a company's sustainability report filled with complex data analyses and technical jargon derived from advanced AI algorithms, making it difficult to assess the company's actual sustainability performance.

The use of advanced technologies for data collection and analysis could raise concerns about data privacy and security, especially when overseeing sensitive information. A company using IoT devices to collect environmental data faces a data breach, leading to stakeholders' mistrust in the company's ability to securely manage sensitive information.

Disparity in access to advanced technologies between large corporations and smaller enterprises could widen the gap in the quality and comprehensiveness of sustainability reporting. Smaller companies in developing countries may lack the resources to employ advanced technologies for sustainability reporting, leading to less detailed and frequent reports compared to larger, tech-savvy corporations.

While technology advancements have the potential to enhance the quality, accuracy, and timeliness of non-financial disclosures, they also present challenges that need to be carefully managed to ensure these advancements contribute positively to the evolution of sustainability reporting.

Navigating the Evolving Landscape of Sustainability and Financial Reporting Standard

The integration of sustainability into financial reporting has become a crucial focus for both businesses and regulators. As the global regulatory environment evolves, it is essential for companies to stay informed about new guidelines issued by bodies such as the European Union, the US Securities and Exchange Commission (SEC), and other regional authorities. These regulators play a key role in shaping the frameworks for environmental, social, and governance (ESG) reporting. Standards set by organizations like the IFRS Foundation, which oversees the International Financial Reporting Standards (IFRS), and the International Sustainability Standards Board (ISSB) are expected to have a significant impact on global ESG disclosure practices.

ESG reporting requirements can vary widely across industries, making it critical for companies to understand the specific standards relevant to their sector. As regulatory expectations evolve, businesses must stay current to remain compliant and strategically position themselves for success in an increasingly complex environment.

Technological advancements, particularly in data analytics and reporting tools, are set to play a key role in enabling companies to meet the growing demands for detailed and transparent ESG disclosures. It is crucial for organizations to explore and leverage emerging technologies that facilitate efficient and standardized reporting. For example, the development of IFRS digital taxonomies provides a standardized, computer-readable format for financial data, helping to streamline

ESG reporting in line with IFRS standards.[28] Companies should actively engage with such innovations, testing them early and often, to ensure competitiveness and compliance with evolving reporting requirements.

Engagement in global dialogues around ESG reporting standards and best practices is also vital. Discussions held by international bodies such as the World Economic Forum (WEF) and the International Organization of Securities Commissions (IOSCO) play a pivotal role in shaping the regulatory framework for ESG issues. By staying informed and participating in these conversations, companies can ensure alignment with global efforts toward standardized and transparent ESG disclosures.

Lastly, businesses must remain agile in responding to the evolving expectations of stakeholders, including investors, consumers, and employees, who are increasingly demanding higher levels of ESG transparency and accountability. The incorporation of ESG disclosures into international accounting standards marks a significant shift in corporate reporting and investment decision-making. For companies, staying attuned to these changes not only ensures regulatory compliance but also presents opportunities for sustainable growth, bolstering long-term value and trust with stakeholders.

As corporate reporting continues to evolve, organizations must prioritize ongoing education, technological adoption, and active participation in global discussions. By doing so, they can stay ahead of the curve and position themselves as leaders in responsible business practices.

Non-Financial Disclosure Management Strategies

Managing and overcoming challenges associated with ESG factors in murky or fast changing regulatory environments is a complex task that requires a strategic and multifaceted approach. Here are some well-researched perspectives and strategies based on insights from various sources.

Aligning Stakeholders and Clarifying ESG Objectives
Financial services firms, like banks, should align stakeholders on key ESG issues like decarbonization and transition finance priorities. This alignment is crucial for setting clear ESG objectives and strategies (Bain & Company, 2023). Companies need to develop long-term strategies that respond to ESG imperatives. This involves balancing defensive and offensive approaches to ESG issues, such as managing downside risks and capturing upside opportunities.

---

[28] For updated IFRS Sustainability Disclosure Taxonomy 2024, visit https://www.ifrs.org/issued-standards/ifrs-sustainability-standards-navigator/

Establishing Robust Internal Controls for ESG Reporting

Firms should strengthen internal controls for ESG reporting, like controls used for financial reporting. This includes addressing pitfalls like inadequate risk assessment, disconnect between ESG targets and business strategies, and insufficient commitment from leadership (KPMG, 2022). And, conducting a materiality assessment to determine ESG focus areas based on regulatory requirements and stakeholder priorities is essential. This helps in establishing an effective control environment for ESG reporting and compliance.

Preparing for Regulatory Changes and Emerging Risks

Businesses must stay informed and adapt to changing ESG regulations globally. This includes understanding and preparing for new climate regulations and integrating ESG into strategic business planning (AON Insights, 2022). Companies should consider ESG risks as part of their enterprise risk management strategy. This includes assessing climate-related risks and opportunities, internal carbon pricing, and capital deployment toward climate-related risks (Deloitte, 2023).

Decision-Making and Trade-Offs in ESG

Firms should identify their "high jumps" (basic ESG standards to meet) and "long jumps" (areas where they can take a leadership role in ESG). This involves leveraging a company's unique strengths and addressing societal demands such as living wages, net-zero emissions, and diversity (McKinsey, 2023). Forward-looking companies should think systematically about ESG trade-offs, balancing the benefits and costs, including the costs of inaction. This involves clear decision-making about investing in various ESG initiatives and their potential impact on different stakeholders.

Measuring and Assessing ESG Performance

Companies need to approach ESG performance management effectively, focusing on meaningful KPIs and objectives that tie directly to their business model. Regular assessment with robust data analytics is crucial for maintaining agility in rapidly changing circumstances.

Integrating ESG into Operations

This easy decision could prove to be the hardest in most modern organizations. ESG should be operationalized throughout the organization. Companies should act purposefully to align ESG initiatives with their operations, ensuring that these initiatives are not just a "tick box" exercise but contribute to improving the business and creating positive impacts.

Managing ESG in uncertain regulatory environments requires a balanced approach that involves aligning stakeholders, strengthening internal controls, adapting to regulatory changes, making informed trade-offs, and systematically integrating ESG into business operations. Firms should focus on both managing risks and capturing opportunities presented by ESG, while ensuring transparency and accountability in their ESG reporting and strategies.

Conclusion

The development of non-financial disclosures has marked a significant evolution in how companies report on sustainability and broader societal impacts. This chapter examined the frameworks that underpin these disclosures, highlighting their role in helping firms align with growing stakeholder expectations and regulatory requirements. While challenges remain, particularly around standardization and comparability, the progress made underscores the importance of integrating these practices into corporate reporting.

This discussion transitions seamlessly into Chapter 5, which focuses specifically on greenhouse gas (GHG) emissions reporting at the firm level. Building on the foundations laid here, Chapter 5 delves into the Greenhouse Gas Protocol and its application across Scopes 1, 2, and 3. It explores the complexities and sector-specific nuances of emissions measurement, offering a closer look at how companies are navigating these challenges in practice.

Together, these chapters trace the trajectory from broad sustainability frameworks to the detailed, actionable disclosures necessary for understanding and managing climate-related risks and opportunities. This progression highlights the role of corporate reporting in bridging high-level commitments with measurable actions, setting the stage for the detailed examination of firm-level emissions in the following chapter.

Key Takeaways

1. **Rising Importance of Non-Financial Disclosures (NFDs)**
   NFDs have evolved from voluntary initiatives to essential components of corporate reporting, reflecting growing stakeholder demands for transparency on environmental, social, and governance (ESG) issues.
2. **Standardization and Fragmentation**
   Frameworks such as the Task Force on Climate-related Financial Disclosures (TCFD) and the Global Reporting Initiative (GRI) have driven standardization in sustainability reporting, though the coexistence of multiple frameworks continues to create challenges for consistency and comparability.
3. **Regulatory and Market Drivers**
   Increasing regulatory requirements, such as the EU's Corporate Sustainability Reporting Directive (CSRD), and growing investor emphasis on ESG factors are compelling firms to enhance the scope and quality of their disclosures.
4. **Challenges in NFD Implementation**
   Firms face hurdles such as data collection, verification costs, and adapting to evolving reporting standards. Addressing greenwashing and ensuring the reliability of disclosed information remain critical concerns.
5. **Role of Technology**

Technological advancements, including artificial intelligence and data analytics, are improving the collection, analysis, and dissemination of non-financial data, enabling more actionable insights.

6. **Strategic Benefits of NFDs**

   Beyond compliance, NFDs can uncover operational efficiencies, identify risks and opportunities, and strengthen stakeholder trust by demonstrating a firm's commitment to long-term sustainability.

7. **Link to Firm-Level Disclosures**

   This chapter lays the groundwork for Chapter 5 by connecting the broad evolution of NFDs with the specific challenges of greenhouse gas emissions reporting, particularly in high-impact sectors.

8. **Broader Impact on Corporate Finance**

   The integration of non-financial metrics into corporate reporting reflects their increasing relevance to financial performance, risk management, and investment decisions in a changing global economy.

Questions

1. Why have non-financial disclosures (NFDs) become increasingly important in corporate reporting?
2. What are some key frameworks for non-financial disclosures, and how do they contribute to standardization?
3. Discuss the challenges companies face in implementing effective NFD practices.
4. How have technological advancements supported the evolution of non-financial disclosures?
5. What are the strategic benefits of integrating NFDs into corporate reporting?
6. Explain the relationship between non-financial disclosures and regulatory developments such as the CSRD.
7. How does Chapter 4 transition into the focus on firm-level greenhouse gas emissions reporting in Chapter 5?

References

AON. (2022). *2022 Weather, Climate & Catastrophe Insight: 2022 Annual Report*. AON. Retrieved from https://www.aon.com

Aluchna, M., Roszkowska-Menkes, M., & Kamiński, B. (2023). From talk to action: The effects of the non-financial reporting directive on ESG performance. *Meditari Accountancy Research, 31*(7), 1–25. https://doi.org/10.1108/MEDAR-12-2021-1530

Andorfer, V. A., & Liebe, U. (2012). Research on fair trade consumption: A review. *Journal of Business Ethics, 106*, 415–435. https://doi.org/10.1007/s10551-011-1008-5

Bartolacci, F., Bellucci, M., Corsi, K., & Soverchia, M. (2022). A systematic literature review of theories underpinning sustainability reporting in non-financial disclosure. In Cinquini, L., & De Luca, F. (Eds.), *Non-financial disclosure and integrated reporting* (pp. 50–72). Springer. https://doi.org/10.1007/978-3-030-90355-8_4

Bain & Company. (2023). *Climate and ESG: Driving business value through sustainability.* Bain & Company Insights. Retrieved from https://www.bain.com

Boelders, F. B. (2020). *The relationship between firm characteristics and greenwashing in the European energy sector: An NLP approach* (Utrecht University Master Thesis). https://studentheses.uu.nl/handle/20.500.12932/38034

Borghei, Z. (2021). Carbon disclosure: A systematic literature review. *Accounting & Finance, 61*(4), 5255–5280. https://doi.org/10.1111/acfi.12757

Bose, S. (2020). Evolution of ESG reporting frameworks. In Esty, D. C., & Cort, T. (Eds.), *Values at Work.* Palgrave Macmillan. https://doi.org/10.1007/978-3-030-55613-6_2

Deloitte. (2023). *Sustainability and climate change: Insights and strategies.* Deloitte. Retrieved from https://www2.deloitte.com

Fiedler, T., Pitman, A. J., & Mackenzie, K. (2021). Business risk and the emergence of climate analytics. *Nature Climate Change, 11,* 87–94. https://doi.org/10.1038/s41558-020-00984-6

Hales, J. (2018, July 16). The future of accounting is now. *The CPA Journal.* https://www.cpajournal.com/2018/07/16/the

Hira, A., & Ferrie, J. (2006). Fair trade: Three key challenges for reaching the mainstream. *Journal of Business Ethics, 63,* 107–118. https://doi.org/10.1007/s10551-005-3041-8

Halkos, G., & Nomikos, S. (2021). Corporate social responsibility: Trends in global reporting initiative standards. *Economic Analysis and Policy, 69,* 106–117. https://doi.org/10.1016/j.eap.2020.11.008

Ioannou, I., & Serafeim, G. (2017). *The consequences of mandatory corporate sustainability reporting.* (Harvard Business School Research Working Paper No. 11–100). https://ssrn.com/abstract=1799589

Koster, M., Vos, B., & Van der Valk, W. (2019). Drivers and barriers for adoption of a leading social management standard (SA8000) in developing economies. *International Journal of Physical Distribution & Logistics Management, 49*(5), 534–551. https://doi.org/10.1108/IJPDLM-01-2018-0037

KPMG. (2022). *KPMG Survey of Sustainability Reporting.* KPMG Insights. Retrieved from https://home.kpmg

La Torre, M., Sabelfeld, S., Blomkvist, M., Tarquinio, L., & Dumay, J. (2018). Harmonising non-financial reporting regulation in Europe: Practical forces and projections for future research. *Meditari Accountancy Research, 26*(4), 598–621.

Llach, J., Marimon, F., del Alonso-Almeida, M., & M. (2015). Social Accountability 8000 standard certification: Analysis of worldwide diffusion. *Journal of Cleaner Production, 93,* 288–298. https://doi.org/10.1016/j.jclepro.2015.01.044

Macve, R., & Chen, X. (2010). The "equator principles": A success for voluntary codes? *Accounting, Auditing & Accountability Journal, 23*(7), 890–919. https://doi.org/10.1108/09513571011080171

McKinsey & Company. (2023). *The net-zero transition: What it would cost, what it could bring.* McKinsey Global Institute. Retrieved from https://www.mckinsey.com

Pelsmacker, P. D., Driesen, L., & Rayp, G. (2003). *Are fair trade labels good business? Ethics and coffee buying intentions.* (University of Gent Working Paper, 2003/165).

Pizzi, S., Principale, S., & De Nuccio, E. (2023). Material sustainability information and reporting standards: Exploring the differences between GRI and SASB. *Meditari Accountancy Research, 31*(6), 1654–1674. https://doi.org/10.1108/MEDAR-11-2021-1486

Raynolds, L. T., Murray, D. L., & Wilkinson, J. (2007). *Fair trade: The challenges of transforming globalization.* Routledge.

Reed, D. (2009). What do corporations have to do with fair trade? Positive and normative analysis from a value chain perspective. *Journal of Business Ethics, 86*(Suppl 1), 3–26. https://doi.org/10.1007/s10551-008-9757-5

Rogers, J., & Herz, R. (2013, December). Corporate disclosure of material information: The evolution—and the need to evolve again. *Journal of Applied Corporate Finance, 23.* https://doi.org/10.1111/jacf.12028

Runhaar, H., & Lafferty, H. (2009). Governing corporate social responsibility: An assessment of the contribution of the UN Global Compact to CSR strategies in the telecommunications industry. *Journal of Business Ethics, 84*, 479–495. https://doi.org/10.1007/s10551-008-9720-5

Stolowy, H., & Paugam, L. (2023). Sustainability reporting: Is convergence possible? *Accounting in Europe, 20*(2), 139–165. https://doi.org/10.1080/17449480.2023.2189016

Szadziewska, A., & Kujawski, J. (2022). Environmental disclosures in the non-financial reporting of energy companies. Creating a reliable business image or impression management? *Zeszyty Teoretyczne Rachunkowości, 46*(2), 157–194.

Vallone, C. (2022). The role and expectations of stakeholders in the new non-financial disclosure regulations. In Cinquini, L., & De Luca, F. (Eds.), *Non-financial Disclosure and Integrated Reporting*. Springer. https://doi.org/10.1007/978-3-030-90355-8_22

Zaid, M. A., & Issa, A. (2023). A roadmap for triggering the convergence of global ESG disclosure standards: lessons from the IFRS foundation and stakeholder engagement. *Corporate Governance: The International Journal of Business in Society, 23*(7), 1648–1669.sx.

Measuring and Disclosing Greenhouse Gas Emissions

5

Introduction

In the previous chapter, we explored the motivations for corporate disclosure of greenhouse gas (GHG) emissions, emphasizing both external pressures and internal benefits. External drivers, such as regulatory requirements and investor expectations, compel firms to disclose emissions for compliance and to attract capital. Internally, disclosure enables firms to identify cost-saving opportunities, enhance operational efficiency, and build reputational value.

However, these benefits come with challenges. Companies face significant costs, including investments in data collection systems, compliance with reporting frameworks, third-party verification, and capacity building. Measuring Scope 3 emissions—those linked to value chain activities—is particularly resource intensive due to its complexity. Additionally, firms must navigate evolving regulatory requirements, such as the European Union's Corporate Sustainability Reporting Directive (CSRD), and manage reputational risks associated with inadequate or misleading disclosures.

This chapter shifts to the practical aspects of emissions reporting, focusing on the Greenhouse Gas Protocol, the leading framework for categorizing and measuring emissions across Scopes 1, 2, and 3. We explain the distinctions across the three scopes and drill down on the food sector to illustrate the application of Scope 3 emissions in more detail. We then expand the discussion to examine GHG emissions across major economic sectors, highlighting the challenges firms face in reducing emissions and the progress made toward mitigation. These efforts align with the **Paris Agreement's dual goal** of keeping global temperature rise well below 2 °C while striving to limit it to 1.5 °C (UNFCCC, 2015). In fact, the **1.5 °C target** has become central to climate policy, as IPCC reports increasingly emphasize the severe risks of exceeding this threshold (IPCC, 2021). This last report card sets the stage for the following chapter that describes pathways to a decarbonized future as well as what could happen if the world fails in sufficiently containing the climate crisis.

5.1   Measuring Emissions: The GHG Protocol

Greenhouse Gas (GHG) Emissions and the GHG Protocol

To effectively reduce greenhouse gas emissions, it is essential to measure both current emissions and projected future emissions. A standardized measurement methodology is needed across firms or other units of interest to ensure comparability and to observe emissions trends over time. The Greenhouse Gas Protocol,[1,2] is the most widely used reporting standard for this purpose, with ninety percent of Fortune 500 firms that disclose emissions to the CDP (formerly the Carbon Disclosure Project) utilizing the GHG Protocol. Various sustainability and climate accounting standards, such as the Task Force on Climate-related Financial Disclosures (TCFD), recommend the use of the GHG Protocol to measure emissions, and advise that firms provide historical emissions data.

Defining Emissions Scopes

The GHG Protocol formalizes how a firm's carbon footprint should be calculated and categorized into three distinct scopes: Scope 1 (direct emissions), Scope 2 (indirect energy emissions), and Scope 3 (indirect emissions not counted under Scope 2):

- Scope 1: Emissions that a company owns or controls through direct operations. Scope 1 emissions are straight forward to calculate.
- Scope 2: Emissions associated with the use of purchased energy to run all facets of the business. These are "indirect" emissions but focus solely on the energy-use component of the firm.
- Scope 3: Emissions associated with customers using company products and emissions associated with suppliers making products used by the company. These emissions capture all indirect emissions except those associated with energy, which are recorded as Scope 2.

---

[1] See: Greenhouse Gas Protocol, https://ghgprotocol.org/.
[2] World Resources Institute (WRI) & World Business Council for Sustainable Development (WBCSD) (2011), Corporate Value Chain (Scope 3) Accounting and Reporting Standard (PDF), https://ghgprotocol.org/sites/default/files/standards/Corporate-Value-Chain-Accounting-Rep oring-Standard_041613_2.pdf, page 5.

Measuring and Interpreting Emissions Scopes

Scope 1

Scope 1 emissions are considered relatively straightforward to calculate because they represent direct GHG emissions from sources that are owned or controlled by a company. These include emissions from combustion in company-owned or controlled boilers, furnaces, vehicles, and chemical production in manufacturing processes. Since these emissions are generated directly by a firm's operations, they are typically easier to measure accurately compared to Scope 2 and Scope 3 emissions, which rely on indirect factors and external data (Greenhouse Gas Protocol, n.d.).

For example, in industries such as manufacturing or logistics, Scope 1 emissions come directly from burning fossil fuels in equipment or transportation fleets. Firms can often use fuel consumption data and emission factors to calculate these emissions. The direct control over the processes generating these emissions makes measurement more reliable, as there are fewer dependencies on external systems or third parties (EPA, 2023).

However, while Scope 1 emissions are simpler to measure, relative to Scopes 2 and 3, the challenge lies in ensuring that all relevant sources are accounted for, particularly in large organizations with diverse operations. Misclassification or overlooking smaller emission sources can still lead to inaccuracies. Furthermore, for certain sectors, such as oil and gas, direct emissions may include methane leaks or flaring, which require more advanced monitoring systems to quantify accurately (IEA, 2021).

Scope 2

Scope 2 emissions are the indirect energy emissions of the company. Firms have discretion in how Scope 2 emissions are reported: location based or market based.

Location-based approach: This approach relies on average grid emissions factors based on the energy generation mix in a specific geographic location where electricity is consumed. It assumes that the emissions associated with the electricity consumed are representative of the average emissions from the electricity grid.

Market-based approach: This approach allows companies to use contractual instruments, such as purchase power agreements (PPAs) and renewable energy certificates (RECs), to claim the environmental benefits of renewable energy. PPAs and RECs reduce market-based Scope 2 emissions by allowing companies to claim the environmental benefits of renewable energy generation. Through a PPA, a company supports renewable energy projects and typically receives RECs, which certify that the energy was generated from renewable sources. These RECs enable the company to offset the emissions associated with its electricity consumption, regardless of the actual carbon intensity of the grid supplying its operations. As a result, market-based disclosures reflect the renewable energy purchased, not the physical energy consumed, leading to lower reported emissions.

Although the GHG Protocol provides flexibility for companies to choose either the location-based or market-based approach for reporting Scope 2 emissions, it recommends the use of location-based measures as the default method. This is because location-based measures provide a consistent and standardized approach that aligns with the emissions associated with the average electricity grid in a specific location.

Practically speaking, there was likely not a substantial difference in overall emissions disclosed between the location-based and market-based methods for Scope 2 reporting. This was largely because electricity grids in most regions heavily relied on fossil fuels, resulting in high emissions regardless of the methodology. Both approaches, whether reflecting the actual grid mix (location based) or adjusted for renewable energy purchases (market based), produced similar results due to the limited availability of renewable energy and the dominance of traditional energy sources.

However, this has changed in recent years as renewable energy adoption has accelerated and companies increasingly rely on the market-based method to report lower emissions. By using PPAs and RECs, firms can claim to offset their energy use with renewable sources, even when their actual electricity consumption is tied to fossil fuel-heavy grids. This importance of measurement of Scope 2 is particularly apparent in energy-intensive sectors like tech. In the Fall of 2024, The Guardian shone a spotlight on this issue explaining that companies such as Google, Microsoft, and Meta are underreporting their Scope 2 emission. Using market-based reporting, they claim lower emissions due to renewable energy purchases, but the location-based figures—reflecting their actual grid mix—often tell a different story. For example, The Guardian estimated that the emissions from these companies' data centers are 662% higher than officially reported, driven by the increasing energy demands of artificial intelligence (AI) workloads (O'Brian, 2024).

Scope 3

Scope 3 emissions span the firm's entire value chain—from suppliers through to customers. Thus, Scope 3 emissions capture all indirect emissions except those associated with energy, which are recorded as Scope 2.

Scope 3 emissions are the most difficult to measure and, unfortunately, often the most relevant in terms of global GHG emissions. The UN Global Compact estimates that on average, 70% of a firm's carbon footprint resides in Scope 3 emissions (United Nations Global Compact UK, n.d.). Ideally, firms report their emissions within each category; however, only a relatively small number of firms report their Scope 3 emissions. Scopes 1 and 2 emissions are generally comparable across years and companies, while Scope 3 emissions are less comparable due to variations in activities, boundaries, and methodologies.

The Scope 3 Standard under the GHG Protocol is the only internationally accepted method for companies to account for these types of value chain emissions. Currently, many companies that disclose GHG emissions limit their disclosure to Scopes 1 and 2. When firms are required to report GHG emissions (e.g., under Cap-and-Trade schemes), mandated emissions disclosure typically covers only Scope 1 and Scope 2 emissions. However, this is changing with new legislation in the EU. As part of the EU's European Green Deal, businesses will be required to comply with the Corporate Sustainability Reporting Directive (CSRD) and disclose Scope 3 emissions. The CSRD phase-in began on January 1, 2024, for certain large EU and EU-listed companies, and will eventually apply (by January 1, 2028) to a broader group, including non-EU companies with significant EU revenues or significant EU operations, such as an EU branch or subsidiary. Additionally, under a new set of SEC proposals, large companies, and eventually many publicly traded companies, will be required to disclose Scope 3 emissions.

Scope 3: Double Counting?

Within a firm, Scopes 1 and 2 are uniquely identified, preventing double counting of emissions, and therefore comparable across firms. However, Scope 3 emissions present complexities due to their nature as indirect emissions that overlap along the value chain. For example, the direct emissions (Scope 1) of a car manufacturer are the indirect emissions (Scope 2) of a steel supplier, which also become part of the upstream emissions (Scope 3) for a vehicle dealership. Each company within this chain has unique opportunities to reduce emissions: the car manufacturer might use materials with lower carbon footprints, the steel supplier could adopt energy-efficient processes, and the dealership could prioritize vehicles with higher fuel efficiency or electric models. Measuring Scope 3 emissions across a firm's value chain highlights essential areas for reducing greenhouse gases, despite some degree of double counting. Examining emissions by scope reveals targeted opportunities for mitigation and can significantly impact global emissions, as Scope 3 often constitutes a substantial part of a company's overall carbon footprint.

Emissions Scopes Across the Value Chain

In Fig. 5.1, you can see a picture of how a product (or service) moves across the value chain going upstream and downstream, both catching Scope 3 emissions. Together, Scopes 1, 2, and 3 form the comprehensive carbon footprint of a company, capturing the full range of emissions associated with its operations, supply chain, and product life cycle.

Defining the emissions scope is one thing, and implementing measurement is quite another. To illustrate the breadth of Scope 3 emissions, we will examine sectoral emissions in the food and beverage sector: from farm to fork.

Graphic by Stacy Smedley, 2021

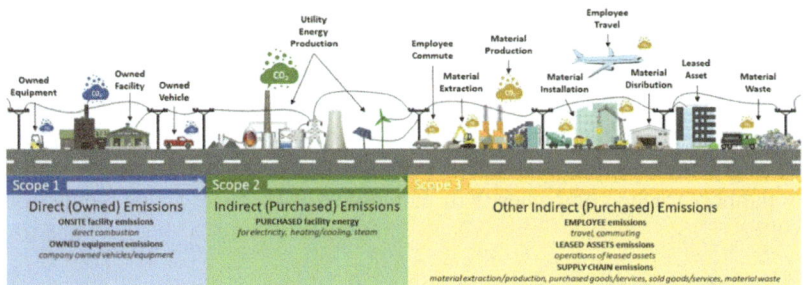

Fig. 5.1 Understanding Scope 1, 2, and 3 emissions[3]

Illustrating Scope 3 Emissions: From Farm to Fork

The two largest sources of greenhouse gas emissions globally are energy and food production. While much attention is given to transitioning away from fossil fuels, this focus alone is not enough to address climate change. Even if we were to halt all emissions from fossil fuels immediately, the emissions from food production alone would still push us beyond the carbon budget needed to limit global warming to 1.5 °C.

The food system is responsible for an estimated 25–33% of global GHG emissions, which are generated at multiple stages in a complex supply chain where Scope 3 emissions dominate. Figure 5.2 illustrates the value chain for food and the associated emissions at each stage. Typically, food production goes through seven key stages: land-use change, farm, animal feed, processing, transport, retail, and packaging. Notably, the land use and farm stages account for roughly 80% of GHG emissions across all types of food. This highlights the immense impact of activities such as deforestation, soil management, fertilizer use, and methane emissions from livestock.

Most of the emissions sources shown in Fig. 5.2 require no further explanation apart from "loss." "**Loss**" represents greenhouse gas (GHG) emissions from food waste and spoilage at various stages of the supply chain, including production, transportation, processing, retail, and consumption. These losses occur when food that could have been consumed is wasted, yet the emissions from its production, such as farming, energy use, and transportation, have already been generated. Losses also contribute additional emissions when wasted food decomposes in landfills, releasing methane (CH_4), a potent GHG. Despite being discarded, food waste

---

[3] Graphic by Stacy Smedley. (2021). Accessed November 4, 2021, at: https://climateeverything.com/climategifs/scopeonetwoandthreeemissions.

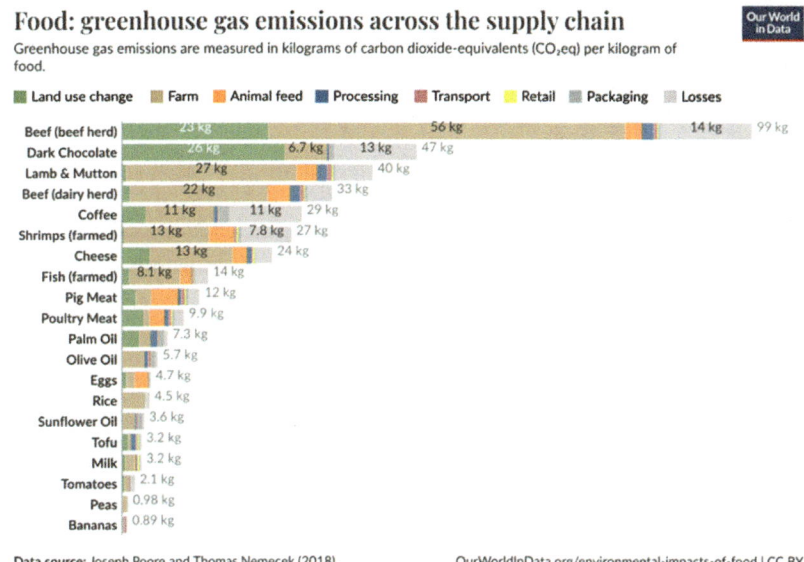

Fig. 5.2 GHG emissions across the food supply chain (*Data source* Joseph Poore and Thomas Nemecek [2018])

carries the environmental cost of its entire lifecycle, from fertilizers and fuel used in farming to refrigeration and packaging during distribution.

Figure 5.2 illustrates the significant carbon footprint of beef production, driven by factors such as animal feed production, deforestation for pastureland, and methane emissions from cattle. Beef production is particularly resource intensive, accounting for approximately **40% of land use** associated with domestic livestock in the United States (Eshel et al., 2014). These factors contribute to beef's outsized greenhouse gas (GHG) emissions compared to other food types.

Interestingly, transportation plays a relatively small role in the overall GHG emissions of food, contributing an average of **10% of total emissions** across food categories and just **0.5% for beef**. This indicates that efforts to reduce food-related GHG emissions through local sourcing may have limited impact for high-carbon foods like beef, as most emissions arise earlier in the supply chain—specifically during feed production, animal rearing, and land-use changes.

Figure 5.2 also highlights substantial variability in emissions by food type. Animal-based foods, such as beef, lamb, and cheese, exhibit significantly higher emissions per kilogram than plant-based foods. For example, beef from dedicated beef herds emits approximately **60 kg CO_2-equivalents per kilogram** of product, with methane emissions and land-use changes being the primary contributors. By contrast, plant-based foods such as grains, vegetables, and nuts produce far fewer emissions. Root vegetables, for instance, emit less than **1 kg CO_2-equivalent per kilogram**, making them a much lower-impact dietary choice.

Table 5.1 GHG emissions by firm (metric tons CO_2e, reported in millions)

Firm	Scope 1	Scope 2	Scope 3	Total	% Scope 1	% Scope 2	% Scope 3
ADM	12.6	2.03	107.5	122.13	10.32%	1.66%	88.02%
PepsiCo	3.4	0.3	54	57.7	5.89%	0.52%	93.59%
Nestle	3.16	0.31	84.08	87.55	3.61%	0.35%	96.04%
Tyson	3.56	2.2	N/D	N/D	N/D	N/D	N/D
JBS	3.46	1.55	151.53	156.54	2.21%	0.99%	96.80%

Source ADM (2024), PepsiCo (n.d.), Nestle (2023), Tyson Foods (2022) and JBS (n.d.)

Big Farma???[4]

The food and beverage industry, particularly segments like meat processing, ready-to-eat cereals, and retail grocery, is dominated by large companies with extensive global supply chains. These companies operate through numerous subsidiaries and rely on complex international networks, making their climate impact broad and significant. A critical aspect of assessing their greenhouse gas (GHG) emissions lies in distinguishing between direct emissions (Scope 1) from production processes and indirect emissions (Scopes 2 and 3), which encompass energy consumption, agriculture, land use, and broader supply chain activities.

Table 5.1 highlights emissions data for selected prominent firms in the food and beverage industry. These companies represent key sectors, including agricultural products, food and beverage production, and meat processing. The data are categorized into Scope 1 (direct emissions), Scope 2 (indirect emissions from energy use), and Scope 3 (indirect emissions from supply chains and land-use changes).

Table 5.1 confirms what we saw from Fig. 5.2: the lion's share of GHG emissions for the food sector reside in Scope 3 emissions, with Scope 2 emissions quite irrelevant. However, what we also see is that Tyson is very much the laggard among industry peers. The most recent emissions information we were able to find for Tyson was from a report on the company website published in 2022. There has been significant media criticism of Tyson Foods for their poor disclosure practices as well.

We have already seen in Fig. 5.1 that the beef sector is by far the larger GHG emitter in the food and beverage sector. This suggests that firms in this sector will probably experience significant scrutiny from investors, the media, and regulators; and this is exactly what happened. Tyson Foods, the world's second-largest meat processor, and JBS, the largest, have faced allegations of "greenwashing" misleading consumers about their environmental commitments. In September 2024, the Environmental Working Group (EWG) sued Tyson Foods, claiming the company lacked a concrete strategy to achieve its net-zero emissions goal by 2050 and that its "climate-friendly" beef marketing was deceptive (Gibson, CBS News,

---

[4] We couldn't resist the "corny" reference, nor our "play on words" footnote!

2024). Similarly, in February 2024, New York Attorney General Letitia James filed a lawsuit against JBS, accusing the company of making unsubstantiated claims about reaching net-zero emissions by 2040 without a feasible plan (Plumer, Vox, 2024). These legal actions highlight concerns that major meat processors are making environmental promises without implementing the necessary measures to fulfill them.

These legal actions underscore the critical importance of aligning emissions measurement and disclosure with actionable strategies to achieve climate goals. They also highlight the unique challenges faced by the **food and beverage sector**, where Scope 3 emissions—arising from agricultural practices, land use, and supply chains—dominate the carbon footprint. In the next section, we will continue our sectoral focus, identifying two more sectors: heavy industry and energy which have significant GHG emissions, often residing in Scope1, that are hard to abate.

5.2 Measuring GHG Emissions

The concept of a carbon budget—the maximum amount of carbon dioxide (CO_2) that can be emitted while limiting global warming to a specific target—is a stark reminder of the urgency of climate action. Scientists estimate that to keep global temperature rise below 1.5 °C, humanity has a finite budget of about four hundred gigatons of CO_2, a figure projected to be exhausted within the next few decades at current emissions rates (IPCC, 2021). Every ton emitted today reduces the amount left in this budget, intensifying the need to address emissions growth and focus on targeted reductions across sectors (Rogelj et al., 2019). Yet, global emissions continue to rise despite widespread recognition of climate risks, largely because certain sectors are harder to decarbonize than others. Recognizing these sectoral differences is critical; it informs where to focus innovation and policy efforts for maximum impact.

Overview of GHG Emissions by Sector

Understanding global greenhouse gas emissions on a sectoral basis provides essential insights into the sources of climate change and the challenges each sector faces in decarbonizing (Rogelj et al., 2019). This section provides an overview of the major sectors contributing to GHG emissions, also drawing on data that indicate the continuing upward trend in emissions despite efforts to curb them. There is no silver bullet for tackling climate change; achieving the Paris Agreement goal of net-zero emissions by 2050 will require innovation and collaboration across

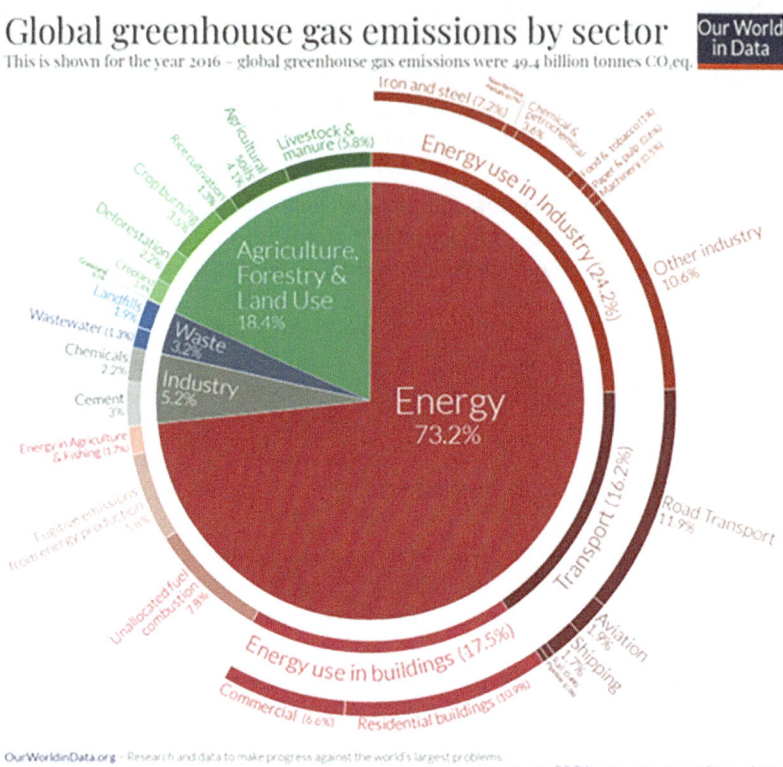

Fig. 5.3 Global greenhouse gas emissions by sector *Source* Climate Watch the World Resource Institute (2020)

diverse areas of the economy, from energy to industry, agriculture, and beyond. Figure 5.3 illustrates global emissions by sector in 2016.[5]

Interpretation of Fig. 5.3

There is no escaping the dominance of the energy sector in contributing to global GHG emission. Of course, the energy sector is the fuel for the rest of the economy. If you look at the next ring you can see that nearly two-thirds of total energy lands in buildings (17.5%); transportation (16.2%) and energy use in industry (24.2%). The last outer band breaks down each of the former categories to ever more targeted sub-sectors. For example, look at energy use in industry and you will see that chemicals sit at 3.6%. We point this out because if you zoom to the other

___

[5] Figure 5.1 is from Our World in Data, based on a 2020 analysis. This is the most up-to-date depiction we could find.

side of the chart, you will see industry at 5.2% taking a chunk of the pie and then within industry you will see chemicals at 2.2%. So, we can infer that chemicals are contributing 2.2% to global emissions through their processes (these are direct Scope 1 emissions) and Scope 2 (at 3.6%) through energy use.

With your guide to Fig. 5.3 in hand, we will begin now to look through sector by sector, starting with energy so it will be a little repetitive as we get off ground. We will cover the four pieces of the pie to start: energy, agriculture, industry, and waste, examining possible strategies for reducing emissions while providing an overview of why it is so hard to accomplish!

Energy (73.2% of Global Emissions)

The energy sector is the largest contributor to global GHG emissions, responsible for nearly three-quarters of total emissions. This sector includes emissions from electricity and heat generation, transportation, and energy use in buildings and industrial activities:

- Electricity and Heat Production: Fossil fuel combustion for electricity and heat is a major source of CO_2 emissions. Transitioning to renewable energy sources, such as wind, solar, and hydro, is critical but faces challenges in scalability and grid integration.
- Transportation: Emissions from road transport, aviation, shipping, and rail contribute significantly, with road transport alone accounting for a large portion. Shifting to electric vehicles and alternative fuels is essential, yet infrastructure and technology development remain barriers.
- Buildings: Both residential and commercial buildings emit GHGs through heating, cooling, and electricity use. Improving energy efficiency and integrating low-carbon technologies can mitigate these emissions, but retrofitting existing buildings is often costly and complex.

Agriculture, Forestry, and Land Use (18.4% of Global Emissions)

Agriculture and land-use change contribute to nearly a fifth of global GHG emissions. We discussed in Sect. 5.1 that about 80% of emissions reside in land-use change and farming: methane from livestock, nitrous oxide from fertilizers, and CO_2 from deforestation.

- Livestock and Manure: Methane emissions from livestock digestion and manure management are significant, particularly in beef and dairy production. Innovative feeding strategies and manure management practices could reduce these emissions.
- Deforestation and Crop Production: Clearing forests for cropland releases stored carbon, while activities like rice cultivation produce methane. Sustainable land management and reforestation are essential but face competing demands for agricultural land and food production.

- Soil Management: Fertilizer use and soil disturbances emit nitrous oxide, a potent greenhouse gas. Reducing fertilizer usage and adopting regenerative agriculture practices can help, though adoption varies across regions.

Waste (3.2% of Global Emissions)
The waste sector's emissions primarily arise from landfills and wastewater treatment. Organic waste in landfills decomposes anaerobically, releasing methane, while wastewater treatment produces both methane and nitrous oxide:

- Landfills: Methane capture technologies can mitigate emissions, but widespread implementation is needed.
- Wastewater Treatment: Emissions can be reduced through improved treatment methods, though costs and infrastructure needs vary by region.

Industry (5.2% of Global Emissions)
The industrial sector includes emissions from cement production, chemicals, metals, and other manufacturing activities. High energy demand and process emissions make it one of the hardest sectors to decarbonize:

- Cement: Cement production is particularly emissions intensive due to the calcination process, which releases CO_2 from limestone. Alternatives like low-carbon cement and Carbon Capture, Utilization, and Storage (CCUS) are being explored but face significant scalability challenges.
- Chemicals: The chemical sector emits CO_2 and other gases from high-temperature processes and energy use. Transitioning to low-carbon feedstocks and electrifying processes can help, yet technological and economic barriers remain.
- Steel and Metals: Steel production requires enormous amounts of energy, primarily from coal in traditional blast furnaces. Low-carbon technologies, such as hydrogen-based steelmaking, are emerging but still in early stages.

Each of these sectors contributes to global GHG emissions in distinct ways, and the solutions to mitigate emissions are similarly diverse. Figure 5.4 illustrates that emissions across key sectors are still on an upward trajectory from 1990 to 2020. This trend persists despite global efforts to reduce emissions, highlighting the challenges these sectors face in achieving meaningful reductions as well as their well-earned descriptor as "hard-to-abate" sectors. If we put it in perspective, the Paris Agreement goal is for emissions to reach net zero by 2050—that's only twenty-five years away—and Fig. 5.4 is showing emissions growth in hard-to-abate sectors over a period only five years more (1990–2020).

Greenhouse gas emissions by sector, World, 1990 to 2020

Greenhouse gas emissions[1] are measured in tonnes of carbon dioxide-equivalents[2] over a 100-year timescale.
Land-use change emissions are not included.

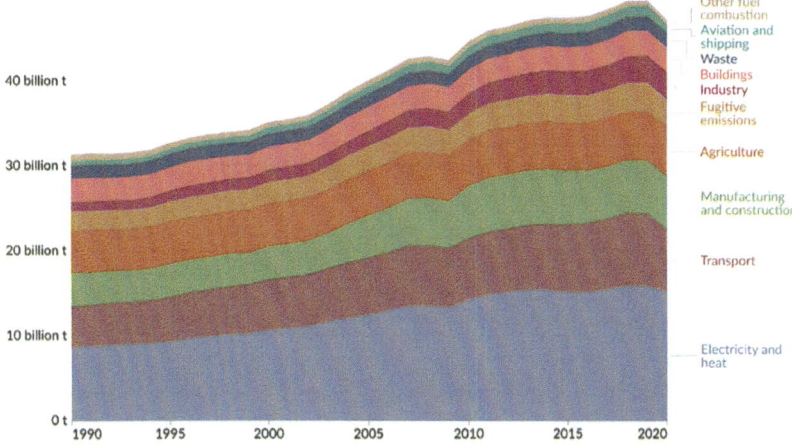

OurWorldinData.org/co2-and-greenhouse-gas-emissions | CC BY

1. **Greenhouse gas emissions**: A greenhouse gas (GHG) is a gas that causes the atmosphere to warm by absorbing and emitting radiant energy. Greenhouse gases absorb radiation that is radiated by Earth, preventing this heat from escaping to space. Carbon dioxide (CO_2) is the most well-known greenhouse gas, but there are others including methane, nitrous oxide, and in fact, water vapor. Human-made emissions of greenhouse gases from fossil fuels, industry, and agriculture are the leading cause of global climate change. Greenhouse gas emissions measure the total amount of all greenhouse gases that are emitted. These are often quantified in carbon dioxide equivalents (CO_2eq) which take account of the amount of warming that each molecule of different gases creates.

2. **Carbon dioxide equivalents (CO_2eq)**: Carbon dioxide is the most important greenhouse gas, but not the only one. To capture all greenhouse gas emissions, researchers express them in "carbon dioxide equivalents" (CO_2eq). This takes all greenhouse gases into account, not just CO_2. To express all greenhouse gases in carbon dioxide equivalents (CO_2eq), each one is weighted by its global warming potential (GWP) value. GWP measures the amount of warming a gas creates compared to CO_2. CO_2 is given a GWP value of one. If a gas had a GWP of 10 then one kilogram of that gas would generate ten times the warming effect as one kilogram of CO_2. Carbon dioxide equivalents are calculated for each gas by multiplying the mass of emissions of a specific greenhouse gas by its GWP factor. This warming can be stated over different timescales. To calculate CO_2eq over 100 years, we'd multiply each gas by its GWP over a 100-year timescale (GWP100). Total greenhouse gas emissions – measured in CO_2eq – are then calculated by summing each gas' CO_2eq value.

Fig. 5.4 Greenhouse gas emissions by sector, World, 1990 to 2020 *Data source* Climate Watch (2023)

5.3   Examples of Sector Challenges: The Second Tier

In this section, we will take a closer look at GHG emissions in some "hard-to-abate" sectors, expanding to look at some of the sub-sectors contained in the energy slice and the heavy industry slice.

Energy Sector

The contribution of the energy sector to overall global emissions is primarily driven by the combustion of fossil fuels—coal, oil, and natural gas—for electricity generation, heating, transportation, and industrial processes. Despite significant efforts to expand renewable energy sources, fossil fuels still dominate, supplying over 80% of the world's energy.

Within energy production, different sub-sectors contribute to emissions in various ways:

- Electricity Generation: Coal-fired power plants are among the most carbon-intensive sources of energy. Natural gas, while producing less CO_2 per unit of energy than coal, still contributes heavily to emissions. Meanwhile, renewables like wind and solar provide a low-emission alternative but currently lack sufficient scale to meet global demand alone.
- Transportation: The transportation sector relies primarily on oil-based fuels (e.g., gasoline and diesel), which are major sources of carbon dioxide. This sector includes emissions from cars, trucks, planes, and ships, each with unique energy needs and challenges for decarbonization. Electric vehicles (EVs) and alternative fuels like Sustainable Aviation Fuel (SAF) are promising solutions, but their adoption and infrastructure need rapid scaling to reduce emissions meaningfully.
- Heating and Industrial Processes: Energy for industrial processes, including the direct use of fossil fuels to generate high heat, is essential in heavy industries like steel, cement, and chemicals. Industrial heat generation is particularly challenging to decarbonize due to the high temperatures required, which renewables struggle to achieve economically.

The shift to cleaner energy sources faces several key challenges:

- Intermittency of Renewable Sources: Solar and wind power, though clean, depend on weather conditions, creating gaps in energy supply. Energy storage technologies (like batteries) are essential but currently expensive and limited in capacity.
- High Transition Costs: Retrofitting existing infrastructure, decommissioning fossil fuel assets, and building new renewable capacity require significant upfront investments.
- Developing Regions: In many developing countries, coal remains a vital energy source due to its low cost and availability, making the transition to renewables slower.

Heavy Industry

In looking at Figs. 5.3 and 5.4, the contribution of industry to global emissions seems modest relative to emissions in energy and agriculture: 5.2% versus 73.2% in energy and 18.4% for agriculture, forestry, and land use. It is tempting to ask, "Why worry so much?"

Heavy industry manufactures products that are essential materials in our daily lives—chemicals, steel, and cement, serving various purposes from infrastructure to consumer products. The demand for these materials has significantly increased in recent years due to a growing population and economic development. For

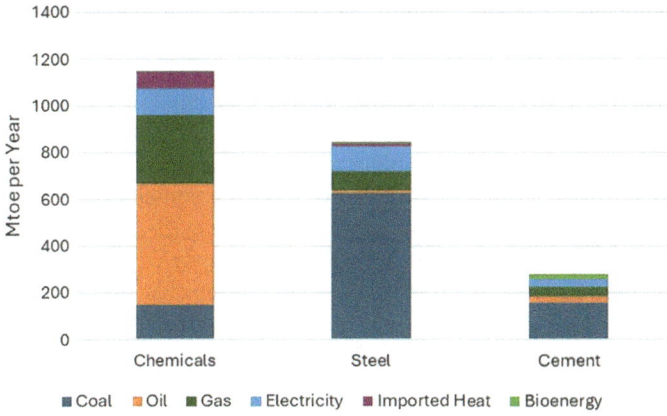

Fig. 5.5   Final energy demand of chemicals, steel, and cement by fuel, 2019 (*Source* IEA [2020])

example, the production of cement and steel has more than doubled since the millennium, while plastics, a crucial product of the chemical industry, have seen a 90% increase.

The wrinkle in heavy industries, however, is that while chemicals, steel, and cement fulfill our current needs, they also play a vital role in enabling a sustainable transition in the energy sector. Analysis by the IEA reveals that in a Sustainable Development Scenario, steel demand for renewable energy technologies like wind turbines is projected to be nearly three times higher by 2070 compared to baseline projections (IEA, 2020).

Plastics and cement are also essential for clean energy applications such as electric vehicles, wind turbines, and solar panels. So, it would seem that to fix the emissions problem we must create more emissions. The production of these materials requires vast amounts of energy, approximately 2,300 million tons of oil equivalent (Mtoe)[6] in 2019, which is roughly equivalent to the total primary energy demand of the United States. Think about this: **"If the cement industry were a country, it would be the third largest emitter in the world"** (after China and the United States (Timperley, 2018)). Various fuels are used in these industries, including fossil fuels (1900 Mtoe), electricity (250 Mtoe), bioenergy (30 Mtoe), and imported heat (85 Mtoe). These energy sources are utilized for generating heat, providing raw materials, sustaining chemical reactions, and powering mechanical equipment. Figure 5.5 shows the fuel sources used in these three sub-sectors.

The review of the energy and heavy industry sectors, alongside the earlier analysis of the food and beverage sector, emphasized why emissions in these sectors

---

[6] Mtoe stands for million tons of oil equivalent. It is a unit of energy commonly used to compare different energy sources or quantify energy consumption. Mtoe provides a standardized measurement by equating the energy content of different fuels to that of one million metric tons of oil.

are so challenging to abate. While both the energy and heavy industry sectors are primary contributors to global emissions, they are also essential for producing materials critical to renewable energy infrastructure. This paradox—relying on fossil fuels and heavy industry to enable the clean energy transition—underscores the complexities and trade-offs inherent in the journey to a low-carbon economy.

The food industry presents a unique challenge. Unlike heavy industry, where the paradox lies in needing more emissions to reduce emissions, the food sector's complexities stem from its diverse environmental impacts. Balancing the nutritional needs of a growing population with the necessity of lowering greenhouse gas (GHG) emissions requires systemic change. Sustainable agricultural practices, reducing food waste, and aligning with dietary shifts, such as plant-based diets, are critical steps for firms aiming to reduce emissions without compromising food security.

The sector also holds potential to support decarbonization in other industries. Agricultural feedstocks, such as energy crops and food waste, are increasingly used in Sustainable Aviation Fuel (SAF), a key strategy for reducing emissions in aviation. This dual role—as both a contributor to and a solution for emissions—underscores its significance in the climate crisis.

Ultimately, the food industry must navigate these complexities by embracing innovation and sustainable practices, ensuring it contributes to a low-carbon economy.

Firms in carbon-intensive sectors face significant policy risks. Measures aimed at curbing emissions growth and slowing the pace of global warming disproportionately target these industries. For these firms, identifying emissions is not optional—it is typically mandated by national regulators. Moreover, firms with substantial emissions are increasingly scrutinized by financial institutions, as discussed in Chapter 3, because this sector's emissions pose a threat to financial stability. From an investor perspective, high-emission firms are perceived as risky, leading to higher costs of equity and debt. In essence, these sectors are under pressure from all directions.

To address this, developing a strategy to mitigate GHG emissions has become a necessity, which must be taken in tandem with emissions disclosures. Such strategies might involve adopting green hydrogen for iron and steel production, implementing regenerative agriculture in the food sector, addressing stranded assets by exiting parts of the business, or investing in mitigation technologies. Whatever the approach, the pathway begins with measuring and disclosing emissions, progresses through identifying areas of significant impact, and culminates in a clear commitment to emissions reduction—ultimately aiming for net-zero emissions (NZE).

With accurate emissions measurement, targeted disclosures, and a commitment to reduction across the value chain, firms in hard-to-abate sectors can navigate the challenges of the transition and position themselves not merely to survive but to thrive in a decarbonized economy.

5.4   Sector-Level Progress Toward Net-Zero Emissions

Net zero cannot be achieved unless business initiates and implements solid plans to attain NZE in line with the Paris Agreement. Business interest in documenting NZE strategies accelerated in 2018 following research published by the IPCC that highlighted the urgency surrounding climate change (Axios, 2018). The report presents compelling evidence and arguments to support a global warming limit of less than 1.5 °C—effectively moving down the previous target range of warming of less than 2 °C by 2100. The IPCC report received widespread media attention and spurred investor interest and activism to have companies produce their own net-zero commitments. In response, many of the world's largest corporations have published NZE commitments and the prevailing wisdom is increasingly tilted toward corporate-level commitments to NZE if their sustainability initiatives are to be taken seriously.

As we saw in the previous section, different sectors face different challenges and opportunities in achieving net-zero emissions, depending on their level of dependence on fossil fuels, their exposure to carbon pricing and regulation, their potential for innovation and efficiency, and their stakeholder expectations and demands. While firms in some sectors, cement for example, have homogeneous emissions-mitigation landscapes, others like the transportation sector exhibit far greater heterogeneity, especially in terms of mitigation opportunities (e.g., aviation versus auto sector). As we close out this chapter, we will take a last look at how the decarbonization in hard-to-abate sectors is progressing.[7]

Decarbonization in Hard-to-Abate Sectors: Insights from the Systems Change Lab

The Systems Change Lab is an organization that monitors global progress toward climate and sustainability goals.[8] Through its comprehensive Dashboard, the Lab provides information on a wide range of indicators across multiple sectors, each important for achieving net-zero emissions by 2050. By tracking progress in areas such as energy, industry, food systems, and transport, the Systems Change Lab identifies where advancements are on track or off track, highlighting specific barriers and requirements for making necessary transformations. With numerous indicators, the Lab offers a detailed view of the key factors influencing progress toward a sustainable future. Where available, interim targets are also included to offer a more immediate look at progress relative to shorter-term milestones. However, most assessments are based on whether sectors are likely to meet or miss the broader 2050 goals of the Paris Agreement, signaling areas in urgent need of

---

[7] There is no shortage of resources that discuss sectoral decarbonization guidelines including the World Economic Forum, SBTI, Transition Pathways Initiative, and Systems Change Lab. Scenario providers and private sector consultancies also will include sectoral examples.
[8] https://systemschangelab.org/.

change. In Tables 5.2, 5.3 and 5.4, we present the report cards for segments of heavy industry, power generation in the energy sector, and the food sector. The assessments are dismal: only in "progress toward renewable energy" and "better access to energy in developing countries" does the Systems Change Lab give a passing grade of "some progress."

Each sub-sector within heavy industry, shown in Table 5.2, has specific decarbonization hurdles. For steel, processes such as green hydrogen steel making and molten oxide electrolysis show promise but need rapid scaling. The chemicals industry requires breakthroughs in low-carbon production methods, particularly for plastics, while the cement sector is focusing on adopting carbon capture and alternative materials. Currently, each of these sectors is off track in meeting the necessary 2050 goals.

Table 5.2 Heavy industry sector: steel, chemicals, and cement

Sub-sector	Current emissions impact	Key issues	Progress and innovations	On track/off track
Steel	~7% of global CO_2 emissions	Carbon-intensive production processes	Green hydrogen steelmaking, molten oxide electrolysis	Off track
Chemicals	Major industrial CO_2 source	High emissions from plastics and production	Low-carbon chemical processes and alternative materials in R&D	Off track
Cement	Significant CO_2 contribution	Emissions from calcination and energy use	CCUS technology, alternative materials	Off track

Source Systems Change Lab

Table 5.3 Energy sector: power generation and coal phase-out

Sub-sector	Current emissions impact	Key issues	Progress and innovations	On track/off track
Renewable energy	Reduced emissions with increased usage	Slow deployment in some regions	Solar, wind, and hydropower expansions	Some progress
Coal phase-out	High emissions from coal power plants	Economic dependence on coal in many regions	Declining coal usage, but needs sevenfold acceleration	Off track
Energy access	Critical for equity and just transition	Lack of infrastructure in developing areas	Growing investment in clean and affordable energy access	Some progress

Source Systems Change Lab

Table 5.4 Food sector: agriculture and food systems

Sub-sector	Current emissions impact	Key issues	Progress and innovations	On track/off track
Agriculture	Large GHG source due to livestock, land use	Deforestation, methane emissions from livestock	Sustainable agricultural practices	Off track
Food Waste Reduction	Contributes significantly to emissions if unmanaged	Inefficiencies across the food supply chain	Waste reduction initiatives, circular economy	Off track
Dietary Shifts	Diets impact land and resource use	Slow transition to low-emission food sources	Promotion of plant-based diets	Off track

Table 5.3 indicates that renewable energy sources are increasing, and they are not yet scaling at the pace required for NZE by 2050. Coal phase-out efforts are especially lagging, with progress needed to accelerate dramatically to meet climate targets. Ensuring equitable energy access remains another challenge, as infrastructure limitations in certain regions create barriers to a just energy transition.

The food sector faces considerable challenges in reducing emissions, with significant contributions from livestock, deforestation, and food waste. Efforts to adopt sustainable agricultural practices, minimize waste, and promote dietary shifts toward plant-based foods are advancing but remain insufficient to meet the goals. Currently, this sector is largely off track, with accelerated action needed across each area.

5.5   Linking Firms to Sectors: The Emissions Landscape of Firms in Hard-to-Abate Sectors

Before leaving this chapter, we will circle back to the original focus of this chapter, measuring and disclosing firm-level GHG emissions. The analyses of patterns of GHG emissions across sectors allowed us to see why some sectors exhibit stubbornly challenging climate risk contexts. At the firm level, firms that can correctly identify where they can be most effective in reducing their emissions come hand in glove with identifying the source of these emissions. Table 5.5 shows the emissions of select firms in hard-to-abate sectors for their most recent disclosures (2022 or 2023).

These tables illustrate the heterogeneity of emissions sources across sectors by providing examples of firm-level emissions. We can clearly see how for firms in both the industry sector and the transportation sector, their direct emissions are the most significant component of their overall emissions. To reduce their

Table 5.5 GHG emissions in hard-to-abate sectors (metric tons CO_2e, reported in millions)

Company	Total emissions (Mt CO_2e)	Scope 1 (%)	Scope 2 (%)	Scope 3 (%)	% Scope 1	%Scope 2	%Scope 3
ArcelorMittalSteel	296.64	223.43	6.16	67.05	75.32%	2.08%	22.60%
LafargeHolcim Cement	127	75	5	47	59.06%	3.94%	37.01%
Maersk Shipping	34.1	34.1	0	0.04	100.00%	0.00%	0.00%
ExxonMobil Oil & Gas	600	92	7	n.d.	n.d.	n.d.	n.d.
Dow Inc. Chemicals	109.01	26.48	2.89	79.64	24.29%	2.65%	73.06%
United Airlines Aviation	49.31	36.5	0.14	12.67	74.02%	0.28%	25.69%
PepsiCo Food & Beverage	57.7	3.4	0.3	54	5.89%	0.52%	93.59%

GHG emissions, these firms will have to fundamentally adjust their manufacturing processes in the case of ArcelorMittal and LafargeHolcim. United Airlines and Maersk shipping must adjust how they face the future by looking at ways to replace conventional fuels in their operations with sustainable non-fossil fuel-based alternatives.

Oil and gas majors, like ExxonMobil, have the majority of emissions sitting firmly in Scope 3, as they lie behind the fuel consumed by the sheer number of vehicles, industrial facilities, and power plants using their products multiplying their direct emissions many times over.

Other firms, like PepsiCo and Dow Chemical, also exhibit significant Scope 3 emissions, but for different reasons. For PepsiCo, Scope 3 emissions arise largely from the agricultural supply chain, packaging, and transportation of its products, as well as consumer use and disposal of items such as single-use bottles. Meanwhile, for Dow Chemical, Scope 3 emissions are driven by its production of petrochemical products, including plastics, synthetic materials, and specialty chemicals. These emissions stem from the upstream extraction and transportation of fossil fuel feedstocks, as well as downstream use and disposal, such as the incineration of plastics or emissions from chemical-intensive applications in agriculture and manufacturing. These examples highlight the diversity of emissions profiles across sectors and the unique challenges each firm faces in addressing their carbon footprint.

Conclusion
The diverse sources and magnitude of GHG emissions across sectors highlight the challenges firms face in navigating the climate crisis. Accurate emissions measurement, as outlined in this chapter, is foundational for firms

to understand their carbon footprint and design strategies for reduction. This process is not only about compliance but also a necessary step for aligning business operations with global climate goals.

In the following chapter, we explain how scenario analysis can be used to highlight potential pathways to a decarbonized future and evaluate the risks and opportunities posed by different climate scenarios. As we will show, the firm's carbon emissions will depend upon the future—a hothouse world with unbridled emissions or a decarbonized net-zero world?

Key Takeaways

1. **The Importance of Emissions Measurement**

Accurate measurement of GHG emissions is foundational for firms to understand their carbon footprint and take actionable steps toward decarbonization.

The Greenhouse Gas Protocol is the primary framework used by firms to categorize emissions into Scope 1 (direct), Scope 2 (indirect energy), and Scope 3 (value chain emissions), ensuring consistency and comparability across disclosures.

2. **Scope 3 Dominance and Complexity**

For many firms, Scope 3 emissions constitute the majority of their carbon footprint, especially in sectors like food, beverages, chemicals, and oil and gas.

Measuring Scope 3 emissions is complex due to their indirect nature, requiring extensive value chain data and collaboration with suppliers and customers.

3. **Sectoral Differences in Emissions Profiles**

Emissions sources vary significantly by sector:

- **Heavy industry (e.g., steel, cement)**: Dominated by Scope 1 emissions due to energy-intensive production processes.
- **Transportation (e.g., shipping, aviation)**: Primarily Scope 1 emissions from fuel use.
- **Food & beverage (e.g., PepsiCo)**: Scope 3 emissions dominate, arising from agriculture, packaging, and product disposal.
- **Chemicals (e.g., Dow)**: Significant Scope 3 emissions from the lifecycle of petrochemical products.

4. **Drivers and Benefits of GHG Disclosure**

Firms disclose emissions due to regulatory requirements, investor pressure, and reputational concerns.

Beyond compliance, measuring emissions allows firms to identify efficiency opportunities, reduce costs, and align with global climate goals.

5. Challenges in GHG Disclosure

Measuring Scope 3 emissions is resource intensive and involves navigating data gaps, potential double counting, and evolving reporting standards (e.g., EU's Corporate Sustainability Reporting Directive).

Balancing accuracy, consistency, and transparency is critical, especially as stakeholders increasingly scrutinize emissions data.

6. Progress Toward Net Zero

Firms in hard-to-abate sectors face unique challenges in reducing emissions, requiring significant innovation, investment in low-carbon technologies, and alignment with global climate targets.

Questions

1. What are the three scopes of GHG emissions as defined by the Greenhouse Gas Protocol?
2. Why are Scope 3 emissions often the largest component of a firm's carbon footprint?
3. What are the key challenges firms face in measuring Scope 3 emissions?
4. How do emissions profiles differ across sectors such as food and beverages, chemicals, and heavy industry?
5. How does measuring emissions benefit firms beyond regulatory compliance?
6. What role does the Greenhouse Gas Protocol play in standardizing GHG emissions reporting?
7. How do firms in hard-to-abate sectors like steel or cement address their emissions challenges?

References

Archer Daniels Midland (ADM). (2024). *2023 Corporate Sustainability Report*. Retrieved November 4, 2024 from https://www.adm.com/globalassets/sustainability/sustainability-reports/final_archer-daniels-midland-adm_2023-corporate-sustainability-report_51424.pdf

A.P. Moller-Maersk. (n.d.). *Climate change*. Retrieved November 24, 2024, from https://www.maersk.com/sustainability/our-esg-priorities/climate-change

Axios. (2018, October 8). UN details massive changes needed to slow global warming. https://www.axios.com/2018/10/08/ipcc-report-massive-energy-changes

Dow Inc. (2024). *Greenhouse Gas (GHG) Protocol Disclosure Report*. Retrieved November 4, 2024, from https://corporate.dow.com/content/dam/corp/documents/about/066-00475-01-2023-intersections-report-disclosure-ghg.pdf

Environmental Protection Agency (EPA). (2023). *Understanding scope 1, 2, and 3 emissions*. Retrieved February 19, 2025, from https://www.epa.gov/climateleadership/scopes-1-2-and-3-emissions-inventorying-and-guidance

Eshel, G., Shepon, A., Makov, T., & Milo, R. (2014). Land, irrigation water, greenhouse gas, and reactive nitrogen burdens of meat, eggs, and dairy production in the United States. *Proceedings of the National Academy of Sciences, 111*(33), 11996–12001. Retrieved November 4, 2024, from https://doi.org/10.1073/pnas.1402183111

ExxonMobil. (2024). *Metrics and data*. Retrieved November 4, 2024, from https://corporate.exxonmobil.com/sustainability-and-reports/metrics-and-data#Greenhousegasemissionsperformancedata

Gibson, K. (2024, September 19). *Tyson Foods misleads shoppers about its carbon emissions, climate group says*. CBS News. Retrieved November 4, 2024, from https://www.cbsnews.com/news/tyson-foods-greenwashing-climate-change-jbs/

Greenhouse Gas Protocol. (n.d.). *A corporate accounting and reporting standard: Revised edition*. Retrieved November 23, 2024, from https://ghgprotocol.org/sites/default/files/standards/ghg-protocol-revised.pdf

Holcim. (2024). *Climate report 2023*. Retrieved November 4, 2024, from https://www.holcim.com/sites/holcim/files/2024-04/28022024-holcim-climate-report-2023.pdf

IEA. (2020). *Final energy demand of selected heavy industry sectors by fuel, 2019, IEA*. Retrieved November 4, 2024, from https://www.iea.org/data-and-statistics/charts/final-energy-demand-of-selected-heavy-industry-sectors-by-fuel-2019

International Energy Agency (IEA). (2021). *Methane emissions from oil and gas*. Retrieved November 4, 2024, from https://www.iea.org/reports/methane-tracker-2021

Intergovernmental Panel on Climate Change (IPCC). (2021). *Climate change 2021: The physical science basis*. Cambridge University Press.

JBS. (n.d.). *Climate*. Retrieved November 23, 2024, from https://jbsesg.com/our-environment/climate/#:~:text=In%20addition%2C%20we%20are%20currently,to%20JBS

Nestle. (2023). *Creating Shared Value and Sustainability Report 2023: Advancing regenerative food systems at scale*. Retrieved November 4, 2024 from https://www.nestle.com/sustainability/reports/csv-sustainability-report-2023

O'Brien, I. (2024, September 15). Data center emissions probably 662% higher than big tech claims. Can it keep up the ruse? *The Guardian*. Retrieved November 4, 2024, from https://www.theguardian.com/technology/2024/sep/15/data-center-gas-emissions-tech

PepsiCo. (n.d.). *Climate change*. Retrieved November 23, 2024, from https://www.pepsico.com/our-impact/esg-topics-a-z/climate-change

Plumer, B. (2024, March 8). *Why New York is suing the world's biggest meat company*. Vox. Retrieved November 4, 2024, from https://www.vox.com/future-perfect/2024/3/8/24093774/big-meat-jbs-lawsuit-greenwashing-climate-new-york

Poore, J., & Nemecek, T. (2018). Reducing food's environmental impacts through producers and consumers. *Science, 360*(6392), 987–992. https://doi.org/10.1196/science.aaq0216

Rogelj, J., Forster, P. M., Kriegler, E., Smith, C. J., & Séférian, R. (2019). Estimating and tracking the remaining carbon budget for stringent climate targets. *Nature, 571*(7765), 335–342. https://doi.org/10.1038/s41586-019-1368-z

Stern, N. (2007). *The economics of climate change: The Stern review.* Cambridge University Press.

Timperley, J. (2018, September 13). Q&A: Why cement emissions matter for climate change. Carbon Brief. Accessed November 4, 2024 at https://www.carbonbrief.org/qa-why-cement-emissions-matter-for-climate-change/

Tyson Foods. (2022). *Growing a more sustainable future: Sustainability report 2022.* Retrieved November 4, 2024, from https://www.tysonfoods.com

United Airlines. (2024). *Environmental sustainability data and insights.* Retrieved November 4, 2024, from https://crreport.united.com/documents/united-environment-data.pdf?v=20241001

United Nations Framework for Climate Change (UNFCCC). (2015). Adoption of the Paris Agreement. FCC/CP/2015/L.9/Rev.1). Retrieved November 4, 2024 from https://unfccc.int/resources/docs/2015/cop21/eng/109r01.pdf

United Nations Global Compact UK. (n.d.). *Scope 3 emissions.* Retrieved November 23, 2024, from https://www.unglobalcompact.org.uk/scope-3-emissions/

World Resources Institute, & World Business Council for Sustainable Development. (2011). Corporate value chain (Scope 3) accounting and reporting standard: Supplement to the GHG Protocol corporate accounting and reporting standard. Greenhouse Gas Protocol. https://ghgprotocol.org/sites/default/files/standards/Corporate-Value-Chain-Accounting-Reporing-Standard_041613_2.pdf

World Resources Institute. (2020). Climate Watch: Historical greenhouse gas emissions by sector. https://www.climatewatchdata.org/ghg-emissions

Unraveling the Pathways: Exploring the Role of Scenarios in Projecting Climate Futures

6

Introduction

In 2008, the global financial crisis revealed the fragility of interconnected systems, shaking assumptions about stability and resilience. In 2020, the COVID-19 pandemic demonstrated how sudden disruptions can reshape economies and societies almost overnight. Now imagine a crisis that unfolds over decades, affects every nation, and leaves no sector untouched. Climate change is that crisis—a slow-moving but deeply transformative force requiring urgent, coordinated responses.

Understanding and preparing for climate change demands more than numbers and projections—it calls for stories. Climate scenarios, especially the Intergovernmental Panel on Climate Change (IPCC) Shared Socioeconomic Pathways (SSPs), are a way of telling these stories. Each scenario paints a picture of a possible future, shaped by decisions on policy, technology, and human behavior. Will we prioritize sustainability and global cooperation, or will we fragment into regional rivalries with uneven progress? Scenarios give us the tools to imagine these futures, not to predict them, but to explore what might happen—and what should happen.

In this chapter, we unravel the pathways offered by leading scenario frameworks. We start with the foundational tools of Integrated Assessment Models (IAMs) and the IPCC's SSPs, which weave narratives of socioeconomic development and climate response. We then examine how the International Energy Agency (IEA) and the Network for Greening the Financial System (NGFS) refine these narratives to offer actionable insights for energy transitions and financial stability.

Finally, we bring these ideas to life through the story of Air New Zealand. The airline began with SSPs to map risks and opportunities but shifted its emphasis as it integrated IEA and NGFS insights into its strategy. This real-world example shows how scenarios are more than abstract frameworks—they are powerful tools for decision-making in a world of uncertainty.

6.1   IAMs and Scenarios

Integrated Assessment Models (IAMs) lie at the intersection of three critical pillars: climate science, energy systems, and economics. By combining insights from these disciplines, IAMs provide an integrated framework to explore the complexities of climate change. They help quantify the relationships between greenhouse gas emissions, climate impacts, and economic outcomes, enabling the evaluation of mitigation and adaptation strategies. IAMs are essential tools in painting possible futures, estimating the social cost of carbon (SCC), and shaping net-zero emissions (NZE) strategies for governments and businesses. However, their complexity and reliance on assumptions invite scrutiny, particularly regarding loss and damages[1] from climate change.

The Three Pillars of the IAM

Climate Science
IAMs incorporate climate models to estimate how greenhouse gas emissions affect the Earth's climate system. They translate emissions trajectories into changes in temperature, precipitation patterns, sea level, and the frequency of extreme weather events. By integrating physical science data, IAMs provide insights into the likely impacts of different emissions pathways on ecosystems, human health, and infrastructure. These projections are critical for evaluating the potential consequences of climate inaction and for designing effective mitigation and adaptation strategies.

Energy Systems
Energy systems are central to IAMs because they account for the majority of global greenhouse gas emissions. IAMs model energy supply and demand dynamics, exploring how transitions to renewable energy, improvements in energy efficiency, and the phase-out of fossil fuels influence emissions trajectories. They also assess the potential of emerging technologies such as carbon capture and

---

[1] Loss and Damage refers to the negative impacts of climate change that go beyond what communities can adapt to, including economic and non-economic losses caused by extreme weather events and slow-onset changes such as sea-level rise. The concept has gained prominence in international climate negotiations, particularly in relation to financial support for vulnerable countries (IPCC, 2022).

storage (CCS) and green hydrogen, as well as shifts in energy use across sectors like transportation, industry, and residential consumption. This analysis helps identify cost-effective pathways for decarbonizing energy systems, a critical step in achieving net-zero emissions.

Economics

The economic pillar of IAMs examines how socioeconomic factors drive emissions and influence the costs and benefits of climate action. These models capture underlying economic trends, such as population growth, urbanization, industrialization, and the rise of emerging middle classes, which shape resource demands and emissions trajectories. IAMs also evaluate trade-offs between mitigation investments and the avoided damages from climate impacts, providing key metrics such as the social cost of carbon (SCC). This pillar links economic growth and development pathways with climate outcomes, offering insights into the feasibility of various strategies and their implications for global equity and economic stability.

By integrating these three pillars, IAMs create a comprehensive framework that connects the physical science of climate change with human behavior and policy responses. This confluence enables policymakers, businesses, and researchers to evaluate the implications of different climate pathways and make informed decisions about mitigation and adaptation strategies.

Painting Possible Futures

IAMs are indispensable for exploring potential futures under different levels of climate action. These models generate scenarios that highlight the consequences of policy choices, technological innovation, and societal behavior. For example, they help answer questions such as: What would happen if we delayed mitigation efforts? How might ambitious policies accelerate the transition to a low-carbon economy?

By translating narratives into quantitative outputs, IAMs provide a structured way to evaluate the trade-offs associated with different climate strategies. Their results inform global and regional policies, shaping international agreements like the Paris Accord and guiding corporate NZE strategies (Rogelj et al., 2018).

Estimating the Social Cost of Carbon

IAMs play a central role in estimating the SCC, a critical metric for policymaking. The SCC reflects the monetary value of the damages caused by emitting one additional ton of CO_2, encompassing effects on health, agriculture, and ecosystems. Models like DICE (Dynamic Integrated Climate-Economy) and PAGE (Policy Analysis of the Greenhouse Effect) calculate the SCC by integrating emissions pathways with economic and climate damage functions (Nordhaus, 2017).

While SCC estimates are invaluable, their accuracy depends on assumptions about future climate impacts, discount rates, and societal preferences. Critics have pointed out that these estimates often fail to account for low-probability, high-impact events or non-market damages like biodiversity loss, potentially undervaluing the true costs of emissions (Pindyck, 2013).

Shaping Net-Zero Strategies

IAMs are equally critical in guiding NZE strategies for governments and firms. By simulating the implications of various interventions—such as renewable energy adoption, carbon capture and storage (CCS), and land-use changes—IAMs help identify cost-effective pathways to carbon neutrality. For firms, IAM outputs provide a basis for aligning business strategies with global climate goals (Rogelj et al., 2018).

IAMs also highlight risks, such as stranded assets in high-carbon sectors, and opportunities, such as the growth potential in clean technologies. However, their outputs are highly sensitive to assumptions, requiring careful interpretation.

Dynamic Feedback Loops

One of the key complexities of Integrated Assessment Models (IAMs) lies in the presence of dynamic feedback loops, which reflect the interconnected nature of mitigation, adaptation, economic development, and societal change. These feedback mechanisms are crucial for accurately representing the long-term impacts of climate policies but also significantly complicate the modeling process.

Feedback Between Mitigation and Adaptation

Aggressive mitigation efforts, such as transitioning to renewable energy or implementing large-scale reforestation, can significantly reduce future climate impacts. This, in turn, lowers the costs and urgency of adaptation measures, such as building seawalls or developing drought-resistant crops. For instance, early investments in low-carbon technologies might stabilize global temperatures, reducing the risks of catastrophic climate events that would otherwise necessitate costly adaptation measures.

Conversely, inadequate or delayed mitigation exacerbates the severity of climate impacts, leading to escalating adaptation costs. These costs are not linear; as climate impacts intensify, adaptation may become prohibitively expensive or even ineffective in some cases. For example, rising sea levels could surpass the height of existing seawalls, forcing communities to either retreat or invest in more extreme measures.

Economic Feedback Loops

Mitigation itself reshapes economic systems in ways that feed back into emissions trajectories. For instance, transitioning to renewable energy reduces reliance on fossil fuels, which can lower energy prices and drive further shifts in consumption and production patterns. This feedback loop accelerates the adoption of cleaner technologies and amplifies the overall mitigation effect. However, such transitions also require substantial upfront investments, which may temporarily increase emissions in sectors like construction and manufacturing.

Economic development, particularly in emerging markets, introduces additional layers of complexity. Many developing countries are experiencing rapid population growth and the emergence of a middle class, which drives increased demand for energy, transportation, and consumer goods. IAMs must account for this growth and its implications for emissions. For instance, a growing middle class in regions like Southeast Asia or sub-Saharan Africa may lead to higher emissions in the short term, even as global mitigation efforts intensify. These shifts highlight the challenge of balancing economic growth with sustainability.

Population Growth and Resource Demand

Population growth is another critical driver that IAMs must consider. In scenarios with high population growth, the demand for resources—such as food, water, and energy—increases significantly. This can strain ecosystems, drive deforestation, and increase greenhouse gas emissions. IAMs integrate demographic trends to evaluate how population growth interacts with mitigation and adaptation efforts. For example, higher population scenarios may require more ambitious mitigation strategies to offset increased emissions, as well as more adaptive capacity to manage climate impacts.

Technological Innovation and Adoption

IAMs must also account for the role of technological innovation and adoption, which can create positive feedback loops. Advances in renewable energy technologies, energy storage, and carbon capture and storage (CCS) can accelerate the transition to a low-carbon economy. Once these technologies reach a tipping point of affordability and scalability, their adoption often grows exponentially, reducing costs further and driving broader societal shifts.

However, the pace of innovation and adoption is highly uncertain, and IAMs must make assumptions about how quickly these technologies will develop and diffuse. For instance, optimistic scenarios assume rapid breakthroughs in areas like green hydrogen or direct air capture, while pessimistic scenarios emphasize slower progress and higher costs.

Behavioral and Policy Feedback

Behavioral responses and policy feedback are additional factors that complicate IAM dynamics. Policies aimed at reducing emissions, such as carbon taxes or subsidies for renewable energy, influence individual and corporate behavior, which in turn shapes emissions trajectories. For example, higher carbon prices can incentivize energy efficiency and renewable energy adoption, creating a virtuous cycle of emission reductions. However, poorly designed policies may have unintended consequences, such as economic stagnation or inequitable burdens on vulnerable populations, which could undermine global cooperation and delay progress.

Cascading Risks and Nonlinear Effects

Finally, IAMs must grapple with the possibility of cascading risks and nonlinear effects, which can amplify feedback loops. For example, climate-induced disruptions to food or water supplies may trigger social and political instability, leading to reduced economic growth and increased vulnerability to future impacts. These cascading risks are difficult to model but are critical for understanding the full scope of climate impacts and the importance of proactive mitigation.

Interdependencies Between Pillars

The dynamic feedback loops in IAMs highlight the interdependencies between the three pillars of climate science, energy systems, and economics. Mitigation strategies influence energy systems and economic structures, while population growth and technological innovation shape both emissions and adaptive capacity. These interconnections underscore the need for integrated approaches to climate modeling and policymaking.

IAMs provide a structured framework to evaluate these feedback loops, but their complexity also raises challenges for interpretation. Policymakers and stakeholders must recognize that IAM outputs are not precise predictions but scenario-based explorations of potential futures. By examining the interplay of mitigation, adaptation, and societal change, IAMs illuminate the trade-offs and opportunities inherent in addressing climate change, helping society make informed decisions about its path forward.

Challenges and Limitations

Despite their utility, IAMs face limitations. Their reliance on assumptions about population growth, economic development, and technology adoption introduces uncertainty. Furthermore, IAMs often struggle to model nonlinear climate impacts, such as tipping points, which could lead to abrupt and irreversible changes in the Earth system. These limitations underscore the need for IAM outputs to be interpreted as scenarios, not predictions.

6.2   Societal Choices

IAMs are tools to help society navigate the complex choices involved in addressing climate change. They illuminate the trade-offs between different pathways, from ambitious mitigation to adaptation-focused approaches. While IAMs do not prescribe solutions, they provide the analytical foundation for informed decision-making.

The climate decisions we make today as a society will reflect how global warming impacts future generations. What might the future look like? Each generation is experiencing a warmer world than the last. A person born in 1950 entered a world with only a 0.25 °C increase above pre-industrial temperatures. Today, this same person has lived through a total warming of 0.85 °C, averaging around 0.12 °C per decade. By contrast, a person born in 1980 was already in a world 0.4 °C warmer and has since experienced an additional 0.75 °C rise over just 40 years. Putting this another way, people born in 1980 are experiencing nearly 50% faster warming than those born in 1950 (IPCC, 2023).

Projections indicate that limiting warming to 1.5 °C by 2050 requires substantial reductions in greenhouse gas emissions. This requires global efforts to transition from fossil fuels toward renewable energy, expand natural carbon sinks like forests, and decarbonize core industries such as agriculture, transportation, and manufacturing. The IPCC estimates that CO_2 emissions need to decline by 45% from 2010 levels by 2030, reaching net zero by 2050. Turning this projection into a reality will require cross-border collaboration, sector-wide innovation, and significant capital investment in both mitigation and adaptation efforts.

Net-Zero Emissions (NZE) and Nationally Determined Contributions (NDCs)

Nationally Determined Contributions (NDCs) are expressions of societal choices regarding climate action. Submitted to the UNFCCC, countries outline how they plan to reduce greenhouse gas emissions and adapt to climate impacts. While many NDCs aim for net-zero emissions by mid-century, their ambition and detail vary widely based on national circumstances. Current NDCs are insufficient to limit warming to 1.5 °C, with projected increases of 2.4–2.7 °C by 2100 (Climate Action Tracker, 2023). Table 6.1 highlights the NDCs of the four largest GHG emitters—China, the United States, India, and the EU—detailing their targets, sectoral actions, and adequacy ratings. These countries' commitments fall short of achieving NZE or the 1.5 °C target. "Highly Insufficient" ratings (e.g., China, India) signal inadequate alignment, while "Insufficient" ratings (e.g., United States, EU) reflect greater ambition but persistent policy gaps. To close these gaps, both national and corporate actions must align with NDCs. The next section explores how firms can engage with and track progress toward NZE.

Table 6.1 Country NDC report card

Country	Target	Sectoral actions	CAT score
China	Peak carbon dioxide emissions by 2030 and achieve carbon neutrality (NZE) by 2060	Increase non-fossil energy share to 25%; enhance carbon sinks through afforestation and forest stock growth; promote green and low-carbon development in energy, transportation, and industrial sectors	Highly Insufficient
United States	Reduce greenhouse gas emissions by 50–52% below 2005 levels by 2030, Achieve NZE by 2050	Transition to 100% carbon pollution-free electricity by 2035; electrify transportation; promote energy efficiency in buildings; invest in clean manufacturing and industrial decarbonization; implement climate-smart agriculture and conserve forests	Insufficient
India	Reduce emission intensity by 33–35% by 2030 compared to 2005 levels. No NZE target included in NDC	Increase renewable energy capacity to 450 GW; enhance forest cover and carbon sinks; promote energy efficiency in industry and transportation; advance climate resilience in agriculture and infrastructure	Highly Insufficient
European Union	Reduce greenhouse gas emissions by at least 55% by 2030 compared to 1990 levels. Achieve carbon neutrality (NZE) by 2050	Expand renewable energy and improve energy efficiency; reform the emissions trading system (ETS); enhance carbon sinks through forestry and land use; strengthen adaptation and resilience	Insufficient

Source Information for NDCs is obtained from the UNFCC NDC Registry. CAT scores are from the Carbon Action Tracker

Companies Pledging Carbon Neutrality vs. Net-Zero Emissions: United States vs. Europe

While many US companies focus on carbon neutrality pledges, European firms are more likely to commit to achieving net-zero emissions (NZE) or full climate neutrality, as mandated by the European Union's Fit for 55 package. These differences highlight varying levels of ambition and comprehensiveness in addressing greenhouse gas (GHG) emissions.

Defining the Terms

- **Carbon Neutrality**: Carbon neutrality refers to balancing the amount of carbon dioxide (CO_2) emitted with equivalent removals or offsets, focusing exclusively on CO_2 and often excluding other GHGs. Companies may achieve carbon neutrality through measures like investing in offset projects without necessarily reducing operational emissions significantly.
- **Net-Zero Emissions (NZE)**: NZE aims to reduce all GHG emissions, not just CO_2, across a company's entire value chain to as close to zero as possible, with any residual emissions balanced by equivalent removals. NZE requires deep emission cuts and comprehensive strategies that include direct (Scope 1), indirect (Scope 2), and value chain emissions (Scope 3).
- **Climate Neutrality**: Climate neutrality extends the NZE concept to a broader framework. It involves achieving a balance between all GHG emissions produced and removed, while also considering the overall impact on the climate system. Under the EU's Fit for 55 package, climate neutrality is defined as balancing all GHG emissions with equivalent removals to achieve no net impact on the climate by 2050 (European Commission, 2021).

Science Based Target Initiative (SBTi) and Firm-Level Commitments

For firms aiming to align with NZE goals, the Science Based Target initiative (SBTi) offers a robust framework. SBTi helps companies define and pursue emissions reduction targets that align with the Paris Agreement's objective of limiting global warming to 1.5 °C or well below 2 °C. A collaboration between CDP, the UN Global Compact, WRI, and WWF, SBTi ensures:

- **Comparability**: Standardized metrics enable benchmarking across industries.
- **Transparency**: Public reporting of progress fosters accountability.
- **Alignment**: Targets align with the Paris Agreement, ensuring progress toward NZE by 2050.

SBTi's sector-specific guidance outlines clear targets. For example, firms in the iron and steel sector who design an NZE goal in accordance with SBTi guidelines need to indicate the following concrete measures across their value chain:

- **Scope 1 (Direct Emissions)**: A 91% reduction by 2050, targeting processes like blast furnaces.
- **Scope 2 (Indirect Emissions)**: Increased reliance on renewable energy.
- **Scope 3 (Value Chain Emissions)**: Tackling significant upstream and downstream emissions through technologies like green hydrogen and Carbon Capture, Utilization, and Storage (CCUS).

By having their NZE plan approved by the SBTi, firms can enhance their credibility and contribute to global decarbonization efforts.

Comparing US and European Approaches to NZE Goal

An analysis of companies in the S&P 500 index found that among 434 firms responsible for 95.6% of the index's total emissions, most US companies focus on carbon neutrality rather than NZE. This often involves reliance on carbon offsets and excludes Scope 3 emissions, which account for a significant portion of a company's overall climate impact (S&P Global, 2023).

In contrast, European firms are more likely to adopt NZE or climate neutrality strategies. Driven by regulatory frameworks like the EU Emissions Trading System (ETS) and policies such as Fit for 55, European firms align with science-based targets and demonstrate greater transparency in reporting progress.

Challenges in Achieving NZE Commitments

Despite credible frameworks like SBTi, achieving NZE at the firm level remains fraught with challenges:

- **Lack of Standardization**: Firm-level commitments often lack uniform metrics or enforcement, making it difficult to compare or verify progress.
- **"Future Washing"**: Firms may overstate long-term climate strategies. The Volkswagen emissions scandal, for example, highlighted how misleading claims erode trust.
- **Regulatory Gaps**: Without strong oversight, firms can exploit loopholes, undermining the integrity of their climate goals.

Emerging Markets: Commitments to Net-Zero Emissions

In emerging markets, particularly China and India—two of the world's largest greenhouse gas emitters—corporate commitments to net-zero emissions (NZE) are increasing, though they still lag more developed economies.

In China, thirty-seven companies have set climate targets through the Science Based Targets initiative (SBTi), with thirteen having submitted their targets and four committing to achieve net-zero emissions by 2050. Additionally, eleven companies have pledged to use 100% renewable electricity through the RE100 initiative. However, significant challenges, including high costs and limited regulatory mandates, constrain broader adoption of NZE strategies (CDP, 2023).

In India, corporate NZE commitments are also emerging but remain limited. A growing number of Indian companies are setting science-based targets, as reported by SBTi, though the total number of firms with explicit NZE goals is not as prominent. This slower adoption reflects financial constraints and competing developmental priorities that often delay ambitious climate action (Science Based Targets Initiative, 2023).

Emerging legal frameworks aim to curb these practices, ensuring that NZE claims are credible and transparent. However, achieving NZE demands rigorous action, accountability, and innovation. By aligning with frameworks like SBTi and following stricter European standards, firms can enhance their role in global decarbonization efforts.

Closing the Gap: From Ambition to Action

There is considerable disparity between ambitious climate pledges and the actual ability to deliver on those commitments. This issue is particularly evident in the different approaches taken by companies in the United States, Europe, and emerging markets like China and India. While many firms are announcing goals to achieve net-zero emissions (NZE) or carbon neutrality, the structural, financial, and regulatory barriers to meaningful action remain significant. Bridging this gap requires addressing these barriers while tailoring solutions to the specific challenges of each region.

In the United States, companies often focus on carbon neutrality, emphasizing offsets rather than deep reductions across their entire value chains. This approach falls short of the comprehensive changes needed for NZE, especially in addressing Scope 3 emissions. However, recent policies, like the Inflation Reduction Act in the United States, present an opportunity to support a more robust transition. By leveraging these incentives, US firms can pivot from short-term strategies to broader commitments aligned with science-based targets.

Europe leads the way in ambitious climate policies, with frameworks such as Fit for 55 and the Emissions Trading System driving corporate action. Yet even in the EU, challenges persist. Some companies struggle to deploy innovative solutions like green hydrogen and carbon capture technologies at scale, while inconsistencies in how member states implement regulations can undermine progress. Strengthening the alignment of standards across Europe and scaling up the deployment of low-carbon technologies will be critical to closing the gap.

Emerging markets, particularly China and India, present a different set of challenges. While corporate NZE commitments are growing in these regions, the pace of adoption remains slower than in developed economies. Financial constraints, limited regulatory frameworks, and competing developmental priorities often hinder progress. For instance, China has made strides with companies setting science-based targets, but only a fraction has committed to achieving NZE. In India, the number of firms making similar commitments remains relatively small. To close the gap in these regions, increased access to green financing, stronger domestic climate policies, and partnerships with multinational corporations are essential.

Ultimately, closing the gap will require aligning corporate actions with the national climate goals laid out in Nationally Determined Contributions (NDCs) under the Paris Agreement. While the United States and Europe provide regulatory and financial leadership, emerging markets need targeted support to overcome structural barriers. By fostering global collaboration—through financing mechanisms, technology transfers, and consistent international standards—companies worldwide can move beyond ambitious pledges to deliver tangible results. This collective effort is crucial to achieving net-zero emissions and creating a sustainable global economy.

6.3 Climate Scenarios

The climate crisis presents an array of possible futures, from sustainable, decarbonized societies to catastrophic "hothouse" scenarios—depending on our choices as a society. While addressing the gaps in corporate and national commitments (NDCs) to decarbonization is critical for achieving net-zero emissions (NZE), the question remains: what does the future hold if these efforts succeed—or fail? The path to a sustainable future is shaped not only by actions taken today but also by the uncertainties of tomorrow. Climate scenarios, therefore, play a critical role in envisioning potential outcomes of the climate crisis, helping stakeholders anticipate challenges and opportunities.

To understand how the climate crisis might unfold, this section examines scenarios from the IPCC, NGFS, and IEA. It bears repeating that these frameworks are not predictive; rather, they outline plausible futures shaped by various trajectories of emissions, socioeconomic shifts, and policy responses. Nonetheless, by exploring these pathways, decision-makers can better identify risks and opportunities while crafting informed strategies to navigate an uncertain future.

Scenario Brands

Three key scenario providers are widely used in business and policy:

- **IPCC**: The authoritative source on climate science, offering comprehensive insights into physical impacts and socioeconomic pathways in its Sixth Assessment Report (AR6, 2021).
- **NGFS**: Focuses on the macroeconomic and financial implications of climate change, highlighting transition risks for the financial sector.
- **IEA**: Delivers granular analyses of the energy sector, emphasizing pathways for reducing GHG emissions, transitioning to renewables, and ensuring energy security.

Together, these scenarios provide diverse perspectives, and using a combination is often essential to fully capture the complexities of climate risks.

6.4   IPCC Scenarios

The IPCC offers a "menu" of possible futures by combining physical and socioeconomic projections using Representative Concentration Pathways (RCPs) and Shared Socioeconomic Pathways (SSPs). RCPs represent trajectories of greenhouse gas concentrations and their associated radiative forcing, while SSPs focus on socioeconomic factors such as population growth, technological development, and policy priorities. Together, these frameworks provide integrated scenarios that depict a wide spectrum of outcomes, from effective decarbonization to severe global warming, highlighting the challenges of both mitigating and adapting to climate change.

Representative Concentration Pathways (RCPs): Emissions and Radiative Forcing

RCPs quantify the pathways of greenhouse gas (GHG) emissions and their impact on radiative forcing—the net energy imbalance in the atmosphere—expressed in watts per square meter (W/m^2). First introduced in the IPCC's Fifth Assessment Report (AR5), RCPs form the backbone of physical climate projections (van Vuuren et al., 2011):

- **RCP2.6**: Represents a low-emission pathway, requiring rapid reductions in GHG emissions and achieving net-zero carbon by mid-century. It limits warming to approximately 1.5 °C–2 °C by 2100.
- **RCP4.5** and **RCP6.0**: Stabilization pathways that assume moderate mitigation, resulting in mid-range warming between 2 °C and 3 °C.
- **RCP8.5**: A high-emission trajectory characterized by unrestrained fossil fuel use and minimal climate action, leading to warming above 4 °C by 2100.

While RCPs provide valuable insights into emissions and radiative forcing, they lack detail about the socioeconomic drivers influencing these trajectories.

Shared Socioeconomic Pathways: Mapping Climate Futures

To address this gap, the IPCC introduced SSPs in its Sixth Assessment Report (AR6). SSPs provide socioeconomic narratives that describe global development trends, policy choices, and their impact on emissions, mitigation, and adaptation. They complement RCPs by adding the human dimension, exploring how factors such as population growth, economic development, and technological innovation shape climate futures (O'Neill et al., 2017; Riahi et al., 2017).

SSP1: Sustainability (The Green Road)

SSP1 describes a sustainable world that prioritizes environmental stewardship, equity, and global cooperation. Population growth is low, supported by high investment in education and health. Strong international collaboration drives efforts to address climate change, while rapid adoption of clean technologies and renewable energy minimizes environmental impacts. Economic development focuses on reducing inequality and improving resource efficiency, resulting in low challenges for both mitigation and adaptation. This pathway envisions a future where proactive policies and green technology enable significant progress in decarbonization, with limited need for adaptation due to effective mitigation efforts.

SSP2: Middle of the Road

SSP2 envisions a world continuing along historical trends, characterized by uneven progress in development and technology. Population growth and economic development are moderate, with mixed success in addressing inequality and sustainability goals. While cleaner energy technologies are adopted, progress is slow and incremental. This scenario presents moderate challenges for both mitigation and adaptation, resulting in outcomes that are neither highly positive nor negative. Climate impacts are manageable, but the potential for transformative progress is limited.

SSP3: Regional Rivalry (A Rocky Road)

SSP3 represents a fragmented world dominated by strong nationalistic tendencies and limited international cooperation. Population growth is high, driven by insufficient investments in education and healthcare. Weak governance and frequent regional conflicts hinder climate action, leading to a reliance on local and traditional energy sources rather than clean alternatives. Inequality remains widespread, with many regions lacking the resources to adapt to climate impacts. This pathway poses severe challenges for both mitigation and adaptation, resulting in high emissions and significant climate-related risks.

SSP4: Inequality (A Divided World)

SSP4 describes a world sharply divided between high-income, technologically advanced regions and low-income, vulnerable areas. Population growth is low in wealthier regions but high in poorer ones, exacerbating global disparities. Advanced economies effectively implement mitigation measures, but poorer regions face severe adaptation challenges due to limited access to education, technology, and financial resources. This scenario reflects a divided world where the wealthy mitigate their emissions successfully, while the poor bear the brunt of escalating climate impacts.

SSP5: Fossil-Fueled Development (Taking the Highway)

SSP5 depicts a world characterized by rapid economic growth driven by intensive fossil fuel use and technological innovation. Population growth is high initially but slows as development advances. The focus is on economic expansion and market-based solutions, often at the expense of environmental policies. Technological advancements aim to address climate impacts, but the continued reliance on fossil fuels leads to significant emissions and warming. Mitigation challenges are high, but adaptation challenges are low, as economic and technological capacity enables effective responses to climate risks. This pathway envisions a future of substantial warming, with adaptation primarily managed in wealthier regions.

SSP-RCP Matrix: Theoretical and Practical Combinations

Riahi et al. (2017) introduced the SSP-RCP matrix as a framework to analyze the interactions between socioeconomic pathways (SSPs) and climate forcing trajectories (RCPs). In theory, this matrix allows for twenty unique combinations, as there are five SSPs and four commonly used RCPs: RCP2.6, RCP4.5, RCP6.0, and RCP8.5. Each combination represents a distinct scenario that connects socioeconomic development to a specific level of radiative forcing by the year 2100.

However, in practice, not all combinations are feasible or meaningful. Some SSP-RCP pairings are inconsistent with the underlying narratives. For instance, RCP2.6, which requires rapid decarbonization, is incompatible with SSP3, which features weak international cooperation and continued reliance on fossil fuels. Similarly, RCP8.5, representing a high-emission, business-as-usual trajectory, conflicts with SSP1, which focuses on sustainability and low-emission development. These inconsistencies result in a smaller set of practically relevant combinations that align logically to represent plausible futures (Table 6.2).

The SSP-RCP matrix provides insights into the dual challenges of mitigation and adaptation across different pathways. Mitigation challenges are most pronounced under RCP2.6, which requires ambitious emissions reductions, and are most consistent with SSPs emphasizing global cooperation and technological innovation, such as SSP1 or SSP5. In contrast, RCP8.5 aligns with SSP3 or SSP5, which feature high fossil fuel reliance and weak climate policy.

Adaptation challenges also vary significantly across SSPs. Scenarios under SSP1 face lower adaptation challenges, as proactive mitigation reduces the severity of climate impacts. SSP4 presents significant disparities, with wealthier regions achieving effective mitigation while poorer regions grapple with severe adaptation challenges. SSP3 reflects the most extreme adaptation pressures due to high population growth, weak institutions, and fragmented efforts to address climate impacts.

By integrating SSPs and RCPs, the matrix provides a comprehensive framework to explore plausible futures, illustrating how socioeconomic pathways shape climate trajectories and the associated mitigation and adaptation challenges.

Table 6.2 SSP-RCP matrix with mitigation and adaptation challenges

SSP	RCP2.6 (1.5–2 °C)	RCP4.5 (2–3 °C)	RCP6.0 (3–4 °C)	RCP8.5 (4 °C+)
SSP1	Low Mitigation, Low Adaptation	Moderate Mitigation, Low Adaptation	Unlikely pairing	Inconsistent pairing
SSP2	Moderate Mitigation, Moderate Adaptation	Moderate Mitigation, Moderate Adaptation	Moderate Mitigation, High Adaptation	High Mitigation, High Adaptation
SSP3	Inconsistent pairing	Unlikely pairing	High Mitigation, High Adaptation	Severe Mitigation, Severe Adaptation
SSP4	Low Mitigation, High Adaptation	Moderate Mitigation, High Adaptation	High Mitigation, Severe Adaptation	Severe Mitigation, Severe Adaptation
SSP5	Unlikely pairing	Moderate Mitigation, Low Adaptation	Moderate Mitigation, Moderate Adaptation	High Mitigation, Low Adaptation

Source Riahi et al. (2017)

The Socioeconomic Context: An Eye to Our Future

The socioeconomic lens of the IPCC scenarios offers a compelling perspective on how the climate crisis might unfold under different global circumstances. This framework invites us to consider not just the physical impacts of climate change but also how societal structures, economic priorities, and global cooperation—or the lack thereof—shape the world's ability to address these challenges.

We cannot help but reflect on the significant global shifts since 2008, which have underscored humanity's vulnerability to systemic shocks. The 2008 financial crisis highlighted the fragility of global economic systems and the interconnectedness of markets, emphasizing the importance of resilience in the face of systemic risks. The COVID-19 pandemic demonstrated the profound societal and economic disruptions that a global crisis can cause, as well as the potential for rapid behavioral, policy, and technological shifts when urgency demands it.

These events underline how quickly socioeconomic contexts can change, affecting the assumptions underlying Shared Socioeconomic Pathways (SSPs). For instance, the pandemic catalyzed advancements in digital technology and reshaped global supply chains, which could accelerate transitions toward low-carbon economies, aligning with SSP1 (Sustainability). Conversely, rising economic inequality and geopolitical tensions post-COVID-19 mirror the fragmented world envisioned in SSP3 (Regional Rivalry), where mitigation and adaptation efforts face severe barriers. Whether we move toward a sustainable, cooperative world like SSP1 or face the stark challenges of SSP3 or SSP4 will depend on the resilience, innovation, and equity of global systems in the coming decades.

The IPCC's scenarios form the bedrock of climate scenario analysis, offering the most comprehensive integration of socioeconomic and physical drivers of climate change. However, specific sectors, such as finance and energy, require more tailored scenarios to address their unique challenges. IEA and NGFS scenarios leverage the insights from IPCC scenarios to provide actionable frameworks for their respective audiences. These scenarios, while narrower in scope, complement the IPCC's work by focusing on critical dimensions such as financial stability and energy system transitions.

6.5   Specialized Scenarios: IEA and NGFS

Building on the IPCC's foundational scenarios, the International Energy Agency and the Network for Greening the Financial System have developed specialized, derivative scenarios. The IEA focuses on energy transitions, providing insights into decarbonization pathways and the technological and policy shifts required to achieve net-zero emissions (IEA, 2024), while the NGFS focuses on scenarios that capture how climate risks can translate into threats to the stability of the financial system.

Both IEA and NGFS scenarios, however, have direct relevance to the corporate sector so firms will often use one or the other, or both to inform their own climate scenarios.

IEA Projections and Scenarios

The IEA focuses on three primary scenarios, each reflecting different levels of policy ambition and technological advancement. Each scenario incorporates varying assumptions about technological progress, consumer behavior, and policy implementation, providing a spectrum of possible energy futures.

Stated Policies Scenario (STEPS)

STEPS reflects a world where current policies and commitments, as of 2024, are implemented without significant new measures. It assumes that governments meet their announced targets but stop short of ambitious additional action to achieve long-term climate goals. This scenario serves as a conservative baseline for global energy trends, showcasing the consequences of limited progress in climate policy.

Under STEPS, global energy demand continues to grow, driven by economic expansion and population growth, particularly in emerging markets. Fossil fuels remain a dominant part of the energy mix, with moderate growth in renewable energy and efficiency improvements. The scenario highlights an "implementation gap," showing how existing policies fall short of the transformative changes needed to align with the Paris Agreement. Carbon emissions plateau in some regions but continue to rise in others, particularly where industrialization is accelerating.

Announced Pledges Scenario (APS)

APS goes beyond STEPS by incorporating all government pledges, including Nationally Determined Contributions (NDCs) and net-zero commitments. It assumes these targets are met in full and on time, providing a more optimistic view of future energy and emissions trajectories. APS bridges the gap between current policies and the more ambitious actions required for achieving global climate goals.

In this scenario, significant investments are directed toward clean energy technologies, such as solar and wind power, hydrogen, and Carbon Capture, Utilization, and Storage (CCUS). Fossil fuel consumption begins to decline in key sectors, although it remains a major contributor to energy supply in regions with slower transitions. APS reveals the "ambition gap," demonstrating that while announced pledges narrow the distance to net-zero emissions, they are insufficient for meeting the 1.5 °C target without further acceleration.

Net-Zero Emissions (NZE) by 2050 Scenario

The NZE scenario presents a normative pathway that aligns with the Paris Agreement's ambition to limit global warming to 1.5 °C. It assumes a rapid and coordinated global transformation of the energy system, requiring unprecedented policy action, technological deployment, and societal change.

In the NZE world, energy systems are fundamentally reshaped to achieve net-zero carbon emissions by 2050. Fossil fuel use is drastically reduced, with oil and gas consumption declining sharply, and coal nearly phased out. Renewable energy dominates the power sector, supported by advances in storage and grid technologies. Electrification accelerates in transport, industry, and buildings, with hydrogen and bioenergy playing critical roles in sectors where direct electrification is challenging. Investments in CCUS and negative emissions technologies scale up to address residual emissions.

Achieving the NZE scenario requires significant behavioral and structural changes, including efficiency improvements, shifts in consumption patterns, and large-scale reforestation. The scenario highlights the scale of global cooperation and resource mobilization needed to meet the 1.5 °C goal.

Informing Corporate Climate Strategies

The IEA scenarios, including the Stated Policies Scenario (STEPS), Announced Pledges Scenario (APS), and net-zero emissions (NZE) by 2050 scenario, provide sector-specific and detailed projections that are particularly valuable for corporate decision-making. While the overarching framework of the SSP-RCP describes global socioeconomic and climate pathways, the IEA scenarios offer a deeper understanding of key factors influencing the energy transition.

Table 6.3 examines emissions pricing, fossil fuel demand, critical minerals, and decarbonization technologies like CCUS and SAF. It shows how businesses can align strategies with policy changes, energy market shifts, and technology adoption to manage risks and seize opportunities in the evolving energy landscape.

Table 6.3 Granular insights from IEA scenarios: informing corporate climate strategies

Aspect	STEPS	APS	NZE	Relevance
Emissions pricing	Limited implementation of carbon pricing; minimal impact on emissions	Moderate carbon prices aligned with pledged commitments	High carbon prices drive decarbonization in energy-intensive sectors	Helps evaluate cost implications and policy-driven market shifts
Oil prices and demand	Sustained demand keeps prices high, driven by transportation and industry	Declining demand leads to moderate price reductions	Drastic reduction in demand; oil prices drop significantly	Assesses transition risks for fossil fuel industries
CCUS deployment	Slow adoption due to weaker policy incentives	Gradual scaling in line with pledged policies	Rapid deployment in hard-to-abate sectors; large-scale investments in infrastructure	Critical for decarbonization in heavy industry and power sectors
Critical minerals	Limited pressure on critical mineral supply chains due to slower renewable adoption	Moderate growth in demand for materials like lithium and cobalt	Rapid demand growth highlights the risk of supply bottlenecks	Supports planning for resource availability and supply chain security
Sustainable Aviation Fuel (SAF)	Marginal penetration due to high costs and limited incentives	Gradual adoption; constrained by infrastructure and investment gaps	Widespread adoption; significant contribution to aviation decarbonization by mid-century	Aids aviation sector in planning for emissions reductions and technology shifts
Regional pathways	Uneven renewable adoption across regions; fossil fuels dominate in many emerging markets	Pledges drive regional variations in clean energy adoption rates	Coordinated global transition accelerates renewables in all regions	Provides targeted insights for regional policy development and investment

Source IEA (2024)

NGFS Projections and Scenarios

NGFS scenarios provide essential tools for evaluating the financial sector risks associated with climate change. Developed with insights from the IPCC's Shared Socioeconomic Pathways, these scenarios align climate risks with financial sector concerns, such as asset exposure, systemic stability, and economic feedback loops. They explore a range of climate action pathways, highlighting the physical and transition risks tied to varying policy coordination and timing. Given their emphasis on how climate risk can impact the financial sector, and most importantly, the stability of the financial system, NGFS scenarios are concerned if the transition to an NZE world will materialize in an orderly manner or a disorderly one (Network for Greening the Financial System, 2022).

Example: Transmission Channel

Financial institutions with significant exposure to fossil fuel assets face considerable risks during the transition to a low-carbon economy, which has profound implications for the corporate sector's ability to secure debt financing. For instance, the Royal Bank of Canada (RBC), given its strong ties to Canada's fossil fuel industries, is particularly vulnerable to stranded assets—those that lose value prematurely due to regulatory, market, or societal changes. Tightening climate policies exacerbate credit risks for high-carbon companies, as declining revenues and rising operating costs, such as those driven by carbon pricing, strain their ability to service debt. These pressures are compounded by market risks, as securities linked to fossil fuel industries lose value, destabilizing financial institutions' portfolios.

This dynamic introduces the potential for a climate-related **Minsky Moment**, as discussed in Chapter 3. In such a scenario, the overvaluation of carbon-intensive assets—combined with abrupt shifts in policy or market sentiment—could trigger a rapid revaluation of financial markets. A disorderly transition, marked by delayed but aggressive policy interventions or sudden technological advancements, could accelerate this process, creating sharp devaluations in high-carbon assets and triggering systemic instability in financial markets.

For the corporate sector, this means rising borrowing costs, reduced debt maturities, or complete exclusion from credit markets, as financial institutions prioritize low-carbon investments to mitigate their own climate risks. The NGFS framework emphasizes these risks, particularly in its disorderly transition scenarios, highlighting how delayed action increases volatility and limits the time available for both financial institutions and corporations to adjust. In contrast, an orderly transition, with early and predictable policy interventions, provides a more stable pathway, reducing the risks of stranded assets and minimizing financial instability.

This evolving financial reality highlights the interconnectedness of the NGFS's work and the corporate sector's need to adapt. As financial institutions align with climate goals, the pressure on corporates to disclose, mitigate, and adapt to climate risks will continue to grow. This alignment not only reshapes the availability of capital but also fundamentally transforms how businesses operate and compete in the global economy.

Orderly and Disorderly Transitions

The NGFS scenarios are divided into two primary transition types: orderly and disorderly transitions, as well as a third pathway that assumes insufficient climate action.

Orderly Transition

In an orderly transition, climate policies are introduced early and ramped up gradually, allowing economies and businesses time to adjust. This scenario minimizes disruptions to financial markets and reduces the risks of stranded assets. For example, carbon-intensive industries, while still facing declining valuations, are able to transition more predictably, providing financial institutions and corporations sufficient time to adapt their investment strategies. This scenario typically assumes that global warming is limited to well below 2 °C, consistent with the goals of the Paris Agreement (NGFS, 2022).

Disorderly Transition

- In contrast, a disorderly transition occurs when policy actions are delayed and then implemented abruptly. "Too Little, Too Late" describes a future where abrupt measures disrupt economies, leaving physical risks. This creates significant volatility in markets and leaves industries with insufficient time to adapt, resulting in a sharp increase in stranded assets and credit defaults. In such a scenario, the corporate sector faces heightened borrowing costs and diminished access to capital, especially for companies in fossil fuel-intensive sectors. The "Divergent Net-Zero" scenario envisions a world where there are regional disparities in climate policy that create uneven transitions and localized vulnerabilities.

Insufficient Climate Action

Hothouse World: A third pathway modeled by NGFS is a hothouse world, where little to no additional climate action is taken. This scenario assumes that global warming exceeds 3 °C, leading to severe physical risks from climate impacts such as rising sea levels, extreme weather events, and disruptions to supply chains. In this pathway, financial institutions and corporates face increasing costs due to physical damage to assets and infrastructure, alongside disruptions to productivity and market stability. The hothouse world scenario highlights the consequences of inaction, with both financial institutions and businesses bearing the brunt of climate-related damages.

The NGFS scenarios are summarized in Table 6.4.

The NGFS scenarios are particularly relevant to the corporate sector, as they directly influence the availability and cost of debt. Under orderly or disorderly transitions, companies with robust climate strategies and lower carbon footprints are likely to retain better access to financing, as financial institutions seek to reduce their own exposure to climate risks. Conversely, firms in high-carbon industries or

Table 6.4 NGFS climate scenarios: descriptions and projected warming outcomes

NGFS scenarios scenario	Description	Projected warming by 2100
Orderly Transition	Early and ambitious climate action with well-coordinated policies that gradually increase in intensity. Transition risks are moderate, as industries and economies have enough time to adapt. Physical risks are minimized through early mitigation efforts	1.5 °C–2 °C
Disorderly: Too Little, Too Late	Delayed action leads to sudden, severe policy changes in response to urgent climate needs. This scenario involves high transition risks due to rushed measures, resulting in economic and financial instability. Physical risks remain high from earlier inaction	Around 2.5 °C or higher
Disorderly: Divergent Net Zero	Fragmented climate policies across regions and sectors result in uneven transition paths. While some areas implement ambitious policies, others lag, causing high transition risks in specific regions or sectors. Physical risks are inconsistently addressed	1.5 °C–3 °C
Insufficient Action: Hothouse World	Limited or ineffective climate policy action leads to high emissions and severe physical risks. Climate change impacts escalate without sufficient mitigation, increasing the frequency and intensity of extreme weather events, causing widespread damage	3 °C–4 °C or higher

Source NGFS Scenarios (2022)

with insufficient transition plans may find themselves excluded from credit markets or facing prohibitively high borrowing costs. The NGFS scenarios emphasize the importance of proactive climate risk management and adaptation strategies for both financial institutions and corporates in navigating the complexities of the low-carbon transition.

NGFS and IEA Scenarios: Relevance for Corporate Sector

The scenarios developed by the IEA and the NGFS are highly relevant for the corporate sector. Firms are increasingly expected to conduct scenario analysis to anticipate risks and opportunities in a changing climate. Both IEA and NGFS scenarios serve as critical inputs for developing tailored corporate climate strategies. Together, these scenarios provide firms with actionable insights into energy transitions, climate risks, and the broader economic impacts of climate change. Thus, corporations can leverage these scenarios to align their strategies with global climate goals, manage risks, and identify opportunities in the low-carbon economy.

- A manufacturing company might use IEA's net-zero emissions by 2050 (NZE2050) scenario to plan its shift to renewable energy while relying on NGFS's Orderly Transition scenario to assess the financial impacts of gradual carbon pricing.

- An energy company could use IEA's projections to align its portfolio with renewable energy targets while using NGFS physical risk data to safeguard its infrastructure from climate impacts.

IEA and NGFS scenarios highlight opportunities for firms to invest in clean energy, green technologies, and sustainable practices. IEA scenarios emphasize the importance of technologies like green hydrogen, carbon capture, and advanced battery storage. Firms can prioritize R&D and adoption of these innovations to stay ahead. NGFS scenarios guide firms on accessing sustainable financing, such as green bonds or climate-linked loans, by demonstrating alignment with credible transition pathways.

- A logistics company might use IEA's SDS to transition to electric fleets while leveraging NGFS insights to secure financing for this investment.

Navigating Energy and Financial Risks: Insights from IEA, NGFS, and IPCC Scenarios

The IEA and NGFS scenarios offer complementary tools for the corporate sector, addressing the dual imperatives of energy transition and financial stability. The IEA scenarios focus on the structural transformation of energy systems, providing granular insights into emissions pricing, fossil fuel demand, renewable energy adoption, and technological pathways like carbon capture and storage (CCUS) or Sustainable Aviation Fuel (SAF). These insights allow companies to plan investments, manage supply chain risks, and align their operations with decarbonization targets.

The NGFS scenarios, on the other hand, emphasize the financial risks of climate change, particularly for institutions and sectors exposed to physical and transition risks. With their focus on orderly versus disorderly transitions, NGFS scenarios illustrate how delayed policy action or abrupt regulatory shifts can affect access to capital, borrowing costs, and the valuation of high-carbon assets. This is particularly relevant for corporations with fossil fuel exposure or those reliant on emissions-intensive supply chains, as the financial sector increasingly prioritizes portfolios aligned with net-zero goals.

The IPCC's SSP-RCP framework complements these tools by providing a broader context for understanding how socioeconomic pathways interact with climate outcomes. The Shared Socioeconomic Pathways (SSPs) describe global development trajectories that shape mitigation and adaptation challenges. For example, SSP1 ("Sustainability") envisions a cooperative, low-emission future, ideal for leveraging orderly transition pathways as described by the NGFS. Conversely, SSP3 ("Regional Rivalry") reflects a fragmented and high-emission world, where disorderly transitions and physical risks dominate. By integrating these global narratives with IEA and NGFS scenarios, corporations can better understand the macroeconomic drivers of climate change and tailor strategies that address both systemic and sectoral risks.

For the corporate sector, this integration of IEA, NGFS, and IPCC frameworks is indispensable. Together, they provide a comprehensive toolkit to navigate the uncertainties of climate change while maintaining competitiveness and compliance. Firms can use the IEA scenarios to anticipate energy market shifts, the NGFS scenarios to evaluate financial exposure, and the IPCC framework to align with broader socioeconomic trends. This multifaceted approach enables companies to design strategies that align with global climate goals, enhance resilience to climate-related disruptions, and seize opportunities in a low-carbon future. As the transition accelerates, leveraging these scenarios will be critical for staying ahead of regulatory demands, market shifts, and stakeholder expectations.

6.6   Applying Scenario Analysis

The practical application of these scenarios—IPCC's SSPs, IEA's energy-focused pathways, and NGFS's financial risk frameworks—highlights their utility in guiding climate strategies. But the real test lies in how organizations integrate these models into their decision-making while navigating external constraints and internal priorities.

Air New Zealand provides a compelling case study of how an ambitious net-zero emissions (NZE) commitment, supported by scenario analyses and science-based targets, can evolve in response to shifting realities. By examining Air New Zealand's journey, we can better understand how these theoretical tools translate into practice—and what happens when external factors, such as government policy changes or operational challenges, cause pathways to shift or unravel.

Air New Zealand: Navigating Climate Pathways and Shifting Realities

Air New Zealand, established in 1965, serves as New Zealand's flag carrier, operating a global network that provides air passenger and cargo services to, from, and within New Zealand. Each year, the airline flies more than fifteen million passengers on its network, offering more than 3400 flights a week to forty-nine domestic and international destinations.

Despite its significant role in the South Pacific, Air New Zealand is relatively smaller compared to major international airlines like Delta Air Lines, United Airlines, and Lufthansa, which operate extensive global networks with larger fleets and passenger volumes.

Net-Zero Emissions (NZE) Goal

In 2022, Air New Zealand committed to achieving net-zero emissions (NZE) by 2050, joining the global movement to decarbonize the aviation sector. This target aligned the airline with the Paris Agreement's goal of limiting global warming to 1.5 °C. To operationalize this ambition, the airline pledged to reduce emissions across all scopes: direct emissions from aircraft operations (Scope 1), indirect emissions from energy use (Scope 2), and value chain emissions such as fuel production (Scope 3).

To realize the NZE ambition, the company identified several concrete focus areas:

- **Operational Efficiency**: Investments in fuel-efficient aircraft to lower carbon intensity.
- **Sustainable Aviation Fuel (SAF)**: Commitment to increasing SAF usage despite challenges in availability and affordability.
- **Technological Innovation**: Exploration of emerging technologies like hydrogen-powered aircraft and carbon capture.
- **Offsets and Nature-Based Solutions**: Plans to offset residual emissions through high-quality carbon credits and reforestation initiatives.

Air New Zealand's net-zero emissions (NZE) goal by 2050 was supported by the Science Based Targets initiative (SBTi), which provided a structured framework for setting near-term emissions reduction targets. The airline's interim 2030 target aimed to **reduce carbon intensity by 28.9%** from 2019 levels, significantly exceeding the SBTi's baseline requirement for the aviation sector of a 10% reduction in carbon intensity over five years.

One driver of Air New Zealand's more ambitious target was the anticipated regulatory landscape, including the global Carbon Offsetting and Reduction Scheme for International Aviation (CORSIA). As a participant in CORSIA, New Zealand is required to offset emissions exceeding 2019 levels through carbon credits. These offsets represent a significant cost, particularly as the program transitions to mandatory phases and offset prices increase. By reducing emissions beyond the minimum requirements, Air New Zealand could limit its reliance on offsets, mitigating future financial burdens.

Additionally, the European Union's ReFuelEU Aviation initiative introduced another layer of regulatory pressure. Starting in 2025, flights departing from EU airports will be required to use a minimum blend of Sustainable Aviation Fuel (SAF). These mandates are set to incrementally increase, reaching 6% by 2030 and 70% by 2050. Although Air New Zealand's routes to Europe are limited, this legislation signals a growing trend of SAF mandates, reinforcing the airline's need to secure access to SAF and integrate it into long-term planning.

Table 6.5 Air New Zealand socioeconomic pathways

Ambitious	Steady	Delayed	Insufficient
1.5 °C SSP1, RCP 1.9	2.7 °C SSP2, RCP 4.5	3.6 °C SSP3, RCP 7.0	4.4 °C SSP5, RCP 8.5
CO_2 emissions reach net zero by 2050 as green energy dominates	CO_2 emissions remain at current levels until 2050 as green energy becomes the majority source	CO_2 emissions double by 2100, with incremental gains in green energy	CO_2 emissions double by 2050, with no gains in green energy
Aotearoa New Zealand[2] invests heavily in green energy and experiences sharply increasing carbon prices	Aotearoa New Zealand experiences a sharp rise in carbon prices and increased investments in green energy	Aotearoa New Zealand experiences a carbon price increase and moderate green energy investment	Aotearoa New Zealand's carbon price remains stable; there is minimal investment in green energy
Aviation decarbonizes rapidly, bolstered by rapid technological advancement, favorable policy settings, and customers prioritizing climate-conscious businesses	Aviation technology advances steadily, with policy settings supporting the industry's transition. Customers prioritize climate-focused businesses if price is comparable	Aviation technology advances slower than anticipated, with limited policy support and limited SAF incentives. Customers prioritize price over climate concerns	Aviation technology advances significantly slower than anticipated, with no policy support or SAF incentives. Customers prioritize price over climate concerns

Source Air New Zealand: Sustainability Report (2023)

Scenario Planning to Support NZE

To ensure resilience in achieving these targets, Air New Zealand integrated scenario analyses into its strategy. In 2023, the airline adopted IPCC's Shared Socioeconomic Pathways (SSPs) to test its NZE roadmap under varying global warming scenarios. The SSP framework allowed the airline to anticipate both transition risks (e.g., SAF adoption, carbon pricing) and physical risks (e.g., increased weather disruptions). Air New Zealand settled on four SSPs tailored to its own operations. These scenarios are presented in Table 6.5.

Air New Zealand then used the SSP framework to gain insight into potential financial impacts of each scenario. This analysis enabled them to identify vulnerabilities and opportunities across varying levels of global climate action and align their strategy with these potential futures. In Table 6.6, we provide a synthesis of these impacts.

---

[2] **Aotearoa New Zealand** is the official Māori name for New Zealand. "Aotearoa" is commonly translated as "Land of the Long White Cloud," which reflects a description used by the Māori people to identify their homeland.

Table 6.6   Linking financial impact to SSPs

Scenario	Physical risk implications	Transition risk implications	Financial impact
Ambitious	Lower physical risks due to effective mitigation efforts limiting extreme weather and sea-level rise	High transition risks in the short term due to accelerated policy changes, technological adoption, and regulatory requirements	Initial financial pressures from investment in low-carbon technologies, but long-term financial benefits through reduced exposure to carbon pricing and stranded assets
Steady	Moderate physical risks as some climate impacts, such as changing weather patterns, remain partially mitigated	Moderate transition risks as gradual changes allow for adaptation but still require shifts in technology and compliance efforts	Moderate financial impact as gradual adjustments allows for manageable costs, though exposure to future carbon pricing remains
Delayed	Increasing physical risks due to delayed mitigation efforts, with heightened exposure to extreme weather events	High transition risks due to abrupt policy and regulatory changes needed to catch up, leading to stranded assets and cost spikes	Significant financial impact from sudden regulatory compliance costs, stranded assets, and increased insurance premiums due to intensified physical risks
Insufficient	Severe physical risks, including more frequent and intense extreme weather, sea-level rise, and operational disruptions	Low short-term transition risks but severe long-term risks due to reputational damage, policy catch-up costs, and market shifts	Severe financial impact due to escalating operational disruptions, rising insurance costs, asset devaluation, and difficulty accessing capital markets

Source Air New Zealand: Sustainability Report (2023)

It seemed that by 2023, Air New Zealand was well equipped to understand and respond to the challenges it faced from climate change. There was an organized framework that considered an overarching socioeconomic context, and from there Air New Zealand could move forward to achieve its NZE ambitious, even though the company was going beyond what the SBTi guidelines required! A lot could change in less than a few years though.

Redrawing Climate Goals: Moving Beyond SBTi

In its 2024 Climate Statement, Air New Zealand acknowledged a pivotal shift in its approach to decarbonization, stating: "We have chosen to step away from our Science Based Targets initiative (SBTi) commitment as we reassess and refine our pathways to achieving our net-zero emissions goal by 2050. This decision reflects the evolving complexities of the aviation industry and the need for more tailored and realistic goals that account for unique operational challenges (Air New Zealand Climate Statement, 2024)."

This marks a significant departure from the standardized framework provided by SBTi, as the airline adapts its strategy to align with evolving regulatory landscapes, advancements in Sustainable Aviation Fuel (SAF), and other decarbonization technologies. The transition also reflects a broader shift to incorporate IEA and NGFS scenarios to evaluate risks and opportunities with greater sector specificity.

Scenario-Based Risk Analysis

Air New Zealand built upon the SSPs developed in 2023 by blending them with insights from the IEA and NGFS frameworks.[3] They undertook this integration to refine their understanding of potential decarbonization pathways and to better align their strategic planning with both global climate goals and industry-specific challenges. The addition of IEA data, such as SAF blending mandates and energy cost projections, and NGFS insights on transition and financial risks illuminated more precise pathways for addressing the operational and market impacts of climate change. This approach enabled Air New Zealand to better adapt the broad scenarios of the SSPs to its unique operational context (Table 6.7).

Moving Ahead

Summing up, integrating the IEA and NGFS scenarios with the earlier SSPs positioned Air New Zealand to identify key challenges, such as those associated with SAF adoption, operational risks, and potentially CCUS as a complementary technology. By using a more tailored approach to scenarios, the airline could balance ambition with feasibility, adapting to global and regional variations in climate policies.

However, two external shocks highlight the fragility of this planning process. The repeal of certain environmental laws, affecting the availability of feedstock for SAF (linked to agricultural policy), and delays in Boeing aircraft deliveries created significant operational and financial setbacks. These disruptions underscore that operational efficiency and SAF adoption remain pivotal to Air New Zealand's ability to achieve its decarbonization targets. Moving forward, the airline will need to adapt dynamically to such uncertainties while maintaining alignment with evolving climate action frameworks.

---

[3] Air New Zealand dropped the "Insufficient" scenario from consideration and focused on the remaining three SSP scenarios identified a year earlier: Ambitious, Steady, and Delayed.

Table 6.7 Refining scenarios with IEA and NGFS input

Aspect	Ambitious scenario (SSP1-1.9/IEA NZE)	Steady scenario (SSP2-4.5/IEA STEPS)	Delayed scenario (SSP3-7.0/NGFS delayed transition)
Scenario context	Strong global climate action; high ambition toward net-zero emissions	Moderate progress; gradual policy implementation and emissions reductions	Fragmented global policies; delayed climate action leads to higher emissions
Technology pathways	Rapid SAF scaling supported by strong policies; SAF blending targets of 10% by 2025 and 50% by 2040	Gradual SAF adoption; blending targets of 5% by 2030 and 20% by 2050	Limited SAF adoption; mandates reach only 2% by 2035, with reliance on traditional fuels
Energy pathways	High investment in renewable energy and green hydrogen reduces costs and accelerates adoption	Moderate investment slows the pace of cost reductions for renewable energy and hydrogen	Limited investment keeps costs high, delaying renewable energy and hydrogen adoption
Macroeconomic risks	Reduced stranded asset risks due to early policy and market alignment	Moderate stranded asset risks as policies are phased in incrementally	High stranded asset risks due to abrupt policy shifts and market disruptions
Physical risks	Lower physical risks from effective global mitigation efforts	Moderate physical risks from partial global climate action	Severe physical risks due to insufficient mitigation, such as extreme weather impacts
Financial impact	Initial financial pressures from SAF and hydrogen investments; long-term financial benefits	Manageable financial risks due to phased investment in SAF and renewables	Severe financial risks from rising operational costs, asset devaluation, and financing challenges
Reputation risks	Enhanced reputation from transparent decarbonization strategies and investor confidence	Moderate reputation preservation; less competitive in ambitious markets	Damaged reputation due to perceived inaction and lack of alignment with global targets

Source Air New Zealand Climate Statement (2024)

Conclusion

This chapter explored the power of climate scenarios to navigate the uncertainties of a changing world. From the foundational stories crafted by the IPCC's Shared Socioeconomic Pathways (SSPs) to the targeted insights provided by the IEA and NGFS, we have seen how scenarios are more than analytical tools—they are frameworks for imagining futures. By modeling diverse trajectories, these scenarios allow us to understand the risks, opportunities, and trade-offs of different choices, highlighting the pathways to sustainability or the consequences of inaction.

The Air New Zealand case study brought these concepts into the real world, demonstrating how businesses can apply scenarios to identify risks and refine strategies. The airline's journey—from using SSPs to assess broad socioeconomic risks to integrating IEA and NGFS insights for sector-specific and financial challenges—showed the value of layering scenario frameworks. Yet, it also revealed the limits of planning when faced with unforeseen disruptions like regulatory reversals and supply chain delays.

One of the key takeaways from this chapter is that while scenarios do not predict the future, they provide a structured approach to preparing for it. They help us explore "what could happen" under different assumptions and inform "what should happen" to achieve sustainable outcomes.

By integrating scenario analysis into decision-making, businesses, policymakers, and financial institutions can better navigate the complexities of climate change and align their strategies with a rapidly evolving landscape. In the next chapter, we examine the various climate risks confronting corporations and how they can respond to these challenges –both risk and response will crucially depend upon how the climate crisis unfolds: which scenario is realized!

Key Takeaways

1. **Scenarios are Exploratory, Not Predictive**

    Climate scenarios, such as those from the IPCC, IEA, and NGFS, do not predict the future. Instead, they offer structured tools to explore a range of possible futures based on differing assumptions about policy, technology, and behavior.

2. **Integrated Frameworks Provide Comprehensive Insights**

    Combining socioeconomic, energy, and financial scenarios helps businesses and policymakers connect global narratives with specific sectoral risks and opportunities.

3. **Adaptation and Mitigation are Interdependent**

    Effective mitigation strategies can reduce long-term adaptation costs and risks, while delayed mitigation leads to higher physical and financial risks.

4. **Sector-Specific Applications Are Crucial**

    The value of scenarios increases when tailored to sector-specific challenges, such as the aviation industry's need for Sustainable Aviation Fuel (SAF) or the energy sector's reliance on carbon pricing and renewable transitions.

5. **Air New Zealand Case Study Highlights Practical Use**

The integration of SSPs, IEA, and NGFS frameworks by Air New Zealand demonstrates how businesses can align high-level climate scenarios with operational decision-making to achieve ambitious decarbonization goals.

6. **Scenario Planning Illuminates Risks and Opportunities**
   Scenarios reveal potential risks (e.g., regulatory changes, stranded assets) and opportunities (e.g., investments in green technologies) that can inform long-term strategies for resilience and growth.

7. **Dynamic Flexibility is Key**
   Companies must be prepared to adjust their strategies as external shocks, such as regulatory shifts or supply chain disruptions, reshape the pathways to achieving net-zero emissions (NZE).

8. **Collaboration is Critical**
   Achieving sustainable outcomes requires coordination between businesses, governments, and financial institutions, emphasizing the need for alignment across global, national, and sectoral levels.

9. **A Transition from Analysis to Action Is Necessary**
   While scenario analysis offers valuable insights, businesses must also develop actionable strategies to mitigate risks and capitalize on opportunities, bridging the gap between understanding and implementation.

10. **Next Steps: Understanding Climate Risks and Opportunities**
    This chapter sets the stage for the next, where a detailed examination of physical, transition, and systemic risks will help identify the tangible impacts of climate change and the opportunities they present for innovation and competitive advantage.

Questions

1. Define the primary purpose of climate scenarios and explain their value in decision-making.
2. What are the key differences between the IPCC's SSPs, IEA scenarios, and NGFS scenarios? Provide examples of their applications.
3. How does the integration of socioeconomic pathways (SSPs) with energy and financial risk frameworks enhance strategic decision-making?
4. In the Air New Zealand case study, what were the key insights gained from blending SSPs, IEA, and NGFS scenarios?
5. What role do physical, transition, and systemic risks play in scenario planning, and how are they addressed in the SSP-RCP matrix?
6. What challenges arise when applying climate scenarios, particularly regarding unforeseen events or regulatory shifts?
7. Describe how businesses can use climate scenarios to identify and seize opportunities in a decarbonizing economy.
8. Explain the significance of Sustainable Aviation Fuel (SAF) in Air New Zealand's net-zero emissions (NZE) strategy.
9. What lessons can be learned from the transition risks highlighted in NGFS's "Disorderly Transition" scenario?

10. How can firms balance ambition and feasibility when setting long-term climate goals?

References

Air New Zealand. (2023). *Sustainability Report 2023*. Retrieved November 4, 2024, from https://www.airnewzealand.co.nz/sustainability-reporting-and-communication

Air New Zealand. (2024). *Climate Statement 2024*. Retrieved November 4, 2024, from https://p-airnz.com/cms/assets/air-new-zealand-2024-climate-statement.pdf

CDP. (2023). *China's net zero vision: An opportunity for business leaders to seize*. Retrieved November 4, 2024, from https://www.cdp.net/en/articles/climate/chinas-net-zero-vision-an-opportunity-for-business-leaders-to-seize

European Commission. (2021). *'Fit for 55': Delivering the EU's 2030 climate target on the way to climate neutrality*. Retrieved November 4, 2024, from https://ec.europa.eu/commission/pressc orner/detail/en/ip_21_3541

IEA. (2024). *Global energy and climate model*. IEA, Paris. Retrieved November 4, 2024, from https://www.iea.org/reports/global-energy-and-climate-model

Intergovernmental Panel on Climate Change (IPCC). (2022). *Climate Change 2022: Impacts, adaptation, and vulnerability*. Contribution of Working Group II to the Sixth Assessment Report of the Intergovernmental Panel on Climate Change. Cambridge University Press. Accessed February 14, 2025, at https://doi.org/10.1017/9781009325844

Network for Greening the Financial System (NGFS). (2022). *NGFS climate scenarios for central banks and supervisors*. Retrieved November 4, 2024, from https://www.ngfs.net/en/ngfs-climate-scenarios-central-banks-and-supervisors-september-2022

Nordhaus, W. D. (2017). Revisiting the social cost of carbon. *Proceedings of the National Academy of Sciences, 114*(7), 1518–1523.

O'Neill, B. C., Kriegler, E., Ebi, K. L., Kemp-Benedict, E., Riahi, K., Rothman, D. S., Van Ruijven, B. J., Van Vuuren, D. P., Birkmann, J., Kok, K., & Levy, M. (2017). The roads ahead: Narratives for shared socioeconomic pathways describing world futures in the 21st century. *Global Environmental Change, 42*, 169–180. https://doi.org/10.1016/j.gloenvcha.2015.01.004

Pindyck, R. S. (2013). Climate change policy: What do the models tell us? *Journal of Economic Literature, 51*(3), 860–872.

Riahi, K., Van Vuuren, D. P., Kriegler, E., Edmonds, J., O'neill, B. C., Fujimori, S., Bauer, N., Calvin, K., Dellink, R., Fricko, O., & Lutz, W. (2017). The shared socioeconomic pathways and their energy, land use, and greenhouse gas emissions implications: An overview. *Global Environmental Change, 42*, 153–168.

Rogelj, J., Shindell, D., Jiang, K., Fifita, S., Forster, P., Ginzburg, V., Handa, C., Kheshgi, H., Kobayashi, S., Kriegler, E., & Mundaca, L. (2018). Mitigation pathways compatible with 1.5 °C in the context of sustainable development. In *Global warming of 1.5 °C*. IPCC.

S&P Global. (2023). *Still the exception for top U.S. companies, not the rule: Net-zero commitments in the S&P 500*. Retrieved November 4, 2021, from https://www.spglobal.com/esg/insights/featured/special-editorial/net-zero-commitments-are-still-the-exception-for-top-us-companies-not-the-rule

Science Based Targets Initiative. (2023). *Companies taking action*. Retrieved November 4, 2024, from https://sciencebasedtargets.org/companies-taking-action

van Vuuren, D. P., Edmonds, J., Kainuma, M., Riahi, K., Thomson, A., Hibbard, K., Hurtt, G. C., Kram, T., Krey, V., Lamarque, J. F., & Masui, T. (2011). The representative concentration pathways: An overview. *Climatic Change, 109*(1–2), 5–31.

Part II
Risks and Opportunities in Climate Change and Financial Implications

Physical and Transition Risk

Introduction

Climate change poses unprecedented risks and opportunities for businesses, reshaping the way firms assess value and manage uncertainty. From rising temperatures and extreme weather events to evolving regulations and shifting consumer preferences, climate risks manifest in ways that directly affect operations, supply chains, and profitability. These risks fall into two primary categories: physical risks, such as hurricanes and water scarcity, and transition risks, driven by policy changes, market dynamics, and technological advancements.

For corporate finance, these risks translate into measurable impacts on firm value. Physical risks can lead to asset impairments, supply chain disruptions, and increased operational costs, while transition risks may create stranded assets, raise capital costs, or necessitate significant investments in new technologies. Yet, these challenges also create opportunities for firms to innovate, enhance resilience, and align with a low-carbon economy.

This chapter explores the tools and frameworks businesses use to navigate climate risks, focusing on materiality, adaptive capacity, and scenario analysis. By integrating climate risk into financial decision-making, firms can mitigate threats, capitalize on opportunities, and meet the expectations of stakeholders, including investors, customers, and regulators. Through real-world examples, such as Danone, we illustrate how a strategic approach to climate risk can enhance firm value and contribute to long-term sustainability.

7.1   Climate Risk Materiality

The goal of corporate climate risk assessment is to identify, quantify (when possible), and manage the potential impacts of climate change on firm value. When thinking about climate risk, what probably immediately jumps out is how much can these risks impact the value of the firm. In other words, are these risks germane or can they be ignored? Regulators, among others, are taking a second look at climate risk materiality, focusing not only on how climate risk affects firm value (Single Materiality) but also how the firm's operations affect the environment and society (Double Materiality).

Materiality: The Case of EcoPop Inc.

To illustrate how the concepts of single and double materiality are applied we will consider the hypothetical example of EcoPop Inc., a global company which operates in the food and beverage sector.

Single Materiality: Financial Impact Assessment

EcoPop Inc. begins by examining how climate change-related factors directly impact the company's financial performance:

- **Operational Costs**: EcoPop evaluates how extreme weather events increase operational costs, such as energy use for cooling during heatwaves or repairing infrastructure after floods.
- **Regulatory Costs**: Compliance with emerging climate-related regulations, such as carbon pricing, poses additional financial risks.
- **Market Dynamics**: Shifts in consumer demand toward sustainable products influence revenue streams.

Quantifiable Thresholds

To prioritize material risks, EcoPop establishes a measurable threshold: any financial impact amounting to or exceeding 1% of net income is deemed material. For instance, with an annual net income of five hundred million dollars, the threshold is set at $5 million. The thresholds and corresponding impact assignment are shown in Table 7.1

Table 7.1 Financial impact thresholds for EcoPop: assessing materiality by net income

Impact level	Net income threshold
Low Impact	≥ 5 M and < 25 M
Medium–low Impact	≥ 25 M and < 50 M
Medium Impact	≥ 50 M and < 125 M
Medium–high Impact	≥ 125 M and < 250 M
High Impact	≥ 250

Double Materiality: Expanding the Scope

EcoPop extends its climate risk assessment to include environmental and social materiality, addressing how its operations affect non-financial stakeholders, such as communities and ecosystems. These areas of impact are identified through stakeholder engagement, including discussions with investors, customers, policymakers, and civil society. EcoPop identifies the following key environmental and social impacts of its operations:

- **Carbon Emissions**: EcoPop measures its greenhouse gas emissions and sets reduction targets, aligning with global climate goals. Lowering emissions mitigates reputational and regulatory risks while reducing EcoPop's environmental footprint.
- **Water Use**: The company evaluates water usage in its operations and supply chain, recognizing that efficient water management is essential for long-term operational stability, especially in water-scarce regions.
- **Regenerative Agriculture**: EcoPop engages in regenerative agricultural practices that improve soil health, sequester carbon, and enhance biodiversity. This approach reduces risks tied to resource scarcity while improving long-term supply chain resilience.

Quantifying Environmental and Social Materiality

EcoPop captures the key social and environmental issues by establishing "Impact Thresholds." Table 7.2 summarizes the analysis of Double Materiality for EcoPop, indicating the social and environmental Impact Category (Column 1); the Quantifiable Impact associated with the Impact Category (Column 2), and finally an estimate of financial implications associated with the key environmental/social category (Column 3).

Summing Up Materiality for EcoPop

By considering both the direct financial impacts of climate risks and the broader societal implications of its operations, EcoPop aligns its priorities with stakeholder expectations and evolving regulatory demands. For example, EcoPop can

Table 7.2 Environmental and social impact thresholds: quantifying double materiality for EcoPop

Impact category	Quantifiable impact	Potential financial implications
Carbon Emissions	150,000 tons CO_2 reduced annually	$10 million in avoided regulatory costs
Water Use	20% reduction in water consumption	$7 million in operational savings
Regenerative Agriculture	5,000 acres converted annually	$5 million in improved supply chain resilience

visualize "Feedback Loops": reducing emissions and improving water management directly lower operational risks while creating long-term financial benefits. Initiatives like regenerative agriculture not only address environmental impacts but also secure long-term supply chain stability, enhancing EcoPop's ability to adapt to climate change.

The EcoPop example highlights the practical application of double materiality in climate risk assessment. While single materiality ensures a focus on financial impacts, the double materiality framework broadens the scope to include environmental and social considerations, many of which have indirect feedback effects on financial outcomes. By addressing both dimensions, EcoPop builds resilience against climate-related risks and positions itself to adapt effectively to evolving stakeholder expectations and regulatory landscapes. ◄

7.2   Adaptive Capacity and Climate Resilience

Identifying material risks is only half the story! The firm must design strategies not only to mitigate climate risks but also to adapt to a changing climate. At some level, we are actually saying: here are the risks and what can we do about them? Unlike our discussion of materiality where we were able to more or less measure the impact on the firm from both single and double materiality perspectives, the "what can we do about" part hinges on the firm's adaptive capacity.

Adaptive capacity reflects a company's ability not only to adjust to climate impacts and mitigate potential damages but also to seize opportunities arising from the transition to a low-carbon economy. Unlike quantifiable metrics, assessing adaptive capacity often requires a qualitative approach. For example, examining corporate governance can reveal if the Board is "climate aware," while evaluating climate disclosure practices shows alignment with sound climate risk management. Assessing whether a firm identifies both risks and opportunities, sets science-based targets, and integrates climate risk into enterprise risk management and capital planning is also vital. Ultimately, adaptive capacity indicates how well a firm can "future-proof" its value—preparing to thrive amid climate uncertainty, rather than simply survive.

7.3   Physical Risks

The tangible impacts of climate change are increasingly hard to ignore. From hurricanes that devastate infrastructure to prolonged droughts that disrupt agricultural output, physical climate risks pose direct and measurable threats to business operations and supply chains. Unlike transition risks, which stem from policy and market shifts, physical risks are grounded in observable environmental changes. As these events become more frequent and severe, businesses must adapt to safeguard assets, manage costs, and maintain continuity in a rapidly changing world.

Physical Risk Categories

The physical risks stemming from climate change fall into two broad categories:

- **Acute Physical Risks**: These are driven by specific weather events or "hazards," such as heatwaves, floods, wildfires, and storms. Although they can cause immediate damage to physical assets and disrupt operations, they are typically associated with short-term disruptions to a firm's operations, assets, or supply chains. However, an acute event like a severe storm can exacerbate chronic risks, such as flooding in already vulnerable areas.
- **Chronic Physical Risks**: These are driven by longer-term shifts in climate patterns, such as rising sea levels and increasing mean temperatures which pose long-term or ongoing threats to the firm from gradual changes in climate-related conditions. For example, rising sea levels can lead to chronic flooding and saltwater intrusion in coastal facilities.

Physical Climate Risk: The Crucial Role of Geography

Geography plays a crucial role in addressing the physical risks of climate change for businesses. Physical risks due to climate events, whether chronic or acute, need to be considered across the firm's value chain—from suppliers through to customers. There are several sources that rank countries by their climate risk. For example, the Climate Risk Index (CRI) is published annually by the environmental organization German Watch (German Watch, 2023), which assesses and ranks countries based on their vulnerability and exposure to climate change. The CRI rankings consider the number of climate-related events, fatalities, economic losses, and the level of exposure and sensitivity of a country's population and infrastructure to these events.

Importantly, climate vulnerability is often more acute in the "Global South," encompassing many emerging markets and developing economies, where higher exposure to climate events intersects with limited resources for adaptation. While these regions have contributed comparatively little to historical greenhouse gas emissions, they often bear a disproportionate burden of climate impacts. This vulnerability in the Global South is not just a remote concern but has immediate and significant consequences for developed economies, given the deep integration of these regions into global supply chains. Consider how interconnected India and China are with their global partners:

- **India**: As a global hub for IT and telecommunications services, particularly for outsourcing by international firms, climate risks in India directly affect developed economies. Severe heatwaves, water shortages, and flooding in cities like Bengaluru and Chennai do not just disrupt local infrastructure—they threaten the continuity of customer service, technical support, and software development relied upon by firms worldwide. With developed economies heavily dependent on India's digital workforce, climate vulnerability in India has essentially become a shared risk, forcing global firms to rethink their resilience strategies and incorporate climate-adaptive approaches, such as distributed workforces or dual-location setups, to ensure continuity (McKinsey, 2023).

- **China**: China's status as the world's manufacturing powerhouse makes its climate vulnerabilities a direct challenge for developed economies. Typhoons, floods, and escalating water scarcity in industrial regions disrupt the production of everything from electronics and medical equipment to renewable energy components, which developed economies depend on for both daily life and ambitious sustainability goals. When manufacturing halts in China due to climate-related events, supply chain delays ripple across the globe, highlighting how climate resilience in China is, in effect, an operational necessity for businesses in developed economies. This has encouraged both foreign investment in China's green infrastructure and diversified supply chain planning among developed economies to mitigate their shared climate exposure (Oxford Institute for Energy Studies, 2023).

In brief, the interconnectedness of the global economy is catalyzing a new approach where businesses and governments in developed countries see climate adaptation and resilience in the Global South as critical to their own stability, making climate vulnerability an immediate concern for corporate strategies, trade policies, and international investment.

Physical Climate Risk and Sectoral Exposure

The degree of physical climate risk also varies across sectors, for example, agriculture and tourism are more sensitive to changes in temperature and precipitation, while energy and transportation are more dependent on reliable infrastructure and networks. Some examples of physical climate risk by geography and sector are:

- **Agriculture in sub-Saharan Africa**: This sector is highly exposed to chronic risks such as higher temperatures, lower rainfall, and soil degradation, which can reduce crop yields and increase pest and disease outbreaks. It is also vulnerable to acute risks such as droughts, floods, and locust swarms, which can cause crop failures and food insecurity. The sector has low adaptive capacity due to limited access to finance, technology, and insurance (Pereira, 2017).
- **Tourism in Small Island Developing States (SIDS)**: This sector is highly exposed to chronic risks such as sea-level rise, coral bleaching, and coastal erosion, which can damage the natural and cultural attractions that draw tourists. It is also vulnerable to acute risks such as hurricanes, storm surges, and landslides, which can disrupt travel and damage infrastructure and facilities. The sector has low adaptive capacity due to high dependence on imports, limited diversification, and weak governance (World Tourism Organization, 2012).

- **Energy in North America**: The energy sector in North America is moderately exposed to chronic risks such as higher temperatures, lower water availability, and permafrost thawing, which can affect the efficiency, reliability, and safety of energy production, transmission, and distribution. It is also vulnerable to acute risks such as wildfires, ice storms, and heat waves, which can cause power outages and damage to equipment and infrastructure. The sector has strong adaptive capacity due to advanced technology, regulation, and innovation (Hicke et al., 2022).

Physical Risk Assessment: Granularity

Physical risk assessment, regardless of sector, must be conducted at a granular level to capture the geography of climate change. An example of firms in four different sectors located in different regions highlights the significance of granularity when evaluating physical climate risk.

Physical Risk Across Sectors

- TechSolutions Inc. is a global IT services provider with data centers across the globe. A detailed analysis reveals that their data center in Southern California faces significant climate risks.
- GreenFarm Co. is an agricultural enterprise operating in Southeast Asia.
- In the manufacturing sector, AutoMaker Ltd. evaluates physical climate risks at its assembly plants around the world. One of their plants in the Midwest United States is particularly susceptible to climate risks.
- PharmaHealth Inc. is a global pharmaceutical firm that operates several manufacturing plants, research facilities, and distribution centers. One of its primary manufacturing plants is in Florida.

Table 7.3 summarizes the key risks facing each of these companies and possible mitigation strategies.

Table 7.3 Overview physical risk assessment by sector

Company	Sector	Key risks	Mitigation actions
TechSolutions Inc.	Technology	Wildfires, heatwaves	Cooling systems, backup power
GreenFarm Co.	Agriculture	Shifting monsoon, salinization	Flood-resistant crops, soil management
AutoMaker Ltd.	Manufacturing	Tornadoes, thunderstorms	Strengthened building codes, supply chain diversification
PharmaHealth Inc.	Pharmaceutical	Hurricanes, flooding	Resilient infrastructure, compliance with regulations

Table 7.4 Climate risk financial impact analysis for PharmaHealth

Category	Description	Cost/Investment
Infrastructure Damage	Repair costs per major hurricane event	$50 million per event
Operational Disruptions	Revenue loss due to operational shutdowns	$30 million per month
Supply Chain Interruptions	Revenue loss due to supply chain disruptions	$20 million per month
Compliance Costs	Annual costs for compliance with climate resilience regulations	$5 million per year
Capital Expenditure	Investment in climate-resilient infrastructure upgrades	$100 million (over five years)
Operational Costs	Annual budget for disaster preparedness and response measures	$10 million per year
Revenue Losses	Projected cumulative revenue losses due to operational and supply chain disruptions	$150 million (over five years)
Insurance Premiums	Additional insurance premiums for climate-related damages	$5 million per year

The examples in Table 7.3 illustrate how geography and sector influence climate risk and necessitating an array of mitigation/adaptation responses.

Financial Impact: PharmaHealth Inc.

PharmaHealth Inc. conducted a detailed impact analysis to understand the financial implications of climate-related risks for one of its primary manufacturing plants in a coastal area prone to hurricanes and flooding. The analysis is summarized in Table 7.4.

The additional financial analysis for PharmaHealth Inc. highlights the tangible costs associated with climate risks, from infrastructure damage and operational disruptions to increased compliance and insurance premiums. Addressing these risks involves more than managing costs like infrastructure repairs or revenue losses from disruptions—it requires a shift toward resilience. Investments in climate-resilient infrastructure and disaster preparedness can help mitigate potential impacts while aligning operations with regulatory expectations. These efforts are not just about avoiding losses but ensuring operational stability and supporting long-term growth in an environment increasingly shaped by climate challenges.◀

Climate Impact Chain

Our discussion of physical risk to this point has focused primarily on "the event" and how physical risk varies across sectors and geographies. However, there is a little more to the story. The event—"hazard" is just one part of physical climate risk. The IPCC defines climate risk as the product of three factors: hazard, exposure, and vulnerability.

The three components of physical climate risk are illustrated in Fig. 7.1 along with their drivers.

- **Hazard** refers to the potential occurrence of a climate event or trend that may cause harm or damage. If a firm is located in a coastal area, the primary climate hazard might be hurricanes or sea-level rise. Inland firms may be more concerned about droughts or heatwaves.
- **Exposure** refers to the extent to which an element (such as a person, asset, or activity) is subject to the hazard. Climate exposure for a real estate development company might involve assessing the proximity of their properties to flood-prone areas or coastal zones. A manufacturer may assess the sensitivity of its production facilities to extreme heat, which could affect machinery and worker productivity.
- **Vulnerability** refers to the susceptibility of the exposed element to experience adverse effects from the hazard. For example, a firm's climate vulnerability assessment could reveal that its supply chain heavily relies on a single source of raw materials located in a region prone to wildfires. If this source is not diversified or protected against fire risks, the firm is considered vulnerable to supply disruptions.

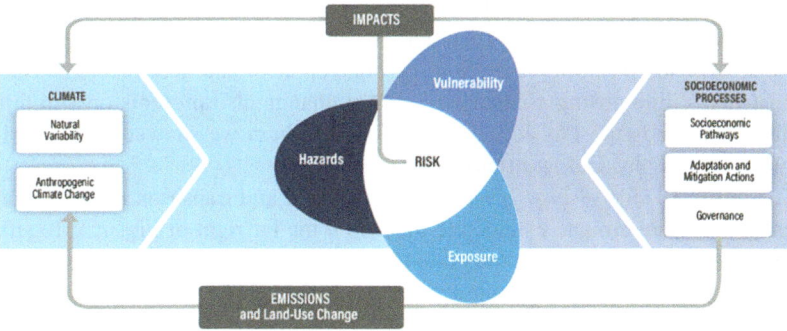

Fig. 7.1 Climate risk = hazard × exposure × vulnerability (*Source* Intergovernmental Panel on Climate Change (IPCC), 2014)

Example: Automobile Manufacture

Consider a hypothetical example of an automobile manufacturer in the United States and another in China. Suppose both firms face the same hazard of heatwaves, which can reduce machinery efficiency, increase cooling and energy costs, and affect worker health and productivity. The US firm has greater exposure to this hazard because it operates in regions with higher average temperatures and more frequent, intense heatwaves than the Chinese firm. However, the US firm experiences a lower impact from this hazard due to investments in resilient, energy-efficient equipment, the installation of cooling systems and ventilation, and the implementation of heat-stress prevention policies for workers. As a result, the US firm faces a lower overall climate risk from heatwaves than the Chinese firm, even though both firms encounter the same hazard, because the US firm has greater adaptive capacity, mitigating the hazard's impact.◄

While physical risks are rooted in tangible environmental changes, the transition to a low-carbon economy brings its own set of challenges. Transition risks, unlike physical risks, are shaped by policy changes, evolving market dynamics, technological advancements, and societal expectations. Together, these forces can reshape industries, disrupt value chains, and redefine competitive landscapes. In the next section, we explore how transition risks complement physical risks, presenting both challenges and opportunities for businesses navigating a climate-conscious future.

7.4   Transition Risk

Extending the Climate Impact Chain to Transition Risk

Transition risk arises from the shift to a low-carbon economy, shaped by climate policy changes and sectoral or regional vulnerabilities. At the firm level, transition risk depends on policy ambition, carbon reliance, and access to low-carbon alternatives. While posing challenges, climate change also presents opportunities. Figure 7.2 extends the Physical Climate Impact Chain we showed in Fig. 7.1 to encompass both physical and transition risk.

The left-hand side of Fig. 7.2 illustrates physical and transition risks. Opportunities arising from climate change are placed on the far right of Fig. 7.2. Together, the risks and opportunities are integrated into the firm's strategic planning and risk management function where their financial impact can be evaluated. As we can see in Fig. 7.2, transition risks go beyond policy changes to encompass a variety of risks arising from the transition to a low-carbon economy.

Fig. 7.2   Climate risk impact chain: physical and transition risk (*Source* Recommendations of the task force on climate-related financial disclosures)

Classifying Transition Risk

Policy Risk: Carbon pricing, including taxes and cap-and-trade systems like the EU Emissions Trading System (EU ETS), introduces a direct cost for greenhouse gas emissions. While the intent is to incentivize emissions reductions, these measures pose significant risks to industries characterized by high emissions. Rising operational costs due to stringent regulations are an immediate concern, but policy stability is equally critical. For example, in an interview with POLITICO following the November 2024 US election, ExxonMobil CEO Darren Woods highlighted the importance of maintaining consistent climate regulations. Woods cautioned that undoing existing policies could create greater challenges in the future, as it would undermine long-term investments in technologies like carbon capture and storage and delay the necessary progress toward reducing emissions (Colman, 2024).

Reputation Risk: Reputation risk involves the potential harm to a company's brand image and stakeholder relationships due to its handling of climate-related issues. Poor environmental performance or perceived inaction on climate matters can result in consumer boycotts, community protests, and decreased sales, all of which have long-lasting impacts on a company's financial performance and market position.

Volkswagen's well-known emissions scandal, in which the company admitted in 2015 to cheating on emissions tests, exemplifies the severity of reputation risk. The scandal, often referred to as "Dieselgate," led to global backlash, legal actions, and a significant loss of consumer trust. By 2019, Volkswagen was still grappling with the consequences, facing billions of dollars in fines and settlements, while its brand image suffered considerable damage.

Market Risk: Market risk stems from evolving consumer preferences, investor sentiment, and changing societal values around climate change and sustainability. Companies could experience reduced demand for carbon-intensive products as consumers shift to eco-friendly alternatives, impacting revenue and profitability.

General Motors was facing declining sales as consumers turned to more fuel-efficient and electric vehicles. In response, in 2020 GM announced plans to invest more in electrification and emission reduction to compete with rivals like Tesla and Toyota, although in 2024 General Motors announced it was scaling back on its EV ambitions (General Motors Press Release, 2020; Wayland, 2024).

Litigation Risk: Firms associated with significant emissions or environmental harm face the risk of lawsuits, fines, or penalties. As regulations tighten, companies increasingly are being held accountable for environmental damages or misleading claims about sustainability ("greenwashing").[1] This risk has grown with recent legislation targeting both misleading environmental claims and inadequate climate strategies.

In 2021, a Dutch court ordered Royal Dutch Shell to reduce emissions by 45% by 2030, as environmental groups argued that Shell's 2050 net-zero target lacked near-term action. This landmark case set a precedent for holding firms accountable for vague or unsubstantiated sustainability promises (Milman & Taylor, 2021). In the period following the decision the stock price fell, not only for Shell but other oil majors (Kolaric, 2023). The Dutch Court of Appeal reversed the 2021 decision against Royal Dutch Shell on November 12, 2024, (Corder, 2024; Reuters, 2024).[2]

Technology Risk: Companies face technology risk as they are pressured to adopt climate-friendly innovations to comply with regulations and meet consumer and investor expectations. Firms that fail to update outdated or carbon-intensive technologies risk falling behind in a market increasingly favoring sustainable practices. For example, rapid advancements in electric and autonomous vehicle technologies are reshaping the automobile sector. Traditional automakers face significant pressure to innovate and adopt climate-friendly technologies to remain competitive, as failure to do so risks obsolescence in a market increasingly driven by regulatory demands and consumer preferences for sustainable mobility solutions (Kuhnert et al., 2017–2018).

Stranded Asset Risk: Stranded asset risk arises when investments in high-carbon assets, such as fossil fuel infrastructure, lose value or become obsolete due to regulatory changes, market shifts, or technological advancements. This can lead to financial losses, asset impairments, and write-downs. One recent study puts the value of untapped fossil fuel output through 2050 in a range from $21.5 trillion to $30.6 trillion, depending on the stringency of climate policies (Chen et al., 2023).

---

[1] Both reputation risk and litigation risk can be linked to greenwashing.

[2] The reversal does not mean that the oil majors will abandon their green investment strategies. However, it does imply that emissions reductions could occur at a slower pace.

Beyond the fossil fuel sector, other industries are also susceptible to stranded asset risk. The automotive industry, for instance, faces challenges as the transition to electric vehicles (EVs) accelerates. Traditional internal combustion engine (ICE) vehicles and related manufacturing facilities may become obsolete, leading to significant financial write-downs (Chaudhary, 2024). The real estate sector is also at risk, particularly concerning properties that do not meet emerging energy efficiency standards. Buildings with poor energy performance may face devaluation or require costly retrofits to comply with new regulations aimed at reducing carbon emissions (Financial Times, 2024).

Transition Risk Across Sectors

The transition risks we just identified will have varied impact on firms depending largely on the sector in which they operate ranging from firms in sectors with emissions that are hard to abate to firms operating in the renewable energy sector and everything in-between.

- **Hard-to-Abate Sectors**: Industries like cement, steel, and aviation have high-carbon footprints and face complex technological challenges in reducing emissions. For these sectors, decarbonization requires substantial R&D, regulatory compliance, and investment in new technologies, making them particularly vulnerable to transition risks.
- **Financial Services**: Financial institutions are exposed through their lending, underwriting, and investment activities. If they have portfolios heavily weighted toward high-carbon industries, they face risks of stranded assets, regulatory pressure, and reputational damage.
- **Consumer Goods and Retail**: These sectors experience transition risks mostly through changing consumer preferences and reputational pressures. Although these risks are less technical than those in heavy industry, companies in consumer-facing sectors must still invest in sustainable sourcing, packaging, and branding to align with climate-conscious customers.
- **Service Sector**: While less energy-intensive, the service sector is still exposed to transition risks through operational energy usage, supply chain sustainability, and customer expectations. For example, firms with high reliance on travel (e.g., consulting or logistics-heavy services) face risks from rising costs associated with aviation and other transport emissions.
- **Renewable Energy Sector**: Renewable energy companies face lower transition risks because they directly benefit from the shift to a low-carbon economy. With continuous advancements in technology, firms in this sector are well-positioned to capitalize on the energy transition, though competition and innovation costs still present challenges.

Financial Implications of Transition Risks

Figure 7.2 implies that transition risks can lead to significant financial impacts, necessitating strategic investments and careful cost management. Even within the same sector, companies may adopt different approaches to reduce greenhouse gas (GHG) emissions and thus exposure to market, legal, reputation and technology risks arising from climate change. For instance, ExxonMobil plans to reduce emissions primarily through carbon capture and storage (CCS) technology, focusing on mitigating the environmental effects of fossil fuel extraction. In contrast, Shell emphasizes investments in green energy, positioning itself toward a renewable-focused transition.

These differing strategies reflect broader challenges for firms in high-emission sectors, where emissions costs, low-carbon capital expenditures (CapEx), and profitability impacts are central considerations.

- **Emissions Costs**: Firms in high-emission sectors face rising costs from carbon pricing and regulatory compliance, which affect operational expenses. The extent of these costs depends on a company's emissions profile and the regulatory landscape in its operating regions.
- **Low-Carbon Capital Expenditures (CapEx)**: To mitigate transition risks, companies are investing in technologies and infrastructure that support low-carbon operations. While these capital expenditures entail significant upfront costs, they are critical for meeting regulatory requirements, achieving net-zero targets, and positioning firms competitively in a low-carbon economy.
- **Profitability Impacts**: Emissions costs and low-carbon investments influence profitability in complex ways. Rising emissions costs can reduce profit margins, while strategic low-carbon investments can unlock new revenue streams, enhance market positioning, and strengthen brand reputation. Companies that proactively invest in sustainable practices often see long-term financial benefits, such as cost savings, access to green financing, and improved customer loyalty.

7.5   Mitigation, Adaptation, and Opportunity[3]

Adaptation and mitigation strategies present pivotal climate opportunities for firms. These strategies enable companies to not only manage and minimize the risks associated with climate change but also to capitalize on the changing physical and policy landscape to generate long-term value for stakeholders (Mendiluce, 2019).

Adaptation strategies involve adjusting business operations and models to better cope with the anticipated impacts of climate change. This could include

---

[3] The World Bank Group provides excellent resources for adaptation and mitigation of climate change.

developing infrastructure that is resilient to extreme weather events, diversifying supply chains, or investing in research and development for climate-resilient products.

Mitigation strategies, on the other hand, focus on reducing greenhouse gas emissions. Companies can achieve this through a variety of means, such as adopting renewable energy solutions, improving energy efficiency, and integrating sustainable practices across their value chains. These efforts help mitigate the adverse effects of climate change and position companies as leaders in the transition to a low-carbon economy.

By implementing both adaptation and mitigation strategies, firms can unlock numerous opportunities. These include entering new markets, developing innovative products and services, and enhancing their corporate reputation among increasingly climate-conscious consumers and investors. Moreover, robust climate strategies can lead to cost savings, improved risk management, and a stronger competitive edge in an evolving regulatory environment. Concentrating on a thorough evaluation of climate risk will reveal opportunities associated with moving toward a low-carbon economy and adapting to the effects of climate change.

Smokestack Ventures and Transition Risk

Smokestack Ventures is a mid-sized company engaged in the extraction, processing, and distribution of coal. Operating several mines across the United States, the firm generates approximately $500 million in annual revenue and employs 1,200 workers. Its client base includes power plants, steel manufacturers, and smaller industrial customers. Smokestack Ventures illustrates the challenges faced by high-emission sectors. With significant emissions, market exposure, and regulatory risks, the company must navigate complex transition risks while seeking to adapt its business model.

Quantitative Analysis

Smokestack Ventures faces significant transition risks as it navigates a rapidly changing regulatory and market landscape. A breakdown of transition risks and the expected financial impact on the firm is shown in Table 7.5.

Qualitative Analysis for Smokestack Ventures

- **Reputation and Competitiveness**:
  - Smokestack Ventures faces growing reputational risks as societal preferences shift toward sustainability. The firm's reliance on coal, a high-carbon commodity, may alienate investors, customers, and partners who prioritize environmental responsibility.
- **Adaptive Capacity**:
  - While the firm has made investments in low-carbon technologies, these efforts are insufficient to offset its transition risks. Significant progress in diversifying its portfolio and adopting greener alternatives is required to reduce long-term exposure.

Table 7.5 Transition risk assessment for smokestack ventures: key impacts and risk level

Factor	Risk level	Explanation
Direct Emissions Costs	High	Direct emissions of 2.5 million metric tons result in $125 million annually in carbon costs
Indirect Emissions Costs	Medium	Indirect emissions of 0.5 million metric tons add $25 million annually to operational expenses
Low-Carbon CapEx	Medium	Annual investment of $20 million (4% of revenue) is significant but insufficient for mitigation
Revenue from Low-Carbon Initiatives	Low	Revenue of $5 million highlights limited diversification and innovation in low-carbon offerings
Market Dynamics	High	Declining global demand for coal due to renewable energy growth poses long-term market risks
Regulatory Compliance	High	Anticipated stricter environmental regulations will increase compliance costs and penalties

- **Sectoral Challenges**:
  Coal is categorized as a "hard-to-abate" sector, meaning that decarbonization pathways are limited and costly. Technologies like carbon capture and storage (CCS) offer potential solutions but remain expensive and underdeveloped, requiring substantial investment and research.

Overall Risk Assessment

Smokestack Ventures is assessed as **High Risk** due to its substantial emissions costs, declining market demand, regulatory pressures, and reputational vulnerabilities.

To address these challenges, the firm must:

- Increase investment in low-carbon technologies.
- Diversify revenue streams to reduce dependence on coal.
- Enhance adaptive capacity through innovative and sustainable business practices.◄

Smokestack Ventures exemplifies the complicated challenges of transition risks, from emissions costs to market dynamics and reputational pressures. Yet, transition risks often intersect with physical risks, amplifying vulnerabilities and creating new layers of complexity. The next section explores how businesses can integrate physical and transition risks into a cohesive framework, enabling a more comprehensive approach to climate risk management.

7.6   Integrating Physical and Transition Risk

Climate risk assessment remains in its early stages for many firms, as they grapple with the complexities of measuring and managing the full spectrum of climate impacts on operations, supply chains, and markets. Typically, companies assess physical and transition risks independently, relying on distinct frameworks and tools. Some, in fact, may focus only on the transition risk aspect and even there, focus more narrowly on carbon pricing. Of course, in following this pathway, insights into the interconnectedness of physical and transition risks are completely missed! Physical risks, such as extreme weather events, can accelerate the urgency of low-carbon transitions, while transition policies, like carbon pricing, can alter firms' exposure to physical hazards by incentivizing investments in mitigation and adaptation solutions. Together, physical and transition risks create a feedback loop that amplifies vulnerabilities and challenges.

PG&E: The First Climate Bankruptcy

The case of PG&E underscores the dangers of an incomplete climate risk strategy. PG&E, California's largest utility company, filed for bankruptcy in 2019 after facing billions of dollars in liabilities from wildfires exacerbated by climate change. While PG&E had invested in clean energy to manage transition risk, it largely ignored physical risks like droughts, heatwaves, and wildfires, failing to integrate climate risk into its corporate governance, risk management, or strategic planning processes. This lack of integration left PG&E unprepared for devastating wildfires in 2017 and 2018, which led to property destruction, loss of life, and prolonged power outages. What was essentially a forensic audit conducted by Ceres (Ceres, 2019) determined that PG&E had not properly assessed its exposure to extreme weather events, nor had it disclosed climate risks transparently to stakeholders. Ultimately, PG&E's lack of a holistic approach to climate risk, combining both physical and transition elements, resulted in financial losses and a dramatic fall in market value.

Climate Risk and Sectoral Impact

Climate risks manifest differently across industries, shaping each sector's revenue potential, cost structure, asset values, and financing needs. For example:

- **Aviation**: The transition to Sustainable Aviation Fuel (SAF) demonstrates how low-carbon demands create both risks and opportunities. Yet, SAF production itself faces physical risks, as it relies on biomass feedstocks vulnerable to drought and pests.

Table 7.6 Major categories of financial impact (non-exhaustive examples)

Category	Description	Steel sector	Airline sector
Revenues	Changes in demand or prices due to economic and environmental shifts	Reduced demand for steel due to economic shifts and adoption of low-carbon alternatives like green steel by end-use sectors	Decrease in demand for air travel due to regulatory restrictions and preference for low-carbon transport options
Expenditures	Increased operating costs from carbon prices, energy expenses, or regulatory changes	Higher costs due to carbon taxes, energy prices, and emissions control technology like CCUS that require additional operating expenses	Increased fuel prices including SAF and costs of carbon offsets
Assets and Liabilities	Potential for asset impairment from physical or regulatory impacts	Stranded assets from outdated equipment in high-emission facilities	Conventional fleet risks stranding, and CapEx associated with fleet renewal and SAF infrastructure
Capital Structure and the Cost of Capital	Changes in the cost or availability of capital due to climate-related credit and reputational risks	Increased borrowing costs or divestment due to high-emission profile	Limited access to capital for non-sustainable practices, with potential growth in green financing
Working Capital	Supply chain impacts on working capital, affecting inventory and cash flow	Disruptions in supply chains Increased inventory holding periods and risk of stranded inventory	Refunds and compensations due to flight interruptions, Increased working capital requirements for SAF investments

- **Steel**: Demand for low-carbon alternatives in steel production heightens transition risks as companies navigate regulatory changes, carbon offset costs, and investments in sustainable technologies.

In Table 7.6 we illustrate how climate risks will manifest in the steel and aviation sectors.

Summarizing Corporate Climate Risk Assessment

A comprehensive approach to corporate climate risk assessment involves multiple interconnected elements. Materiality forms the foundation, identifying which risks and opportunities have the most significant financial, environmental, and social impacts. Adaptive capacity highlights a firm's ability to respond to these risks and capitalize on opportunities, while assessments of physical and transition risks provide insights into the potential disruptions to operations, supply chains, and markets.

Opportunities emerge alongside these risks, enabling firms to innovate, build resilience, and strengthen competitive positioning. Integrated risk assessment combines physical and transition risks into a unified strategy, accounting for how these risks often amplify each other. Finally, the severity of climate risks depends on the evolution of the crisis itself, as shaped by global efforts to achieve net-zero emissions and the pathways defined by the various climate scenarios we saw in the previous chapter.

This framework equips firms to navigate the complexities of climate risk and seize opportunities for long-term sustainability. To illustrate how these elements come together in practice, we turn to Danone, a global leader in the food and beverage sector, which has adopted a holistic approach to climate risk management.

7.7   Danone: Managing Climate Risks

Danone, a global leader in the food and beverage industry, operates across three primary segments: Dairy and Plant-Based Products (e.g., Activia, Alpro), Waters (e.g., Evian, Volvic), and Specialized Nutrition (e.g., Aptamil, Nutricia). The company has set ambitious goals, **including achieving** carbon neutrality by 2050 and **reducing its water** footprint by 60% by 2030. However, its global operations expose it to significant climate risks—both physical and transition related—that could affect its supply chain, markets, and reputation.[4]

Value Chain Risks

Danone's value chain spans over 120 countries, sourcing raw materials globally to ensure quality and availability. While strategically located manufacturing facilities help reduce transportation costs and environmental impact, the complexity of its operations increases its exposure to physical risks like **extreme weather** and **water scarcity**, and transition risks tied to evolving regulations, market dynamics, and consumer expectations.

---

[4] The Danone example relies on publicly available information, most of which is drawn from their website: https://www.danone.com/.

Evaluating Risks and Opportunities[5,6]

Danone employs robust scenario analysis to evaluate climate risks and opportunities under varying climate conditions. In its 2020 TCFD report, the company examined three scenarios aligned with IPCC pathways:

1. 1.5 °C (RCP 2.6):
   - Scenario 1: Consumer-led transition driven by changing preferences.
   - Scenario 2: Policy and technology-driven transition.
2. 3.7 °C (RCP 8.5): A high-emissions scenario with severe physical impacts due to insufficient global action.

Danone's 2023 disclosure expands on this analysis by detailing specific risks, opportunities, and their financial impacts, highlighting how probabilities and financial impacts are evaluated under different scenarios. Table 7.7 summarizes these risks and opportunities.

Physical Risk: Milk Production

Danone identifies milk production as one of the key vulnerabilities in its value chain due to its dependence on water and thermal conditions. In its 2023 disclosure to CDP, the firm reported that water stress affects feed prices, milk volumes, and milk yields, resulting in:

- Long-term financial impact: €42.6 million under the pessimistic IPCC RCP 8.5 scenario.

Adaptation Strategies:

- Farming for Generations Initiative: Focuses on regenerative agricultural practices and farmer support.
- R&D: Invests in sustainable farming techniques to improve soil health and carbon sequestration.

---

[5] Danone is a leader in climate disclosure and analysis. It is among the small minority of firms that disclose not only climate risks and opportunities but also assesses the financial impact of the risks and the costs associated with mitigation and adaptation. It is one of the main motivations for using the Danone example. Within their extensive analyses we have selected two examples: milk production and packaging. The former is largely associated with physical risk while the latter zeroes in on transition risk. The risk and opportunity mapping we show for Danone is itself a summary of more extensive disclosure found in other places for Danone.

[6] Portions of the text in this section "Danone: Managing Climate Risk" is verbatim (or nearly verbatim) from the firm's 2023 CDP disclosure, as made available on Danone's corporate website.

Table 7.7 Danone: climate risks and opportunities

Risk and opportunity categories	Risk and opportunity description	Probability of occurrence 2020–2030	Significance of financial impact of baseline scenario[a]	Significance of financial impact 2030-alternate scenarios[a,b]
Transition	Shift to plant-based alternatives	High	**	***
	Growing consumer engagement in fighting climate change	High	**	***
	Increasing reporting obligations	Medium	*	*
	Carbon pricing and the procurement of packaging and logistics	Medium	**	**/***
	Carbon pricing and the cost of direct operations	Medium	**	**
Physical	Water stress and thermal stress on the milk supply chain	Medium	**	**
	Water stress and thermal stress on agricultural ingredients	Medium	**	**
	Extreme events affecting direct operations	Low	***	***
	Water stress on direct operations	Low	**	**
	Impact of climate change on product use	Low	*	*

Danone Climate Transition Plan, December 2023 and available at danone.com
[a]The significance of the **financial** impact has been assessed on the basis of the reduction in profit margin if the risk occurs
[b]Some risks have two impact assessments because their financial impact differs depending on which climate change scenario is concerned

- Local Projects: Upgraded irrigation systems in South African dairy farms to optimize water usage.
- Total cost of adaptation: €32.9 million.

Transition Risk: Plastic Packaging

Emerging regulations on plastic usage represent a significant transition risk for Danone. These policies will increase the costs of compliance and operations:

- Long-term financial impact: €265 million.

Mitigation Strategy:

- Circular Packaging Commitments:
  - Design 100% reusable, recyclable, or compostable packaging by 2030.
  - Reduce virgin fossil-based plastic by 30% by 2030 and 50% by 2040.
  - Develop and lead effective collection systems to recover as much plastic as is used.
- Total cost of mitigation: €89.3 million.

Danone's Integrated Approach to Climate Risk

Danone's integrated climate risk strategy combines physical and transition risk assessments, scenario analysis, and adaptation initiatives to build resilience across its global operations. The company's key strategies include:

- Setting science-based targets: Aiming for a 50% reduction in GHG emissions by 2030 and carbon neutrality by 2050.
- Scenario analysis: Evaluating climate risks and opportunities under varying pathways and quantifying financial impacts.
- Investment in adaptation: Supporting regenerative agriculture, upgrading infrastructure, and promoting sustainable practices.
- Stakeholder engagement: Collaborating with farmers, suppliers, governments, and NGOs to co-create solutions for a low-carbon future.

Danone's approach demonstrates how a firm can use data-driven insights and targeted investments to navigate the dual challenges of physical and transition risks, while simultaneously capturing opportunities for innovation and sustainability.

Conclusion

A robust climate risk assessment framework enables firms to navigate the complexities of a rapidly changing environment. By addressing materiality, building adaptive capacity, and assessing both physical and transition risks, companies can identify vulnerabilities, seize opportunities, and develop targeted mitigation and adaptation strategies. Integrating these risks into corporate governance ensures that firms are prepared not only to mitigate potential damages but also to capitalize on opportunities in a low-carbon future.

The evolution of the climate crisis will depend heavily on global efforts to achieve net-zero emissions, as illustrated by scenario analyses. Firms must embrace proactive, data-driven strategies to remain resilient in the face of uncertainty. As demonstrated by Danone's integrated approach, combining physical and transition risk management offers a path to resilience, operational stability, and long-term growth. Climate risks are not just challenges; they are catalysts for innovation, adaptation, and competitive advantage in a sustainable economy.

Key Takeaways

1. **Materiality as a Foundation**
   Climate risks must be assessed through both financial (single materiality) and environmental/social (double materiality) lenses to capture their full impact on stakeholders and corporate strategy.

2. **Adaptive Capacity**
   Firms must enhance their ability to respond to climate risks by investing in resilient infrastructure, sustainable practices, and innovative solutions that minimize vulnerabilities and maximize opportunities.

3. **Physical and Transition Risks**
   Physical risks driven by environmental changes like extreme weather and rising temperatures, threaten operations, assets, and supply chains. Transition risks, arising from policy shifts, market evolution, and technological advancements, can disrupt business models and create stranded assets while driving innovation.

4. **Opportunities amid Risks**
   Climate risks also open doors to innovation, market differentiation, and competitive advantage through investments in low-carbon technologies, sustainable products, and circular business models.

5. **Integration of Risks**
   Physical and transition risks are deeply interconnected, often amplifying one another. An integrated risk management approach is essential to ensure comprehensive climate resilience.

6. **Sectoral and Geographic Variation**

The severity and type of climate risks vary widely by sector and geography. Businesses need tailored strategies that account for specific vulnerabilities and regional conditions.

7. **Scenario Analysis**
Evaluating risks under different climate scenarios (e.g., 1.5 °C or 3.7 °C warming) helps firms anticipate future challenges and align strategies with potential policy and market outcomes.

8. **Financial Implications**
Climate risks translate into measurable financial impacts, including higher costs from carbon pricing, stranded assets, operational disruptions, and regulatory compliance. Firms that proactively address these risks can improve financial stability and access to capital.

9. **Case Example—Danone**
Danone's integrated approach demonstrates how firms can combine physical and transition risk assessments, set ambitious climate goals, and implement sustainable practices to align with long-term resilience and growth.

10. **The Role of Proactive Management**
A forward-looking, data-driven climate risk strategy is not just about managing threats; it is about identifying opportunities to innovate and lead in a low-carbon economy.

Questions

1a. What are the two main types of climate risks discussed in the chapter?
1b. How do physical and transition risks differ in their impact on firms?
1c. Provide an example of a sector highly exposed to physical climate risks and explain why.
2. Define "adaptive capacity" and explain its importance for firms facing climate risks.
3. Review physical and transition risks in the aviation sector. What mitigation and adaptation strategies are available to this sector? Using disclosure documents for Lufthansa and Delta, we assess the adaptive capacity of each firm. Refer to sources like the Task Force on Climate-related Financial Disclosures (TCFD) and recent company sustainability reports for more context.
4. Analyze China Shenhua Energy's exposure to both physical and transition climate risks. Provide an overview of coal demand trends and discuss the company's strategies to address these risks. Include references to company reports and recent climate risk assessments where possible.
5. Examine Rio Tinto's exposure to physical and transition climate risks. Discuss the company's strategies to mitigate these risks and how they align with global trends in the mining industry. Refer to Rio Tinto's climate change reports and sustainability disclosures.
6. Analyze how climate risks affecting coffee production in Brazil could disrupt supply chains and impact consumers globally. Explain the potential impacts for

coffee businesses and consumers in cities like London, emphasizing the reality of supply chain risks.

References

Ceres. (2019). *PG&E and climate risk: An assessment of exposure and corporate governance.* Retrieved November 4, 2018, from https://www.ceres.org

Chaudhary, N. (2024). *From stranded assets to assets at risk: Understanding the financial risks of transition.* Retrieved November 4, 2024, from https://www.i4ce.org/wp-content/uploads/2024/06/From-Stranded-Assets-to-Assets-at-Risk.pdf

Chen, Y.-H. H., Landry, E., & Reilly, J. M. (2023). An economy-wide framework for assessing the stranded assets of the energy production sector under climate policies. *Climate Change Economics, 14*(1).

Colman, Z. (2024, November 12). Exxon CEO calls for U.S. climate policy stability. *Politico.* Retrieved November 18, 2024, from https://www.politico.com/news/2024/11/12/exxon-ceo-us-climate-policy-00188927

Corder, M. (2024, November 12). *Dutch appeals court overturns landmark climate ruling against Shell.* AP News. Retrieved November 18, 2024, from https://apnews.com/article/51e84d215df1 3f9c92edf84f301a673f

Financial Times. (2024, October 15). *How big is the real estate 'stranded asset' problem?* Retrieved November 18, 2024, from https://www.ft.com/content/199acbae-f7ef-475b-90ad-985eb25fe6f6

General Motors. (2020, November 19). *General Motors accelerates its transformation to deliver on a commitment to an all-electric future.* Retrieved November 4, 2024, from https://investor.gm.com/static-files/66b83c73-17f3-426a-9618-393cea479c19

Germanwatch. (2023). *Global Climate Risk Index 2023.* Retrieved November 4, 2024, from https://www.germanwatch.org/en/cri

Hicke, J. A., Lucatello, S., Mortsch, L. D., Dawson, J., Domínguez Aguilar, M., Enquist, C. A. F., ... Miller, K. (2022). North America. In H.-O. Pörtner, D. C. Roberts, M. Tignor, E. S. Poloczanska, K. Mintenbeck, A. Alegría, ... B. Rama (Eds.), *Climate Change 2022: Impacts, Adaptation and Vulnerability. Contribution of Working Group II to the Sixth Assessment Report of the Intergovernmental Panel on Climate Change* (pp. 1929–2042). Cambridge University Press. https://doi.org/10.1017/9781009325844.016

Kolaric, S. (2023). The impact of climate litigation and activism on stock prices: The case of oil and gas majors. *Review of Managerial Science, 18*(4), 3141–3172. https://doi.org/10.1007/s11 846-023-00710-4

Kuhnert, F., Stürmer, C., & Koster, A. (2017–2018). *Five trends transforming the automotive industry.* PricewaterhouseCoopers GmbH Wirtschaftsprüfungsgesellschaft. Retrieved November 4, 2024, from https://www.pwc.com/gx/en/industries/automotive/assets/pwc-five-trends-transforming-the-automotive-industry.pdf

McKinsey & Company. (2023). *Will India get too hot to work?* Retrieved November 4, 2024, from https://www.mckinsey.com/capabilities/sustainability/our-insights/will-india-get-too-hot-to-work

Mendiluce, M. (2019). *Business climate resilience: Thriving through transformation.* World Business Council for Sustainable Development (WBCSD). Retrieved November 4, 2024, from https://www.wbcsd.org/news/business-climate-resilience-thriving-through-transformation/

Milman, O., & Taylor, M. (2021, May 26). Dutch court orders Royal Dutch Shell to cut carbon emissions by 45% by 2030. *The Guardian.* Retrieved November 4, 2024, from https://www.the

guardian.com/environment/2021/may/26/dutch-court-orders-royal-dutch-shell-to-cut-carbon-
    emissions-by-45-by-2030
Oxford Institute for Energy Studies. (2023). *Domestic policies: Manufacturing sector in China.*
    Retrieved November 4, 2024, from https://chineseclimatepolicy.oxfordenergy.org/book-con
    tent/domestic-policies/manufacturing-sector/
Pereira, L. (2017, March 29). Climate change impacts on agriculture across Africa. *Oxford
    Research Encyclopedia of Environmental Science.* Retrieved November 15, 2024, from https://
    oxfordre.com/environmentalscience/view/10.1093/acrefore/9780199389414.001.0001/acr
    efore-9780199389414-e-292
Reuters. (2024, November 12). Shell wins appeal against landmark Dutch climate ruling. *Reuters.*
    Retrieved November 18, 2024, from https://www.reuters.com/business/energy/shell-wins-app
    eal-against-landmark-dutch-climate-ruling-2024-11-12/
Wayland, M. (2024, July 23). GM claims its next cycle of EV growth is coming, but when? *Elec-
    trek.* Retrieved November 4, 2024, from https://electrek.co/2024/07/23/gm-claims-next-cycle-
    of-ev-growth-coming-but-when/
World Tourism Organization. (2012). *Challenges and opportunities for tourism development in
    small island developing states.* UNWTO, Madrid. Retrieved November 4, 2018, from https://
    doi.org/10.18111/9789284414550

Additional References for Table 7.6

Airlines for America. (2020). *Analysis of fuel costs and carbon offset requirements for the U.S.
    airline industry.*
Carbon Pricing Leadership Coalition. (2021). *The impact of carbon pricing on global industries.*
    World Bank.
Climate Bonds Initiative. (2020). *Green bonds and capital access for heavy industry.*
Deloitte. (2021). *Managing airline working capital in response to supply chain volatility.*
Environmental Defense Fund (EDF). (2019). *Asset risk and stranded infrastructure in the aviation
    industry.*
International Air Transport Association (IATA). (2021). *Airline industry forecast 2021–2040.*
International Chamber of Commerce (ICC). (2021). *Supply chain resilience in heavy industries.*
International Energy Agency (IEA). (2020). *Energy technology perspectives 2020: The steel
    industry and energy costs.*
Intergovernmental Panel on Climate Change (IPCC). (2014). *Climate Change 2014: Impacts,
    Adaptation, and Vulnerability. Summary for Policymakers. Contribution of Working Group II
    to the Fifth Assessment Report of the Intergovernmental Panel on Climate Change.* Cambridge
    University Press. https://www.ipcc.ch/site/assets/uploads/2018/03/ar5_wgII_spm_en-1.pdf
IATA Economics. (2020). *Financial impacts of COVID-19 on airline working capital.*
McKinsey & Company. (2020). *Decarbonization challenge for steel: An opportunity for collabo-
    ration.* McKinsey & Company.
McKinsey & Company. (2020). *Global steel supply chain disruptions and inventory risk manage-
    ment.*
Moody's Investor Service. (2021). *Climate risk and credit rating implications for the steel sector.*
PwC. (2021). *Climate finance: Transitioning the airline industry with sustainable capital.*
Sustainable Aviation. (2021). *Sustainable aviation fuel: Costs, benefits, and long-term potential.*
Task Force on Climate-related Financial Disclosures. (2017). *Recommendations of the task force
    on climate-related financial disclosures: Final report.* Financial Stability Board. Retrieved
    November 4, 2024, from https://www.fsb-tcfd.org/publications/final-recommendations-report/
Transport & Environment. (2020). *Flying towards a low-carbon future: The impact of COVID-19
    on aviation demand.*
United Nations Environment Programme Finance Initiative (UNEP FI). (2020). *Transition risks in
    aviation.*

World Economic Forum. (2021). *Stranded assets in the industrial sector: Risks and opportunities.*
World Steel Association. (2021). *Sustainable steel: Indicators 2021 and progress.*

Risk, Return, and the Cost of Capital

Introduction

Climate risks are increasingly being incorporated into financial markets across asset prices, discount rates, and investor preferences, though the degree of incorporation varies. Asset prices are already being affected by physical and transition risks, but some long-term risks may still be underpriced. Discount rates are adjusting for climate-related uncertainties, with higher premiums applied to riskier assets, but changes are dynamic as awareness of climate risks does not grow linearly, particularly in the long-term. Investor preferences are shifting toward ESG and green investments, with a growing divestment from high-carbon sectors, but the full integration of climate risks into investor preferences is still developing, with a continued push for better data, disclosures, and clarity around how climate change affects different investments.

The challenge remains that climate risks are inherently long-term, and markets are often short-term focused, which means there may still be significant mispricing of these risks in some asset classes or industries. As climate change impacts become more apparent and policy frameworks evolve, these risks will likely be priced more thoroughly across all three dimensions.

In this chapter, we explore how climate change impacts the cost of equity, the cost of debt, and the weighted average cost of capital (WACC) across different sectors and regions. We examine how climate risk alters systematic risk (beta) and the equity risk premium, affects sovereign and corporate credit ratings, and drives variations in financing costs globally. By bridging theory with real-world examples, we highlight the evolving relationship between climate change and corporate finance, equipping firms with the tools to navigate these emerging challenges.

© The Author(s), under exclusive license to Springer Nature Switzerland AG 2025   187
S. Dow and Y. Shi, *Corporate Finance Under Climate Crisis*,
https://doi.org/10.1007/978-3-031-83487-5_8

8.1   The Weighted Average Cost of Capital (WACC)

The cost of financing for the firm, its weighted average cost of capital (WACC), is a cornerstone that influences a wide array of strategic financial decisions. The cost of capital expresses the expected return, or the minimum required rate, for investing in a company or a project. This expected return is linked with the degree of risk associated with company or project cash flows. The formula for the weighted average cost of capital is given below:

▶

$$WACC = k_e * \frac{E}{(D+E)} + k_d(1-t) * \frac{D}{(D+E)}$$

where:
$k_e =$ the cost of equity
$k_d =$ the cost of debt
$t =$ the corporate tax rate
$D + E =$ the value of the firm's debt (D) and equity (E)
$\frac{E}{(D+E)} =$ the percentage of the firm financed by equity
$\frac{D}{(D+E)} =$ the percentage of the firm financed by debt

We can see two main drivers in the calculation of the cost of capital:

- Component costs of after-tax debt and equity: Component costs must be forward-looking and thus reflect the marginal cost associated with new debt and equity financing.
- Capital Structure: The other main driver of WACC is the capital structure of the firm (the percentage of the firm or project financed by debt).

It makes sense that firms with greater climate risk would have higher costs of capital, reflecting investor demands for compensation for additional risk in equity investments, while creditors for their part will view climate risk as spawning greater default risk. Such shifts in the costs of debt and equity can also lead to adjustments in the firm's target capital structure, in many cases reducing the amount of leverage in the capital structure. We discuss capital structure in the next chapter and focus here on the component costs of debt and equity. We'll begin by looking at how the cost of equity is affected by climate change.

8.2   The Cost of Equity in Climate Crisis

Pinning down the cost of equity, on a good day, is challenging. Adding climate risk to the mix makes the task even more complex. When we are estimating the cost of capital, we need forward-looking estimates. We want to know what the cost of new equity is, and not the past cost of equity. Also, when we use the cost of capital to evaluate projects, the firm's average cost of capital is appropriate only if capital projects carry the same risk as the firm overall. For example, if a firm operating in the oil and gas sector decides to invest in a wind farm, it's certain that the return required on a wind farm will be less than that of drilling a new oil well.

While there are several approaches we could use to determine the cost of equity, we are going to focus on the Capital Asset Pricing Model (CAPM) because it provides us with a clear framework to think about how climate change could affect equity returns. At its core, the cost of equity reflects the return investors require to compensate for the risk of holding a company's stock. However, not all types of risk are relevant in determining this required return. The total risk of an investment, typically measured by the volatility of returns, is composed of two components: unsystematic risk (also known as idiosyncratic or firm-specific risk) and systematic risk (also known as market risk). A foundational principle in finance theory, embedded in the Capital Asset Pricing Model (CAPM), is that investors are only compensated for systematic risk—the risk that cannot be eliminated through diversification. Systematic risk affects the entire market or economy and arises from factors like changes in interest rates, inflation, or economic recessions. Because this type of risk cannot be diversified away, it is priced into the expected returns of assets.

In contrast, unsystematic risk is unique to a specific firm or industry (e.g., management decisions, product failures). Since this risk can be eliminated by holding a well-diversified portfolio, investors cannot receive additional compensation for it. Therefore, the CAPM assumes that unsystematic risk is irrelevant for determining the required return on an investment.

▶   The CAPM formula is:

$$E(R_i) = R_f + \beta_i * \left[E(R_M) - R_f\right]$$

where:
$E(R_i)$ = Expected return of the security
R_f = Risk-free rate
β_i = Beta of the security
$E(R_M)$ = Expected return of the market which includes all assets
$[E(R_M)-R_f]$ = the expected market risk premium
The risk-free rate refers to the theoretical return on an investment with zero risk of economic loss. Government bonds of a stable country, such as US Treasury bonds, are considered free from default risk. The risk-free rate serves as a baseline for evaluating the expected returns of other investments, providing a benchmark against which we can evaluate the additional risks of equities.

▶   If we want to calculate the expected return on a security and we know its systematic risk (β_i) and the market risk premium ($E(R_M)-R_f$) we would be all set to figure out the cost of equity to the firm. To understand how climate risk will impact financial returns we can start by unpacking the elements of the CAPM by rewriting the model in terms of risk premiums.

$$[E(R_i) - R_f] = \beta_i * [E(R_M) - R_f]$$

Now we see that the risk premium on a security, $[E(R_i) - R_f]$, is related to the market risk premium (MRP), $[E(R_M)-R_f]$, via β_i, the systematic risk of the security. We can now focus on how climate risk can reshape overall market expectations (reflected in the MRP) and how specific firms or sectors experience shifts in their sensitivity to market-wide climate risk (reflected in β_i). By initially examining the market risk premium and beta separately we can gain a clearer understanding of how climate risk uniquely affects both the market and individual securities, resulting in a more comprehensive view of its impact on the cost of equity capital.

Market Risk Premium (MRP) and Equity Risk Premium (ERP)

The market portfolio, in theory, includes all assets available in the market, weighted by their market value. This would encompass not only stocks but also bonds, real estate, commodities, and potentially even private equity and other alternative investments. However, in practice, it is impossible to capture every single asset in the market portfolio. So, we need to find a proxy to determine what the risk premium might be. In finance we typically use a couple of methods to come up with a value for the risk premium:

• the survey method and
• estimation drawn from historical data.

The survey method involves collecting expectations about the future market risk premium from a range of financial experts, market participants, and academics. One prominent example of this method is the annual survey conducted by Pablo Fernández, a professor at IESE Business School. Professor Fernández surveys hundreds of finance and economics professors, analysts, and company managers worldwide, asking them for their estimate of the expected market risk premiums. The results provide a comprehensive view of the prevailing sentiment regarding future market conditions.[1,2]

---

[1] Professor Fernandez updates his MRP surveys annually. See, for example, Fernandez, Pablo and Garcia de la Garza, Diego and Fernández Acín, Lucía, Survey: Market Risk Premium and Risk-Free Rate used for 96 countries in 2024 (March 10, 2024). Available at SSRN: https://ssrn.com/abstract=4754347 or http://dx.doi.org/10.2139/ssrn.4754347.

[2] Another source of survey-based estimates comes from financial institutions such as Merrill Lynch and Credit Suisse, which periodically survey their analysts and clients to gauge expectations for the equity risk premium.

Measures of the equity risk premium (ERP) are focused on equity returns and rely on historical data. When using a proxy like the S&P 500, we are examining a broad, diversified index of large-cap US stocks. While the S&P 500 doesn't encompass the entire market portfolio, it serves as a practical and widely accepted approximation for the equity portion of the market. By using this index, we can calculate the ERP—the excess return above the risk-free rate that investors require for holding risky equities.

The simplest approach to estimating the ERP is to look at historical differences between observed risk-free rates and returns on the S&P 500. More nuanced methods estimate an implied discount rate, either through the dividend discount model (DDM), which uses a stream of aggregated dividends, or through free cash flow to equity (FCFE), which focuses on aggregated free cash flows. Professor Damodaran at the Stern School of Business, NYU has compiled extensive data on historical ERPs and provides ERP estimates using both the DDM and FCFE models, offering a comprehensive view of expected equity returns.[3,4]

Climate Crisis and the Risk Premium

Physical and transition risks vary across sectors and regions, and the equity risk premium (ERP) and market risk premium (MRP) also vary across countries. This variance is influenced by multiple factors, including economic stability, market development, political risk, inflation rates, and investor sentiment within each country. Climate risk adds another layer to this variability, as certain regions face heightened physical risks from climate change and the costs of transitioning to a low-carbon economy. Emerging markets often exhibit higher ERPs compared to developed markets due to greater perceived risks in these areas.

For example, countries like Bangladesh, Indonesia, and Brazil not only face economic and political challenges but are also among the most vulnerable to climate-related risks, such as rising sea levels, extreme weather events, and deforestation impacts. These climate risks further elevate the ERP in such markets. The resulting higher cost of capital in emerging markets makes it more challenging to fund climate mitigation and adaptation efforts, creating a vicious cycle: as the risks from climate change increase, so do financing costs, which in turn hinder investment in adaptation.

---

[3] Professor Damodaran updates his MRP analyses annually. See, for example, Damodaran, Aswath, Equity Risk Premiums (ERP): Determinants, Estimation, and Implications – The 2024 Edition (March 5, 2024). Available at SSRN: https://ssrn.com/abstract=4751941 or http://dx.doi.org/10.2139/ssrn.4751941.

[4] Professor Damodaran calculates two methods to determine the MRP based on historical data: the Dividend Discount Model (DDM) and the Free Cashflow to Equity (FCFE) model. These are explained in several places on his website as well as in the annual publication of risk premium data.

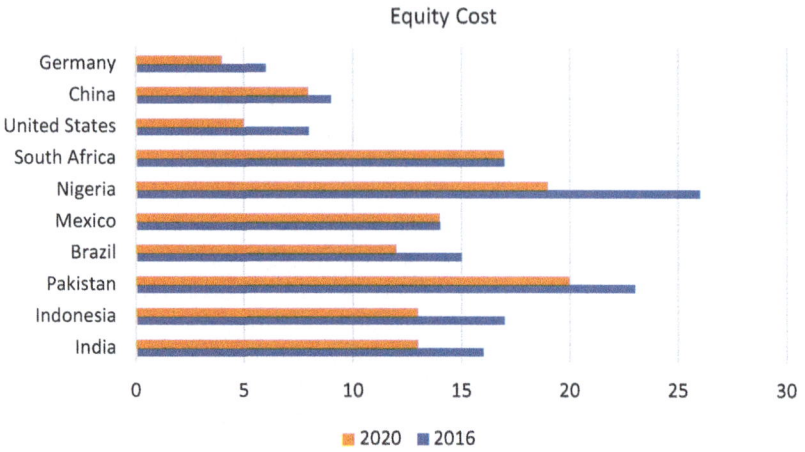

Fig. 8.1 Indicators of economy-wide cost of equity (2016 & 2020)[5] (*Source* IEA [2021])

Figure 8.1 illustrates country differences in cost of equity. This variance often reflects each country's specific economic stability, market development, and exposure to climate-related impacts.

Example: Market Risk Premiums and Climate Change in Europe

Europe is warming faster than any other continent, facing more frequent heatwaves, intense precipitation, droughts, and rising sea levels. These changes pose substantial risks, especially for regions like southern Europe and low-lying coastal areas, which are projected to face severe droughts and flooding. For instance, catastrophic flooding in Belgium and Germany in 2021 caused over €44 billion in damages. A recent report from the European Environmental Agency in 2024 presented the first in-depth analysis of the climate risks facing the EU, concluding that the EU simply wasn't ready to meet these challenges (European Climate Risk Assessment, 2024)!

We decided to look at the evolution of climate disasters in Europe over the past ten years—evoking the "storm of the decade" which is replacing "the storm of the century!" We then wondered if there was any correlation with changes

---

[5] Sources used by IEA: Calculations based on Bloomberg (2021), Bloomberg Terminal; Damodaran (2021), Data: Current; and Refinitiv Eikon (2021), Eikon Data.

Government bond rate = ten-year local currency bond yield. Debt and equity market risk premiums represent the additional return over the risk-free rate (expressed here as ten-year government bond yields) required by investors to invest in the debt and equity securities of a given market.

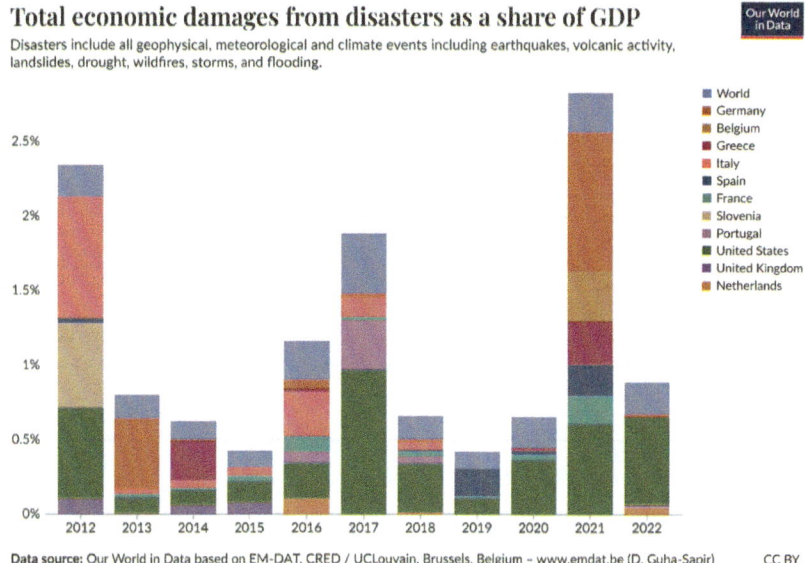

Total economic damages from disasters as a share of GDP

Disasters include all geophysical, meteorological and climate events including earthquakes, volcanic activity, landslides, drought, wildfires, storms, and flooding.

Data source: Our World in Data based on EM-DAT, CRED / UCLouvain, Brussels, Belgium – www.emdat.be (D. Guha-Sapir)   CC BY

Fig. 8.2 Losses due to natural disasters 2012–2022 (Select Countries)

in the ERP over a period roughly corresponding to our time frame. Figure 8.2 shows the economic losses as a percentage of GDP for select EU countries.[6]

Following on from Fig. 8.2, we checked out the equity risk premiums using Professor Damodaran's website to examine the ERPs in 2012 and 2022 for the same set of countries appearing in Fig. 8.2. Table 8.1 shows that the ERP declined over the decade. Admittedly it's a true "back-of-the-envelope" analysis but it supports what a number of researchers have observed: the physical risks of climate change are probably not fully priced (although there is more encouraging support concerning transition risks). In every case the ERP declined over the 2012–2022 period. Of course, there are other reasons for this as well so what we are showing undoubtedly has a good bit of "spurious correlation" in the mix.

---

[6] The data reflect earthquakes, so we are overstating the total climate change disasters. In Greece and Turkey, a 7.0 magnitude earthquake in the Aegean Sea caused damage in Turkey's Izmir province and on the Greek island of Samos on October 30, 2020.

Table 8.1 Equity risk premiums in Europe: 2012 & 2022[7]

Country	ERP 2022	ERP 2013
Germany	4.6	5.8
United Kingdom	5.48	5.8
France	5.32	6.18
Greece	8.26	20.8
Belgium	5.48	6.85
Spain	6.94	8.8
Italy	7.81	8.43
Netherlands	4.6	5.8
Portugal	6.35	10.68
Slovenia	6.35	8.43

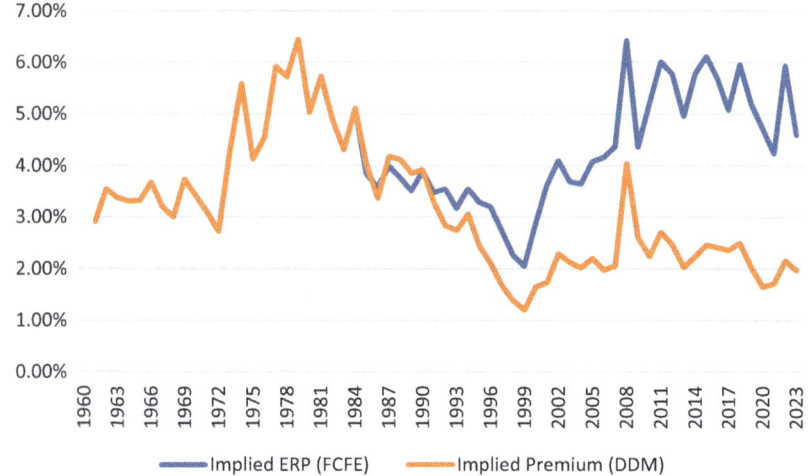

Fig. 8.3 Equity risk premium in the US: 1960–2024

Equity Risk Premiums and Financial Crises

While we have argued that we need forward-looking measures for the market risk premium, there are still lessons to be drawn from the reaction of the equity risk premium to financial crises. Figure 8.3 shows the evolution of the US equity risk premium from 1960 to present day using data from Professor Damodaran's website.

---

[7] Data are from Professor Damodaran's website at: https://pages.stern.nyu.edu/~adamodar/New_Home_Page/dataarchived.html.

  You are encouraged to visit the website to see the method used to determine the ERPs!

Figure 8.3 plots two calculations for the Equity Risk Premium. The blue line shows the Implied ERP from the FCFE approach, and the orange line shows the premium calculated by the Dividend Discount Model.

Taking a closer look at Fig. 8.3 suggests that the market seems to have experience dealing with crises, although nothing as severe as the climate crisis which has been playing out over this century!

- You will see the first spike in our graph corresponds to increases from 1972 to 1974. What was going on then?
- There is another big spike in 2008! What happened?
- There is also a dip caused by "irrational exuberance" in 1999. What was behind the significant reduction in investor risk aversion?

Determinants of the Risk Premium

We cannot predict what the ERP will be in the future under climate change, but relying solely on past data is unlikely to be effective given the unprecedented nature of this crisis. While markets have historically reacted to crises, these have been more "compact" in duration and scope compared to the pervasive, long-term challenge of climate change. However, theoretical insights allow us to outline some fundamental factors affecting the market risk premium, helping us anticipate how climate risks could drive it higher. This effect would be especially pronounced if the transition to a low-carbon economy is disorderly.

Investor Risk Aversion

Climate change introduces substantial physical risks—extreme weather events, rising sea levels, and increased natural disasters—that can disrupt business operations, damage infrastructure, and interrupt supply chains. As these risks become more frequent and severe, investor risk aversion may rise, driving up the ERP. Furthermore, transition risks associated with moving to a low-carbon economy, including regulatory shifts, carbon pricing, and technological changes, create uncertainties that could heighten investors' demand for risk premiums, particularly in high-carbon sectors like energy and transportation.

Economic Growth Expectations and Inflation

Climate thresholds add to economic slowdowns, especially if global warming surpasses critical thresholds, and might even contract economies most exposed to climate risks. This potential for reduced growth and increased economic instability is likely to raise the ERP as investors seek compensation for heightened uncertainty. Additionally, climate impacts can create inflationary pressures through reduced agricultural output, higher energy costs, and disrupted supply chains, further driving up the ERP. For example, rising costs in sectors like energy and transportation, where significant expenditures are required to meet carbon commitments and increase resilience, could elevate the market-wide cost of capital.

Market Inefficiencies and Behavioral Adjustments

Although climate risks are increasingly recognized, they have not yet been fully priced into the market, possibly due to information lags, behavioral biases, and temporary inefficiencies. As these dynamics evolve, the equity risk premium could adjust more clearly to reflect the full impact of climate risks.

Impact on Firms: Sectoral Shifts in Beta and Cost of Equity

As the ERP rises in response to climate risks, the cost of equity for individual firms will vary according to their beta, or systematic risk. Firms in vulnerable sectors, such as fossil fuels and aviation, may see an increase in beta due to heightened exposure to climate risks, raising their cost of equity. Conversely, green sectors might experience lower betas, which could offset a rising ERP, stabilizing or even reducing their cost of equity. Empirical studies support these sectoral shifts, suggesting that climate risks are driving systematic risk adjustments that impact the firm cost of capital differently based on their climate exposure (Addoum et al., 2021; Barnett et al., 2020; Bolton & Kacperczyk, 2021; Engle et al., 2020; Ilhan et al., 2021).

In summary, while climate change is likely to raise the ERP as climate risks are increasingly priced into the market, the effect on firms' cost of equity will depend on how their beta is influenced by these risks. This underscores a broader market adaptation where the equity risk premium reflects climate risk, but the impact on the firm's cost of equity will depend on how the firm's stock moves with this climate-infused ERP. To be clear, this is how things work now but the point we want to get across is that the firm's beta is going to change in climate crisis: it won't be "ceteris paribus!" We will take a closer look at Beta in the next section.

▶   β (Beta) in Climate Crisis
The other main component of the CAPM is Beta, the measure of systematic risk. Beta is simply a measure of how the risk premium on a security is related to the risk premium on the market. In this section we will examine how Beta could behave in climate crisis. To ground this analysis, we will repeat the CAPM formula:

$$E(R_i) - R_f = \beta_i * \left[E(R_M) - R_f \right]$$

where:

$$\beta_{i=} \frac{\sigma_{iM}}{\sigma_M^2}$$

σ_{iM} = the covariance between the return on the stock (R_i) and the return of the market (R_M).[8]

σ_M^2 = the variance of the return on the market.[9]

8.3 Interpretation of Beta

Beta values represent the risk profile of a security relative to the market:

- $\beta = 1$: The returns move in perfect correlation with the market.
- $B > 1$: The returns of the security are more volatile than the market. For example, a beta of 1.5 suggests that the investment is expected to move 1.5 times more than the market.
- $B < 1$: The returns of the security are less volatile than the market. A beta of 0.5 indicates that the investment is expected to move only half as much as the market.
- $B < 0$: The security returns move in the opposite direction to the market. This is rare and typically applies to certain types of hedging instruments.

Finding Beta

We can calculate Beta ex-post by regressing security returns on a market portfolio proxy, like the S&P 500 for US securities or the MSCI (Morgan Stanley Capital International Index) for global analysis. It's a simple exercise which can be done in excel using security returns from different providers (e.g., Yahoo Finance, Google Finance, and a wide number of professional data services). You can also find security Betas on financial websites like Yahoo and Google, where they have already performed the regression using typically five years of monthly returns.

---

[8] Covariance is a statistical measure that indicates the extent to which two variables change together. In the context of beta, it measures the degree to which the return of an investment (R_i) and the return of the market (R_M) move in tandem. A positive covariance indicates that the returns of the investment and the market move in the same direction, while a negative covariance indicates they move in opposite directions.

[9] The measure of covariance on its own is difficult to interpret and compare across firms. However, we can get around this issue if we standardize the measure of covariance of the firm with the market, relative to the variance of the market overall. Variance measures the dispersion of the market returns around the mean. It provides a sense of how much the market returns fluctuate. Higher variance indicates greater volatility, implying that the market returns are spread out over a wider range of values.

Systematic Risk and Climate Change

What we would really like to know is whether Beta incorporates additional risk due to climate change. Do investors price climate risk? The beta for a firm reflects its sensitivity to broader market movements, and it can increase when firms are more exposed to systematic risks driven by climate change. Sectors like fossil fuels, aviation, and heavy manufacturing are more susceptible to regulatory shifts, carbon pricing, and shifts in consumer demand away from carbon-intensive goods, which can raise their systematic risk. As a result, these firms might see an increase in beta alongside a rising ERP, amplifying the rise in their cost of equity.

- The energy sector, particularly oil and gas, is a prime example. As regulators introduce stricter climate policies, such as carbon taxes or emission caps, energy companies face higher operational costs and potential stranded assets. This has been observed empirically in studies showing higher betas for these firms post-Paris Agreement (Bolton & Kacperczyk, 2021). For these firms, both the ERP and beta are likely to increase, resulting in a significantly higher cost of equity.
- Conversely, green sectors—such as renewable energy, electric vehicles, and sustainable technologies—might see the opposite effect. The renewable energy sector has seen increasing investor interest, partly due to its perceived lower exposure to climate risks. A 2020 report by BlackRock Investment Institute shows that renewables exhibit lower beta relative to traditional energy sectors (Blackrock, 2020). For instance, companies involved in solar and wind energy may experience lower systematic risk as the global economy shifts away from fossil fuels, potentially even offsetting the effects of an increased ERP.
- Several empirical studies reinforce the idea that climate change is driving sectoral shifts in beta, impacting the cost of equity. A study published in the Financial Analysts Journal found that companies with lower carbon footprints or those involved in sustainable practices often display lower systematic risk (Andersson et al., 2016). The real estate sector is experiencing this effect as well, particularly in areas vulnerable to physical climate risks where beta and required returns in high-risk areas are increasing relative to less exposed assets (Giglio et al., 2021).

It's fair to conclude that the cost of equity for firms is increasingly influenced by climate change, with high-carbon sectors. In the future we think it is likely that high-carbon sectors will face not only a market-wide increase associated with a higher ERP as well as an elevated beta due to systematic climate risks. In contrast, green sectors with lower exposure to climate-related risks might achieve a lower beta, potentially mitigating the impact of a rising ERP on their cost of equity. This dual impact—where traditional, carbon-intensive firms see rising costs and green firms see potential reductions—could play a substantial role in capital allocation and financing costs as climate risks become more integrated into financial markets. In the next section we look at how Beta will vary by sector and by geography.

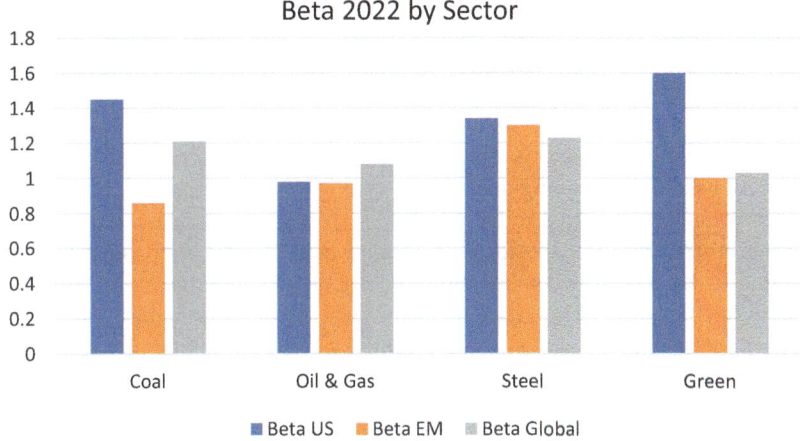

Fig. 8.4 Beta by sector in the US and EMs

Industry Betas

Figure 8.4 shows industry Betas for firms in the US and those in Emerging Markets (EM); as well as a global estimate that includes the entire sample of firms around the world. The data are from Aswath Damodaran's website.[10]

We compared Beta values for the US market and emerging markets, for several carbon-intensive sectors: coal, integrated oil and gas, and steel, with green betas (the green sector encompasses a diverse array of industries from renewable energy and energy efficiency to recycling, waste management, and green building).

It's a one-year snapshot so make of it what you will. If we look at coal the observation is interesting because in emerging markets (where China and India are major players), the systematic risk as captured by Beta is approaching half of what is observed in the US. The other rather remarkable observation is that the Beta in the green sector is much larger in the US compared to emerging markets. The disparity in Beta values between the US and Emerging Markets (EM) for both coal and green sectors could be attributed to several factors:

- In the case of coal, the higher Beta in the US may reflect heightened regulatory and other transition risks. The US has been increasingly implementing more stringent environmental regulations and policies aimed at reducing carbon emissions. This regulatory environment could lead to greater volatility as coal companies face uncertainties related to compliance costs, potential litigation, and shifts in energy policy. Moreover, public sentiment in the US is

---

[10] Data are from Aswath Damodaran's website at https://pages.stern.nyu.edu/~adamodar/New_Home_Page/data.html and click on "Useful Data Sets."

increasingly favoring renewable energy, further enhancing the systematic risk associated with coal investments.

- In emerging markets, coal remains a crucial component of the energy mix, particularly in rapidly industrializing nations like China and India. These countries rely heavily on coal for electricity generation and economic growth, thus potentially stabilizing the sector's Beta. The relative lack of stringent environmental regulations and the ongoing demand for coal in these regions may contribute to lower volatility compared to the US.
- At the other end of the spectrum, the higher Beta in the US green sector could be due to the fast-paced innovation and significant investments in renewable energy technologies. The green sector in the US is experiencing rapid growth, driven by substantial government incentives, venture capital funding, and a strong push toward sustainability. This dynamic environment, while promising, also introduces greater volatility as companies navigate technological advancements, market competition, and policy changes.
- In emerging markets, the green sector is still developing, often with less financial backing and fewer technological breakthroughs. The focus tends to be on scaling up existing technologies rather than pioneering new ones, which may result in a more stable but slower growth trajectory, thereby contributing to a lower Beta. Additionally, the regulatory landscape in EMs might not yet be as supportive or demanding, leading to different risk perceptions.

Climate Change and Beta

Our "back-of-the-envelope" industry betas cannot tell us if the systematic risk is correctly priced. In other words, are the beta estimates right? Whether you use a more sophisticated method as depicted in Fig. 8.3 drawn from Professor Damodaran's website, or you examine Betas from Yahoo for individual firms, you are still relying on past data! This is probably a significant problem when examining the effect of global warming on firm risk. A recurring theme in the discourse on climate change is the uncertainty that envelops not only climate science but the impact that higher temperatures will have on firm value. We can, however, offer some broad ideas about how climate change could affect systematic risk—how security returns might in the future correlate with market movements.

To a rather modest extent we saw this with our example of betas in carbon-intensive sectors versus green industry betas across geographic regions. We'll delve deeper into this relationship now recalling that systematic risk is a measure of risk that cannot be diversified. The question is one of how climate change might increase the "quantity" or level of systemic risk in climate crisis. More succinctly, the issue is frequently expressed as whether global warming is a systematic risk or a firm-specific risk. To do this, we developed a short example of how systematic risk has evolved in the aviation sector.

Risk and Return in the Airline Sector

CORSIA (Carbon Offsetting and Reduction Scheme for International Aviation), announced by the International Civil Aviation Organization (ICAO) on October 6, 2016, is a global regulatory framework designed to limit the growth of carbon emissions from international aviation. Under CORSIA, airlines must offset any increase in CO_2 emissions above 2020 levels by purchasing carbon credits, sourced from projects that capture or reduce emissions, such as reforestation or renewable energy initiatives. This scheme effectively neutralizes the additional emissions produced by the aviation sector.[11]

Importantly, CORSIA applies to both European and non-European carriers operating international flights, making it the first global offsetting mechanism to have worldwide application across borders. The program initially began with a voluntary phase starting on January 1, 2021, but mandatory compliance is set to begin in 2027. This comprehensive, international approach aims to curb aviation's environmental impact without fragmenting regulations by country or region, which could lead to inconsistent and possibly stricter rules.

In our analysis, we wanted to see if firms in the airline sector, a sub-sector of aviation, reacted to ICAO's announcement in 2016. Although the announcement likely did not come as a complete surprise—given ongoing discussions about emissions regulations—airlines may have responded positively to CORSIA's structured pathway, as it provided clarity on future obligations and reduced the risk of unexpected or more stringent regulations.

Our Airline Experiment

We gathered returns for American Airlines (AAL), Delta Airlines (DAL), United Airlines (UAL), Lufthansa (AFLYY), and British Airways (ICAGY) from Yahoo Finance for the period 2011 to 2023. Lufthansa and British Airways are traded in US markets (as ADRs), so this facilitated the comparisons. Figure 8.5 shows the returns for these airlines and the return for the S&P 500.

The airline industry of the early 2010s enjoyed steady growth, riding on the tailwinds of rising global travel demand and low regulatory costs. In this period, airlines like American Airlines (AAL), Delta (DAL), and United Airlines (UAL) closely mirrored the broader market's returns, as seen in the graph. But by October 2016, the industry hit a pivotal moment.

That month, ICAO announced CORSIA. Surprisingly, rather than dragging down returns, the graph shows a peak in airline returns around this announcement. This could be due to several factors. First, markets may have reacted positively to ICAO's proactive approach, as CORSIA provided a structured pathway to sustainable growth, potentially reducing the risk of more stringent regulations down the line. Simply put, CORSIA offered some certainty in the

---

[11] CORSIA applies to international flights (flights between two different countries) from all participating ICAO member states. This includes flights operated by both European and non-European carriers, provided both the departure and arrival countries are participating in CORSIA.

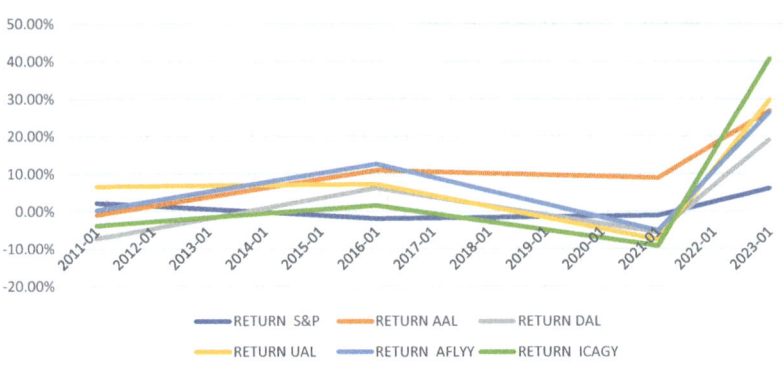

Fig. 8.5 Airline returns 2011–2023

face of many climate-induced uncertainties. Additionally, strong industry fundamentals and a booming travel market could have bolstered airline stocks, temporarily masking any potential concerns about future regulatory costs.

However, as time went on, the realities of compliance with CORSIA began to set in. By **January 1, 2021**, when the voluntary phase of CORSIA officially began, the situation had changed dramatically. The pandemic in 2020 had severely impacted airlines, grounding fleets worldwide and decimating revenues. The **graph in Fig. 8.5 shows a marked drop in returns around the start of CORSIA's voluntary phase**, reflecting the combined weight of pandemic-induced financial strain and the growing awareness of future offset costs. For many airlines, the additional burden of offsetting emissions, even voluntarily, likely seemed daunting amid an already challenging financial environment.

As airlines navigated this turbulent period, they faced mounting pressure to balance recovery efforts with sustainability commitments. The post-2021 period shows gradual improvement as the industry adapted to changing market conditions and started to recover from the pandemic's impact. By 2023, the graph indicates a sharp rise in returns for many airlines, signaling renewed investor optimism as air travel rebounded. Yet, with CORSIA's mandatory phase approaching in 2027, airlines are now more conscious than ever of the challenges—and costs—associated with meeting long-term environmental goals.

Systematic Risk and CORSIA

Let's continue with our CORSIA analysis and see if we can identify an impact on Beta. For this analysis we used simple OLS to estimate the Beta coefficient around the announcement period of October 2016. We looked at a fairly long window of 36 months before and after the CORSIA announcement date. We calculated the excess returns on the S&P 500 and our five airline stocks. To calculate excess returns, we subtracted the risk-free rate (using the yield on 10-year US Treasury bill notes) from the actual returns. Data for the yield

Table 8.2 Systematic risk pre- and post-CORSIA announcement

Firm Ticker	Pre-Announcement 36 months < Oct. 6, 2016	Post-Announcement 36 months > Oct. 6, 2016
AAL	1.21	2.00
DAL	1.04	1.28
UAL	0.65	1.29
AFLYY	0.65	1.28
ICAGY	1.00	1.05

on 10-year treasuries were sourced from the Federal Reserve Economic Data (FRED) database. The results of this exercise are tabulated in Table 8.2.

- **Increased Beta Post-Announcement**: All firms show an increase in beta following the CORSIA announcement, indicating that their stock returns became more sensitive to market fluctuations. This change may suggest that investors viewed the regulatory environment as adding systematic risk to the sector, potentially due to anticipated costs or operational impacts from CORSIA.
- **Firm-Specific Differences**: The degree of increase in beta varies by firm, with American Airlines (AAL) showing the largest jump (from 1.21 to 2.00) and International Consolidated Airlines Group (ICAGY), the parent company of British Airways having the smallest increase (from 1.00 to 1.05). These variations could reflect differences in exposure to international routes, operating structures, or financial resilience to regulatory costs.

This beta analysis around the CORSIA announcement period highlights how sector-wide regulatory changes can shift the market risk profile for firms.◀

Climate Crisis and the Cost of Equity: Summing Up

Understanding how climate change affects the equity risk premium (ERP), and beta allows firms to better gauge the market's perception of climate-related risks. As physical and transition risks grow, many firms can expect higher equity costs, particularly if they operate in high-emission or climate-vulnerable sectors. By analyzing how beta varies across sectors and geographies, firms can identify which investments might face higher capital costs due to climate risks. This information can guide capital allocation decisions, prioritizing projects in sectors or regions with lower exposure to climate risk, or allocating more resources to transition strategies (e.g., renewable energy) that might reduce beta over time. Our analysis highlights the importance of planning for regulatory shifts, carbon pricing, and other transition risks, helping firms stay ahead of compliance requirements and mitigate future financial impacts.

8.4   Climate Change and the Cost of Debt

Imagine a corporation, let's call it GreenWorks Inc., which has long relied on borrowing to finance its growth and sustainability projects. For years, GreenWorks has enjoyed relatively stable interest rates on its debt, thanks to a robust economy and stable market conditions. However, as the world grapples with the escalating impacts of climate change, borrowing costs for corporations like GreenWorks are starting to shift in ways that might not have been anticipated a decade ago. Today, lenders and investors are increasingly aware of climate-related risks—ranging from physical risks, like extreme weather damaging assets, to transition risks, such as regulatory changes aimed at reducing carbon emissions. These risks are changing the financial landscape. Just as a homeowner in a flood-prone area faces higher insurance premiums, a corporation exposed to climate risks may soon find its cost of debt increasing, reflecting the added risk that lenders now see on the horizon.

Estimating GreenWorks' future debt costs is no longer a simple exercise of assuming stable interest rates or straightforward market conditions. Instead, it involves assessing how climate change might impact those costs over time. For example, what if new environmental regulations raise operational costs, affecting GreenWorks' ability to repay its debt? Or, what if extreme weather events damage facilities, increasing the corporation's risk profile in the eyes of lenders? The process of projecting future debt costs now must include a climate lens, acknowledging that the company's environmental strategy, resilience, and ability to adapt to changing conditions will play crucial roles in its borrowing profile. Just as climate change is prompting governments to re-evaluate their risk exposure, corporations like GreenWorks must also factor in these evolving dynamics, preparing for a future where lenders may demand higher premiums in exchange for bearing climate-related risks.

In this context, the cost of debt must be a forward-looking measure, where climate resilience could make the difference between favorable financing and prohibitively expensive borrowing. The journey to estimate future debt costs, therefore, begins with an understanding of how climate change influences corporate risk—and how that risk, in turn, shapes the financial terms companies like GreenWorks will face.

Sovereign Debt and Climate Change

Sovereign debt borrowing costs affect corporate borrowing costs because corporations are often subject to the "sovereign ceiling." This means that companies within a country typically can't achieve a credit rating higher than that of their

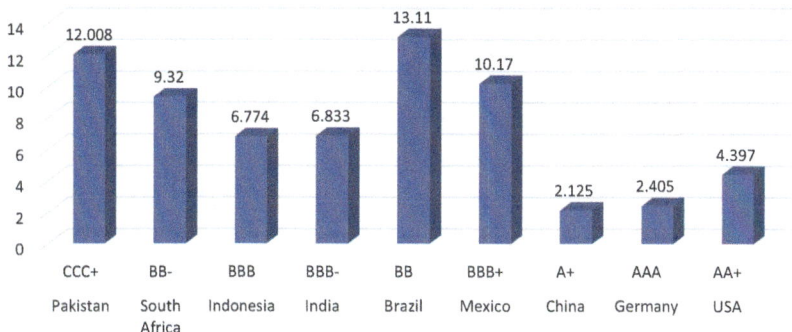

Fig. 8.6 YTM 10-year government bond

government, as the country's fiscal stability influences the perceived risk for businesses operating there. When a country faces higher sovereign borrowing costs due to climate vulnerabilities or other risks, corporate debt costs usually increase as well, since investors demand a risk premium to account for the added uncertainty. Additionally, higher sovereign debt costs can signal broader economic challenges, which may indirectly impact corporate profitability and risk, further raising the cost of corporate borrowing.

A study by the Bank of England (Bank of England, 2021) highlights that physical climate risks increase the probability of default on sovereign debt. Transition risks, including policy changes aimed at reducing carbon emissions, can lead to higher borrowing costs for countries reliant on fossil fuels. The study emphasizes that adverse climate impacts are increasingly being factored into sovereign credit ratings, influencing the terms and availability of debt. Additional analysis provided by The OECD in 2020 indicates that climate change exacerbates fiscal pressures, leading to increased costs for sovereign borrowers (OECD, 2020). Countries with higher climate vulnerability face steeper bond yield spreads, reflecting the higher risk associated with lending to these nations. The report also notes that investing in green infrastructure and technologies can help mitigate long-term debt risk.

Let's take a look at sovereign debt ratings and the yield to maturity (YTM) on 10-year government bonds. We want to use the YTM as it represents the market's current assessment of borrowing costs.[12] Figure 8.6 shows Yields to Maturity (YTMs) and sovereign debt ratings from S&P global ratings for select countries as of November 1, 2024.

What we want to know next is the extent to which the YTMs for sovereign debt translate to corporate borrowing costs. Figure 8.7 shows economy-wide borrowing

---

[12] Sovereign debt ratings are available from multiple sources including major ratings agencies: Standard & Poor's Global Ratings (S&P), Fitch Rating, and Moody's Investor Service. Yields to maturity on sovereign debt are widely available from financial news websites and international financial organizations.

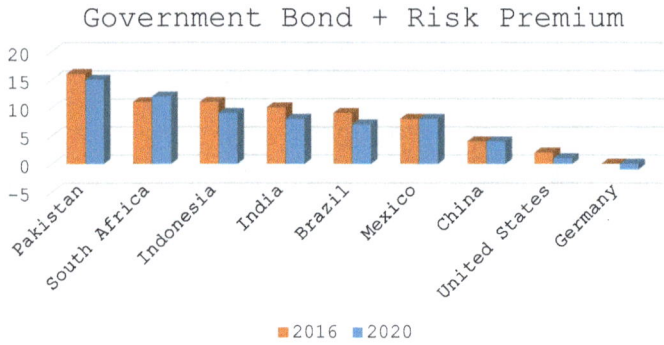

Fig. 8.7 Economy-wide cost of debt (*Source* IEA [2021c])

rates in 2016 and 2020 as reported by the IEA. Even though our time frames differ, we can see that borrowing rates align with the sovereign YTMs in the earlier figure. So, it would seem that the sovereign debt ceiling is intact.

Ex-ante Estimates of Credit Ratings

With much handwringing, we can say that climate change is bad, could get worse, and could be expensive. Unfortunately, we need more information than this to try and figure out (educated guess?) how the cost of corporate debt, and ultimately the cost of capital, will be influenced by climate change, beginning with the impact on sovereign debt. We explained that climate change can influence sovereign debt ratings and increase the cost of sovereign borrowing...but by how much? A very new and innovative article addresses exactly this question (Klusak et al., 2023). The study simulates the effect of climate change on sovereign credit ratings for 109 countries and is able to predict that climate change could induce sovereign downgrades as early as 2030, with the impact increasing in intensity and affecting more countries over the century.[13] The impact on credit ratings and borrowing costs is evaluated over different warming scenarios:

• Strict climate policies aimed at limiting global warming to below 2 °C (RCP 2.6) can almost completely mitigate the negative impact of climate change on sovereign credit ratings. Conversely, under a high-emissions scenario (RCP 8.5), fifty-nine countries could see downgrades by 2030, with an average reduction of

---

[13] The study uses a random forest machine learning model to predict sovereign credit ratings based on macroeconomic indicators and adjusts these inputs to reflect climate impacts under different warming scenarios. See the original article for methodology details.

0.68 notches,[14] increasing to eighty-one countries with an average downgrade of 2.18 notches by 2100.

- Climate-related downgrades in sovereign credit ratings could lead to significantly higher borrowing costs. Even under RCP 2.6, annual interest payments on sovereign debt could increase by $45 billion–$67 billion, and under RCP 8.5, by $135 billion–$203 billion. For corporations, the additional costs could range from $10 billion–$17 billion under RCP 2.6 to $35 billion–$61 billion under RCP 8.5.

Default Risk and the Cost of Corporate Debt

The cost of debt reflects the interest rate firms pay on borrowed funds and is significantly influenced by the perceived risk of default. Lenders assess this risk using a variety of factors, including the borrower's credit rating, which provides a quick indication of creditworthiness. Ratings agencies are also beginning to incorporate climate risk in assessments of creditworthiness. Central banks, for their part, are scrutinizing the portfolios of financial institutions under their supervision to look for climate-induced vulnerabilities. For companies with high-credit ratings, the perceived risk of default is low, resulting in lower interest rates. In contrast, firms operating in volatile sectors or exposed to significant risks—such as those due to climate change—may face higher default risk premiums, increasing their borrowing costs.

Example: Default Risk in the Tourism Sector

Let's assume a central bank rate of 2%. A high-credit company with minimal risk might only pay this base rate or something close to it. However, for a company in a high-risk sector, such as a tourism firm operating in an area vulnerable to climate impacts, lenders might add a significant default risk premium.

Consider Paradise Resorts, a company operating in a coastal region highly vulnerable to climate-related impacts, such as rising sea levels and frequent hurricanes. These environmental risks put the company's physical assets and operations at greater risk, potentially affecting revenue streams and increasing the likelihood of default on its debt.

---

[14] In the context of credit ratings and borrowing costs, "notches" refer to the increments or steps in the credit rating scale used by rating agencies like Standard & Poor's (S&P), Moody's, and Fitch. Each notch represents a different level of creditworthiness and risk. For example, a downgrade by one notch might mean moving from an AA + rating to an AA rating, signaling a slight increase in perceived risk. A more severe downgrade, such as from AA to A, involves multiple notches and indicates a higher risk. The number of notches in a downgrade or upgrade can significantly affect borrowing costs, as investors demand higher returns (interest rates) for increased risk.

Paradise Resorts wants to expand by building a new luxury resort but wonders how financing costs will be affected, especially borrowing costs. Initially, lenders were willing to offer the loan at a 5% interest rate, assuming a low default risk. However, upon reassessing the increased climate-related risks, lenders raise the estimated probability of default from 3 to 8%. In this case, Paradise Resorts will face a borrowing rate of 10%.

Scenario Analysis, Default Risk, and the Cost of Debt

We have made some progress: we see that the costs of borrowing vary across countries, anchored by sovereign debt ratings and we have an idea of how sovereign debt ratings can change in the future due to climate risk. According to an analysis by the Bank of Canada, firms in sectors exposed to significant climate risk are projected to experience larger percentage increases in default probability, relative to less climate-vulnerable firms, particularly when the transition to net-zero emissions occurs with delays (Bank of Canada, 2022). This increased default risk influences both the cost and estimated changes in default risk premiums for various sectors under different climate transition scenarios (Below 2 °C Immediate, Below 2 °C Delayed, and Net-Zero 2050). The Bank of Canada conducted analyses for both Canada and the US. Table 8.3 shows projected changes in default probabilities by sector for US firms.

Table 8.3 is detailed, but not nearly to the extent of careful analysis carried out by the Bank of Canada. In earlier chapters we underscored how the food and beverage sector contributed significantly to overall global emissions, with beef producers having particularly significant emissions. In Table 8.3 we have a glimpse into how firms like Tyson and JBS could see their borrowing costs rise significantly. By 2050, under a Net-Zero pathway, default risk premiums on debt could be 4.5 times greater than they are currently. Also noteworthy is that the increased default risk premiums don't materialize all at once, rather they evolve over time.

ExxonMobil and Default Risk

ExxonMobil, as a leading company in the oil and gas sector, faces significant challenges due to climate transition risks. Table 8.3 suggests that under stringent climate policies, the oil and gas sector could experience **substantial increases in default risk premiums, potentially rising by 250% to 450% by 2050, depending on the scenario.**

Credit rating agencies have already begun to respond to these risks. In 2021, S&P Global Ratings downgraded ExxonMobil's credit rating from AA to AA-, citing increased industry risks from the energy transition and price volatility. ExxonMobil has acknowledged these challenges and is taking steps to mitigate them. The company has announced ambitions to achieve net-zero greenhouse gas emissions by 2050 and is investing in technologies like carbon capture and

Table 8.3 Estimated changes in default risk premiums by sector, year, and scenario[15]

Sector	Year	Below 2 °C Immediate (%)	Below 2 °C Delayed (%)	Net-Zero 2050 (%)
Livestock	2030	−100	50	100
Livestock	2040	50	150	250
Livestock	2050	150	250	350
Crops	2030	0	100	150
Crops	2040	100	200	300
Crops	2050	200	300	400
Coal	2030	100	150	250
Coal	2040	200	300	400
Coal	2050	300	400	500
Oil and Gas	2030	50	100	150
Oil and Gas	2040	150	250	350
Oil and Gas	2050	250	350	450
Refined Oil Products	2030	100	200	300
Refined Oil Products	2040	200	300	400
Refined Oil Products	2050	300	400	500
Electricity	2030	0	50	100
Electricity	2040	50	100	150
Electricity	2050	100	150	200
Energy-Intensive Industries	2030	50	100	150
Energy-Intensive Industries	2040	100	200	300
Energy-Intensive Industries	2050	150	300	400
Commercial Transportation	2030	−50	0	50
Commercial Transportation	2040	0	50	100

(continued)

---

[15] This table summarizes data from Chart A-2 in the Bank of Canada document. It provides a detailed look at the estimated changes in default risk premiums for various sectors in the United States under different climate transition scenarios (Below 2 °C Immediate, Below 2 °C Delayed, and Net-Zero 2050).

Table 8.3 (continued)

Sector	Year	Below 2 °C Immediate (%)	Below 2 °C Delayed (%)	Net-Zero 2050 (%)
Commercial Transportation	2050	50	100	150
Forestry	2030	−100	0	50
Forestry	2040	0	100	150
Forestry	2050	50	150	200

storage. However, the effectiveness of these measures in offsetting the financial risks associated with the global shift toward low-carbon energy remains uncertain.

Applying Scenario Analysis to Exxon's Cost of Debt

We can begin by observing Exxon's current YTM. Of course, the YTM varies depending on the characteristics of outstanding bonds and maturity. We will assume a base rate of 4%, reflecting current yields for high-quality corporate bonds.

Based on Table 8.3 we gathered the default risk premiums for the Oil and Gas sector by scenario and year and present them in Table 8.4. These risk premiums were then applied to the assumed base rate of 4% for ExxonMobil which reflects (by assumption) a base line borrowing rate of 3% in the oil and gas sector and a risk premium of 1% for ExxonMobil. We assumed 1% as the starting risk premium although it is probably more like 1.5% today. However, the simpler number allows us to better see what's going on. Look at the first row of Table 8.5 and then you will be able to see how to proceed through the rest of the table. Under a "Below 2 °C Immediate" Scenario the **change in the risk premium** shown in Table 8.4 is forecast to be150%. In other words, the risk premium in this scenario is forecast to rise from the current 1% level by 150% (and this means the new premium is (1% + (1 + 1.5)); or 3.5%. In the first row of Table 8.5 you can see that we took the base rate of 3% and added the new risk premium of 3.5% to arrive at a projected borrowing cost of 6.5%.

Table 8.5 shows how ExxonMobil's borrowing costs increase steadily but remain relatively moderate if climate policies are implemented gradually. However, if the transition is sudden and steeper due to delay, the cost of borrowing

Table 8.4 Change in default risk premiums for oil and gas by scenario and year

Scenario	2030 Premium	2040 Premium	2050 Premium
Below 2 °C Immediate	150	250	350
Below 2 °C Delayed	200	350	450
Net-Zero 2050	250	400	550

Source Bank of Canada (2022)

Table 8.5 Estimated cost of ExxonMobil debt by scenario

Year	Scenario	Base rate (%)	Default risk premium (%)	Cost of bonds (YTM) (%)
2030	Below 2 °C Immediate	3.0	3.5	6.5
2030	Below 2 °C Delayed	3.0	4.0	7.0
2030	Net-Zero 2050	3.0	4.5	7.5
2040	Below 2 °C Immediate	3.0	4.5	7.5
2040	Below 2 °C Delayed	3.0	5.5	8.5
2040	Net-Zero 2050	3.0	6.0	9.0
2050	Below 2 °C Immediate	3.0	5.5	8.5
2050	Below 2 °C Delayed	3.0	6.5	9.5
2050	Net-Zero 2050	3.0	7.5	10.5

Source Based on data from the Bank of Canada (2022)

for Exxon rises more quickly. The most ambitious climate action, Net-Zero 2050, leads to the highest borrowing costs, as default risks and transition pressures peak. To summarize, if Net-Zero is pursued vigorously, by 2050, Exxon will face borrowing costs of 10.5% assuming the base rate of 3% holds. The borrowing costs could be even greater depending on how different scenarios are reflected in sovereign borrowing rates!◄

Generalizing Default Risk Premiums

The default risk premiums projected by the Bank of Canada for different sectors are likely to be similar across major markets globally, especially as climate policy and transition risks are increasingly coordinated and standardized internationally. Several factors contribute to this alignment:

- Global Climate Policies: Climate commitments, such as the Paris Agreement, set a broad framework that encourages countries to enact similar climate policies. As a result, major economies are progressively adopting comparable carbon pricing, emissions reduction targets, and other regulatory measures, which impact sector risk assessments similarly across regions.
- Industry-Wide Challenges: The oil and gas industry, regardless of geography, faces inherent risks tied to the transition away from fossil fuels. Global demand for oil and gas is projected to decrease in all major climate scenarios, which adds pressure on the sector's profitability and, consequently, on default risk premiums.
- International Financial Standards: Financial institutions and central banks are increasingly integrating climate risk assessments into their regulatory frameworks. Organizations like the Network for Greening the Financial System

(NGFS) are advocating for uniform climate risk assessments, which drive consistency in how default risks are priced for sectors worldwide.

- Cross-Border Investor Sentiment: Large institutional investors and asset managers now often apply consistent climate risk criteria across global portfolios. This sentiment pressures firms in high-risk sectors, like oil and gas, regardless of location, to adopt transition strategies or face higher capital costs and default risk premiums.

8.5   Climate Change and the Weighted Average Cost of Capital

We started this chapter with the WACC formula which we repeat here:

▶

$$WACC = k_e * \frac{E}{(D+E)} + k_d(1-t) * \frac{D}{(D+E)}$$

As companies confront the financial implications of climate change, they face a new question: will their cost of capital inevitably rise, or could some even experience a stable or reduced WACC? This answer isn't straightforward. For some companies, especially those in carbon-intensive sectors like oil and coal, climate change pressures—such as increased regulatory costs, investor scrutiny, and physical climate risks—are likely to drive up both the cost of debt and equity. Investors demand higher returns for the increased risks, and lenders may charge higher interest rates or limit credit availability. For these companies, WACC could rise significantly, reflecting the added costs of operating in a climate-sensitive world.

Yet, this isn't the case for all. Companies in sectors well-positioned to adapt, such as technology or services, might not face such drastic increases. Even within traditionally high-risk sectors, companies proactively adopting green practices can lower their risk profiles. By reducing emissions, committing to transparency, and aligning with climate goals, they might attract investors interested in sustainability, who may accept a lower expected return in exchange for aligning their portfolios with green values. These companies could experience a smaller increase—or, in some cases, even a reduction—in WACC.

Thus, the impact of climate change on WACC is neither uniform nor inevitable. For some, it signals a rising cost of capital. For others, it presents an opportunity to differentiate and potentially stabilize or reduce financing costs by proving their resilience in a changing world. Ultimately, each company's strategy in navigating climate risks and its sector's sensitivity to climate factors will determine whether climate change acts as a financial headwind or, unexpectedly, a tailwind for its capital costs.

Variations in the Cost of Capital across Sectors and Regions

In Figs. 8.1 and 8.7, we saw significant variations in equity and debt costs across countries, which will impact capital structures differently by region and sector. The end result is that we expect the cost of capital to vary across both sectors and regions. To prove our point, we show results of regional variations in the cost of capital between the heavy industry sector (chemicals, steel, and cement) and green projects (utility-scale solar PV).

Emissions Intensive Sectors

As part of a study conducted by the IEA in 2020, data were gathered on capital costs in emissions-intensive industries, specifically the chemical, cement, and iron and steel sectors (IEA). Under IEA's climate-driven scenarios, emission reduction efforts in the next decade will focus on improving industrial equipment efficiency and switching fuels—primarily to electricity and bioenergy, with natural gas as a transition fuel where full-scale clean energy deployment isn't yet feasible.

Meanwhile, heavy industry must also prepare for a large-scale expansion of low-carbon liquids and gases, such as hydrogen, along with carbon capture technologies. These sectors are of particular interest since achieving emissions reductions in these heavy industries will require large-scale investment in innovative technologies. Many of the technologies essential for reaching long-term net-zero goals are still in early market stages, and the typically small transaction sizes make it challenging to secure project finance from banks. Financing costs are critically important. In their study, the IEA found significant differences in capital costs across economies. To illustrate we show the evolution of the cost of capital by region for the Chemical Sector in Fig. 8.8. The results displayed for the chemical sector extend to both cement and iron and steel.

WACC and Green Projects

Stand-alone green projects, such as utility-scale solar PV, often rely heavily on financing conditions and local policy frameworks to remain viable. We've seen that there are important regional differences in the cost of capital for heavy industry, challenging mitigation investments where they are most needed. But what about stand-alone green projects? Fig. 8.9 shows the WACC for Utility-Scale-Solar PV projects in 2019, based on analysis from the IEA. Regions with lower costs of capital, such as Europe and the United States, offer more favorable conditions for these projects, enhancing their bankability and reducing reliance on subsidies. However, in regions like India, where WACC remains high, green projects face greater financial barriers, limiting the scale and speed of green energy deployment.

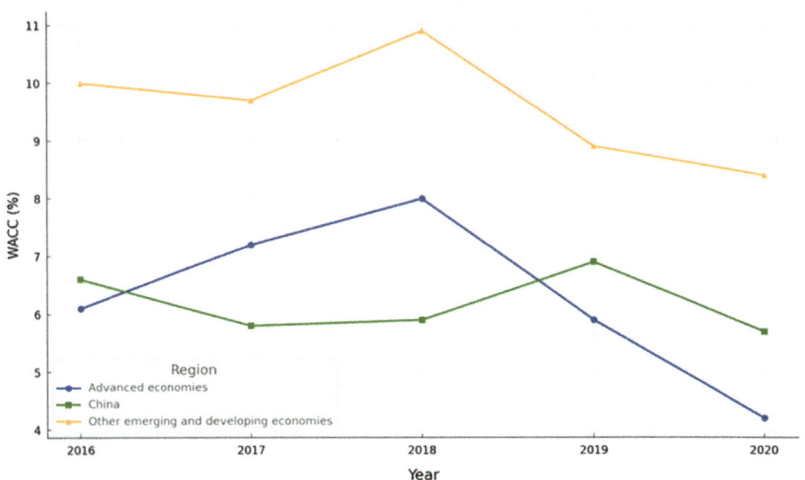

Fig. 8.8 Average WACC of chemical companies by region, 2016–2020 (*Source* IEA [2021b])

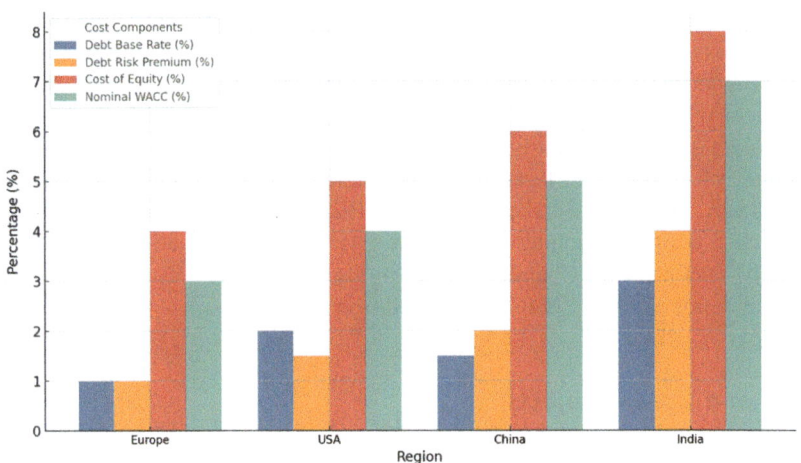

Fig. 8.9 WACC components for uility-scale solar PV projects (2019) (*Source* IEA [2021a])

Examples of WACC Calculations for Carbon-Intensive and Green Firms

Having examined regional variations in the cost of capital, highlighting how financing conditions and policy frameworks impact green energy projects globally, we now turn to firm-level data. By applying the WACC formula to firms within the Oil and Gas sector and those in the green energy sector ("Green firms"), we can observe differences in financing costs that extend beyond regional factors.

Table 8.6 WACC for select oil and gas firms and green firms[16]

Company	D	E	D + E	D/(D + E)	Beta	YTM (%)	k_e	k_d (1-t) (%)	WACC (%)
Total	60.09	131.65	191.74	0.31	0.74	5.22	7.801	3.92	6.58
ExxonMobil	42.55	505.21	547.76	0.08	0.88	4.32	8.445	3.24	8.04
BP	68.49	76.86	145.35	0.47	0.49	3.06	6.651	2.30	4.60
Chevron	23.18	272.92	296.1	0.08	1.08	5.04	9.365	3.78	8.93
Petrobras	331.47	82.51	413.98	0.80	0.94	5.73	8.721	4.30	5.18
Shell	76.61	204.6	281.21	0.27	0.51	5.34	6.743	4.01	6.00
NextEra	82.77	159.06	241.83	0.34	0.55	5.32	6.927	3.99	5.92
Iberdrola	54.84	95.73	150.57	0.36	0.61	5.75	7.203	4.31	6.15
Enel SpA	75.33	77.51	152.84	0.49	0.98	7.75	8.905	5.81	7.38
Orsted	90.37	25.05	115.42	0.78	0.73	4.75	7.755	3.56	4.47

Table 8.6 presents these WACC calculations, offering insight into how sector-specific dynamics influence the cost of capital for individual companies within the broader transition to clean energy. This comparison allows us to see at a more granular level how financing conditions shape investment strategies and operational costs for firms based on their industry alignment.

Table 8.6 shows our estimate of the WACC today. The firm would use this estimate to evaluate capital projects, or an analyst might use these estimates to determine the value of the firm or the value of its equity. These numbers don't tell us whether climate risk is priced, however. Nonetheless we can see from Table 8.6 that for the oil majors in the top half of the table the range for WACC ran from 4.6% (BP) to 8.04% (ExxonMobil). The WACC for Green firms is in a tighter range: 4.4% (Orsted) to 7.38% Enel). If we really tried to push a "story" it is probably that the oil majors generally have far less debt which would explain higher WACCs. It also begs the question: why not use more debt since it is cost advantageous? This is the subject of the following chapter.

---

[16] We used the CAPM to determine the cost of equity (ke), obtaining Betas for the firms from Yahoo Finance and the ERP calculated by Professor Damodaran which for 2023 is 4.60%. In applying the CAPM we used the current (November 2024) rate on US 10-year notes of 4.397. We approximated the cost of debt by obtaining the YTMs for long-dated bonds traded in the OTC (over-the-counter market) and to calculate the after-tax cost of borrowing we assumed a corporate tax rate of 25%. For the debt ratio we calculated the ratio of the book value of debt (D) to the market value of equity (E) which we obtained from Yahoo Finance. It is common to substitute the market value of debt by its book value due to the substantial variation in maturities and classes of debt outstanding. To approximate the cost of new borrowing we used the YTM on long-maturity bonds outstanding. These maturities varied from approximately 15 years (maturing in 2040) to 25 years maturing in 2050. YTMs were obtained from various bond reporting sources, and we asked AI to furnish us with the estimates and sources.

A final look at the cost of capital for BP and Petrobras could be worthwhile. They have the lowest WACCs of the oil majors and also the highest debt ratios. This could result from:

- BP's Transition Efforts: BP has publicly committed to a transition toward renewable energy, possibly attracting investors seeking exposure to both fossil fuels and renewables.
- Market Conditions for Petrobras: While Petrobras operates in the fossil fuel sector, its financing and WACC may benefit from specific market conditions, government support, or investor interest in emerging markets, despite high operational risks.

Forward-Looking WACC and Climate Change

Our previous analyses of the cost of equity and debt indicate that climate change could drive significant increases in both the market price of risk and the cost of equity. As climate-related uncertainties grow, firms facing substantial climate risk may see their borrowing costs climb, influenced by how the climate crisis unfolds and its effects on capital markets. To illustrate, recent findings from the Bank of Canada suggest that ExxonMobil's borrowing costs could range from 6.5% to 10.5%, depending on the year and climate scenario. In our analysis of the cost of capital for ExxonMobil appearing in Table 8.6, we used the yield to maturity (YTM) on Exxon's 2050 bonds, currently indicating a borrowing cost of 4.32%. This comparatively lower rate may reflect Exxon's strategic decision to maintain a relatively low debt level, positioning itself to absorb potential cost increases better than other oil and gas firms with heavier debt burdens.

Given these uncertainties, companies should use a range of WACC estimates when evaluating projects and assessing firm value. Applying varying WACCs can reveal how shifts in the cost of capital, driven by climate change, may impact profitability and the financial viability of long-term projects. Taking this approach will enable more resilient financial planning in an increasingly climate-sensitive future.

Conclusion

In navigating the complexities of climate change, understanding the cost of capital becomes critical for firms. As we've seen, climate risks—both physical and transition—pose unique challenges that elevate the cost of debt and equity, especially for sectors with high emissions or vulnerability to regulatory shifts. The variability in WACC across sectors and regions further underscores the need for firms to assess their financing structures thoughtfully, aligning capital costs with risk exposures. For high-emission industries, increasing investor and lender demand for risk premiums signals a shift in capital allocation, pushing firms toward sustainable and resilient investments. Conversely, green sectors may find new opportunities to attract capital at favorable rates, reflecting the market's evolving priorities and the potential for sustainable projects to yield competitive returns.

Our analysis of the cost of capital highlights the essential role of finance in the NZE transition, underscoring the necessity of forward-looking strategies that incorporate climate risk into long-term planning. Firms that anticipate these trends and adapt accordingly are more likely to access financing on favorable terms, enabling them to stay ahead in a rapidly changing financial landscape.

Building on the insights into how climate risks influence the cost of capital, the next chapter will explore corporate capital structure, dividend policy, and cash holdings under climate uncertainty. As firms confront rising costs of debt and equity, they must also consider how these trends impact decisions around leverage, payout policies, and cash reserves. We'll explore how capital structure adjustments—such as lower leverage or increased cash holdings and the dividend policy of the firm—are altered by the climate crisis.

Key Takeaways

1. Climate Risk and the Cost of Capital

Climate change introduces new layers of risk—physical and transition—that affect both the cost of equity and the cost of debt. These risks are increasingly being priced into financial markets, with significant implications for capital allocation, corporate valuation, and financing costs.

2. Cost of Equity

Systematic risks, reflected in beta, are the primary drivers of the cost of equity under the Capital Asset Pricing Model (CAPM). Climate change can increase beta for carbon-intensive sectors while potentially lowering it for green sectors, leading to diverging costs of equity. The equity risk premium (ERP) varies across regions and sectors, influenced by economic stability, climate vulnerability, and investor sentiment.

3. Cost of Debt

Climate risks are impacting corporate credit ratings and default risk premiums, driving up borrowing costs for vulnerable sectors. Sovereign risk plays a significant role in determining corporate debt costs, particularly in emerging markets where climate vulnerabilities elevate sovereign borrowing costs.

4. Weighted Average Cost of Capital (WACC)

WACC reflects the combined impact of climate-related risks on a firm's debt and equity costs. Emissions-intensive sectors are likely to see higher WACC due to elevated risks, while green sectors may benefit from lower WACC driven by investor preferences for sustainable projects.

5. Sectoral and Regional Variations

Capital costs vary widely across sectors and regions. Developed economies often benefit from lower WACC, whereas emerging markets face higher financing barriers, particularly for green projects. Heavy industries like chemicals, cement, and steel face unique challenges in securing financing for decarbonization due to high upfront costs and nascent technologies.

6. Scenario Analysis and Forward-Looking Metrics

Forward-looking assessments are critical for understanding how different climate scenarios (e.g., 2 °C pathways, delayed transitions) might impact the cost of capital. Scenario-based planning helps firms prepare for varying financial outcomes, enabling better strategic decisions under climate uncertainty.

7. Beta Adjustments and Sector Shifts

Empirical evidence shows sectoral shifts in beta under climate stress. For example, fossil fuel companies face rising betas, reflecting heightened systematic risks, while renewable energy firms often exhibit lower beta values. Firms must monitor beta changes to understand how market perceptions of climate risk influence their cost of equity.

8. Implications for Corporate Strategy

Companies in high-risk sectors need to consider how climate risks affect their financing strategies, from adjusting capital structures to exploring green financing options. Transitioning to sustainable practices not only reduces long-term risk but can also attract green investors, potentially lowering WACC.

9. Investor Preferences and Greenium

Growing investor interest in sustainability creates opportunities for green firms to secure capital at favorable rates, sometimes referred to as the "greenium." This trend underscores the financial benefits of aligning with global climate goals and adopting transparent ESG practices.

10. Importance of Resilience

Firms that proactively address climate risks—through emissions reductions, adaptation measures, and transparent disclosures—are better positioned to secure capital on favorable terms and maintain competitive advantage in a climate-conscious market.

Questions and Problems

1. Access data from Professor Damodaran's website and compare the equity risk premia for emerging markets versus developed economies. Explain how climate change is likely to impact the observed equity risk premia going forward. In thinking about your explanation, refer to physical and transition risk measures.
2. Using the table of pre- and post-announcement betas, answer the following:
   a. Identify which airline experienced the largest increase in beta after the CORSIA announcement. Explain what a higher beta could indicate about the perceived risk of this airline relative to the market.
   b. Compare the beta changes for UAL and ICAGY. Discuss potential reasons why one might have experienced a greater change in beta than the other.
3. Consider the data you have on betas and the potential regulatory impact of CORSIA:
   a. Explain why it may be useful for firms to conduct an event study around major regulatory announcements, using the CORSIA announcement as an example.
   b. Describe how you might expand this study to analyze returns around other events, such as the implementation of the EU Emissions Trading System (ETS) for airlines. Outline what additional data and steps you would need.
4. Using the concept of the CAPM:
   a. Calculate the expected return for each airline both pre- and post-announcement, assuming an ERP of 6% and a rate of 4.5% on 10-year US government bond. Use the provided beta values for each period.
   b. Analyze how the change in expected returns reflects the market's perception of systematic risk for each airline after the CORSIA announcement.
   c. How do your answers compare with the realized returns over the period?
5. Explain how physical and transition risks might affect the cost of debt for a company operating in a high-risk sector such as oil and gas. Describe how a company's cost of debt could be affected if lenders increasingly factor in climate resilience as a criterion for lending.
6. Sovereign credit ratings influence corporate debt costs, especially in emerging markets where climate risks may elevate sovereign debt yields.
   a. Explain the concept of the "sovereign ceiling" and how it impacts the debt cost of corporations within a country.
   b. Describe how increasing sovereign debt costs due to climate vulnerabilities might affect a company's access to financing and the cost of debt in an emerging market.

7. Consider a corporation, "EcoFuel Ltd.," that operates in the fossil fuel sector. Use the following data for this problem:

   - Central Bank Base Rate: 3%
   - Default Risk Premium:
   - Low Climate Action (Below 2 °C Immediate): 1.5%
   - Moderate Climate Action (Net-Zero 2050): 3.5%
   - High Climate Action (RCP 8.5, Delayed Action): 5.0%
     a. Calculate the cost of debt for EcoFuel Ltd. under each climate scenario.
     b. Discuss how the default risk premium changes across scenarios and what it reflects about the impact of delayed climate action on corporate borrowing costs.

8. Imagine you are assessing two companies in the tourism sector: "Beachfront Resorts" and "Mountain Escape Co." Both companies are seeking to issue 10-year bonds, but Beachfront Resorts is highly exposed to rising sea levels.
   a. How might lenders adjust the bond yield for Beachfront Resorts compared to Mountain Escape Co., and why?
   b. Describe one way Beachfront Resorts might mitigate the added cost of debt due to its climate exposure.

9. The text discusses how scenario-based forecasting can help firms prepare for future borrowing costs under different climate transition paths. Consider the following scenarios for "GreenChem Corp.":

   - **Base Rate**: 4%
   - **Default Risk Premiums**:
     - Immediate Action (Below 2 °C): 2%
     - Gradual Transition (Net-Zero 2050): 3%
     - Delayed Action (High-Emissions Scenario): 4%
   - Calculate the cost of debt GreenChem Corp. would face under each scenario.
   - Explain why GreenChem might choose to conduct scenario-based forecasting for its debt financing in a climate-sensitive environment.

10. Explain why green firms have a lower Weighted Average Cost of Capital (WACC) might generally have compared to carbon-intensive firms.

11. Discuss how investor preferences for sustainability might influence the WACC of green firms in the future.

12. Describe two factors that could cause the cost of capital to differ between a developed region like Europe and an emerging market.

13. Suppose a firm in an emerging market wants to attract international investors. What steps could it take to reduce its WACC?

14. In the context of the CAPM, the cost of equity is calculated using the risk-free rate, beta, and market risk premium. Climate risk is expected to influence these components over time.

a. Explain how an increase in systematic climate risk could affect the cost of equity for a high-emission company.
b. Suppose the market risk premium increases due to heightened climate concerns. How might this impact green firms versus carbon-intensive firms?

References

Addoum, J. M., Ng, D. T., & Ortiz-Bobea, A. (2021). Temperature shocks and earnings news. *The Review of Financial Studies, 34*(3), 1463–1502. https://doi.org/10.1093/rfs/hhaa108

Andersson, M., Bolton, P., & Samama, F. (2016). Hedging Climate Risk. *Financial Analysts Journal, 72*(3), 13–32.

Bank of Canada. (2022). *Using scenario analysis to assess climate transition risk.* Retrieved November 4, 2024, from: https://publications.gc.ca/collections/collection_2022/banque-bank-canada/FB4-29-2022-eng.pdf

Bank of England. (2021). *Climate change and financial stability.* Retrieved November 4, 2024, from: https://www.bankofengland.co.uk/stress-testing/2021/key-elements-2021-bie nnial-exploratory-scenario-financial-risks-climate-change

Barnett, M., Brock, W. A., & Hansen, L. P. (2020). Pricing uncertainty induced by climate change. *The Review of Financial Studies, 33*(3), 1024–1066. https://doi.org/10.1093/rfs/hhz144

BlackRock Investment Institute. (2020). Sustainability: The tectonic shift transforming investing. *BlackRock.* Accessed November 4, 2024, from: https://www.blackrock.com/institutions/en-us/insights/blackrock-investment-institute/sustainability-in-portfolio-construction

Bolton, P., & Kacperczyk, M. (2021). Do investors care about carbon risk? *Journal of Financial Economics, 142*(1), 517–549.

Damodaran, A. (2024, March 5). Equity Risk Premiums (ERP): Determinants, estimation, and implications – The 2024 Edition. Accessed November 4, 2024, from: https://ssrn.com/abstract=4751941 or https://doi.org/10.2139/ssrn.4751941

Engle, R. F., Giglio, S., Kelly, B., Lee, H., & Stroebel, J. (2020). Hedging climate change news. *The Review of Financial Studies, 33*(3), 1184–1216. https://doi.org/10.1093/rfs/hhz072

Giglio, S., Kelly, B., & Stroebel, J. (2021). Climate finance. *Annual Review of Financial Economics, 13*, 15–36.

Ilhan, E., Krueger, P., & Sautner, Z. (2021). Carbon disclosure, emission levels, and the cost of capital. *Journal of Banking & Finance, 124*, 106041. https://doi.org/10.1016/j.jbankfin.2020.106041

International Energy Agency (IEA). (2021a). The cost of capital in clean energy transitions. *IEA.* https://www.iea.org/articles/the-cost-of-capital-in-clean-energy-transitions

IEA. (2021b). Average weighted average cost of capitals of chemicals companies by region, 2016–2020. *IEA*, Paris. https://www.iea.org/data-and-statistics/charts/average-weighted-ave rage-cost-of-capitals-of-chemicals-companies-by-region-2016-2020, Licence: CC BY 4.

IEA. (2021c), Indicators of economy-wide cost of capital for debt (government bond + debt risk premium), nominal values, 2016 and 2020. *IEA*, Paris. https://www.iea.org/data-and-statistics/charts/indicators-of-economy-wide-cost-of-capital-for-debt-government-bond-debt-risk-pre mium-nominal-values-2016-and-2020, Licence: CC BY 4.0

Klusak, P., Agarwala, M., Burke, M., Kraemer, M., & Mohaddessin, K. (2023). Rising tempera-tures, falling ratings: The effect of climate change on sovereign creditworthiness. *Management Science.* https://doi.org/10.1287/mnsc.2023.4869

OECD. (2020). Climate change and sovereign risk. Asian Development Bank. Accessed November 4, 202, from: https://www.adb.org/publications/climate-change-and-sovereign-risk

How Climate Change Shapes Capital Structure and Corporate Payout Decisions

Introduction

This chapter extends the discussion of climate risks outlined in Chap. 7 and the impact of these risks on the Weighted Average Cost of Capital (WACC) explored in Chap. 8. While earlier chapters examined the components of financing costs, this chapter considers how climate change influences decisions about a firm's mix of debt and equity financing and its approaches to shareholder returns.

Climate change presents challenges and opportunities that alter the assumptions underlying traditional capital structure theories. Firms in carbon-intensive sectors face constraints on their ability to use debt due to heightened credit risks and regulatory pressures, while green firms may attract favorable financing terms. These shifts affect how firms set their target capital structures and adapt to changing market conditions.

The analysis also considers how payout decisions, such as dividends and share buybacks, are shaped by the need for financial flexibility in an environment of increased climate uncertainty. Companies must balance meeting shareholder expectations with preserving the resources necessary for investments in sustainability and resilience.

By connecting the financial impacts of climate risks to broader corporate decisions, this chapter demonstrates how firms are rethinking their financing strategies in response to the pressures and possibilities of a low-carbon economy.

S. Dow and Y. Shi, *Corporate Finance Under Climate Crisis*,
https://doi.org/10.1007/978-3-031-83487-5_9

9.1   Capital Structure Theory

In the context of climate change, the optimal capital structure can change as the costs and benefits of debt and equity change. For example, if climate change causes increases in the cost of debt more than the cost of equity, firms may reduce their leverage and rely more on equity financing including internally generated funds to finance new investment. This is especially the case if the firm faces significant financial distress costs or low tax benefits from debt. As we work through this chapter, we will view the various factors underlying the capital structure decision, beginning with examining the concept of a target capital structure.

Target Capital Structure and Theoretical Foundations

The concept of target capital structure refers to the optimal combination of debt and equity financing that minimizes a firm's cost of capital and maximizes its value. Establishing a target capital structure requires balancing the benefits and costs of debt and equity financing while accounting for factors such as tax advantages, bankruptcy costs, agency conflicts, and market dynamics. Theoretical frameworks have emerged to provide valuable insights into these decisions, helping to explain variations in leverage across firms and industries.

However, all these theories were developed before the climate crisis became a defining global issue. They neither foresaw nor considered the structural shifts in business operations and investment priorities necessitated by a decarbonized economy. Nonetheless, these frameworks remain essential, offering a foundation for subsequent analyses of how climate change may reshape capital structure decisions. Below, we outline the key theories that have shaped our understanding of capital structure.

Modigliani & Miller (MM)

Modigliani and Miller (1958) introduced the foundational principles of modern capital structure theory, demonstrating that, under idealized conditions, a firm's value and its cost of capital are independent of its financing mix. Known as the Modigliani and Miller (MM) theorem, the theory is built on two key propositions.

The first proposition states that, in perfect capital markets, the value of a levered firm equals that of an unlevered firm, as the benefits of debt are offset by an increase in the cost of equity due to higher financial risk. The second proposition asserts that the cost of equity in a levered firm equals the cost of equity in an unlevered firm plus a premium proportional to the debt-equity ratio, reflecting the additional risk borne by equity holders.

While MM assumed no taxes, bankruptcy costs, or information asymmetry, their model established a benchmark for understanding financing decisions. Later refinements addressed these limitations. Modigliani and Miller (1963) introduced corporate taxes into their framework, demonstrating that the tax-deductibility of interest creates a "tax shield" that increases firm value with greater leverage. Miller (1977) extended the model further, incorporating personal taxes to show how differential tax treatments influence leverage decisions.

The Trade-off Theory

The trade-off theory builds on the MM framework by incorporating the costs and benefits of debt in real-world conditions. Durand (1952) first alluded to the idea of balancing financing trade-offs, suggesting that firms might weigh benefits and costs when choosing their capital structure, though his work lacked formal modeling. Robichek and Myers (1965) expanded this idea, emphasizing the role of financial distress costs in offsetting the tax benefits of debt. Kraus and Litzenberger (1973) formally developed the theory into a mathematical framework, showing that firms maximize value at a debt-equity mix where the marginal tax benefit of debt equals the marginal cost of financial distress.

The theory argues that while debt provides a tax shield, excessive leverage increases the risk of financial distress, leading to costs such as legal fees, operational disruptions, and reputational damage. Firms with stable cash flows and tangible assets, such as utilities or manufacturing companies, are more likely to use higher levels of debt, as they face lower bankruptcy risks. Conversely, firms with volatile earnings or intangible assets, such as technology or pharmaceutical companies, tend to rely more on equity to avoid financial distress.

While MM assumed no taxes, bankruptcy costs, or information asymmetry, their model established a benchmark for understanding capital structure decisions. Subsequent refinements incorporated real-world complexities. In 1963, Modigliani and Miller introduced corporate taxes into their framework, demonstrating that the tax-deductibility of interest creates a "tax shield" that increases firm value with greater leverage. Later, Miller (1977) extended the model to include personal taxes, highlighting how differences in tax rates on interest, dividends, and capital gains influence leverage decisions.

The Pecking Order Theory

The **pecking order theory**, introduced by Myers and Majluf (1984), emphasizes the role of **asymmetric information** in financing decisions. Managers often have better information about the firm's prospects than external investors, which creates a preference for financing sources that minimize information-related costs. According to this theory, firms prioritize financing as follows: first, they use internal funds, such as retained earnings, because these do not involve external scrutiny or transaction costs. If internal funds are insufficient, firms turn to debt, which is less affected by information asymmetry. Equity is used as a last resort because its issuance often signals negative information to the market, leading to a decline in the firm's share price.

The pecking order theory implies that firms with significant retained earnings are less reliant on external financing. It also highlights the signaling effects of financing decisions, as equity issuance may be interpreted by investors as a sign of overvaluation or financial distress.

Signaling Theory

The **signaling theory** extends the insights of the pecking order theory by focusing explicitly on the informational content of financing decisions. The signaling theory, as articulated by Ross (1977) and Leland and Pyle (1977), emphasizes that financing decisions convey information to the market in the presence of asymmetric information. Managers can use debt as a positive signal of confidence in the firm's future cash flows, as higher leverage commits the firm to fixed payments and suggests strong performance expectations. Conversely, issuing equity often signals uncertainty or overvaluation, leading to negative market reactions. Ross highlights that high-quality firms may take on more debt to distinguish themselves from lower-quality firms, creating a separating equilibrium. Leland & Pyle add that managerial retention of equity signals confidence in the firm's profitability, while equity dilution can indicate weaker future prospects. These theories demonstrate that debt can enhance credibility but increases financial risk, while equity issuance must be carefully managed to avoid adverse signals to investors.

Dynamic Trade-off Theory

The **dynamic trade-off theory**, introduced by Fischer, Heinkel, and Zechner (1989), refines the traditional trade-off theory by accounting for adjustment costs and recognizing that firms gradually move toward their optimal capital structure over time. Unlike the static view, this theory assumes a **target leverage range** rather than a fixed ratio, allowing for temporary deviations due to transaction costs and market frictions. Firms only rebalance when the benefits of adjustment outweigh the costs, leading to observed leverage ratios that drift within tolerable bounds. Leland (1994) extended this framework by incorporating default risk and tax shield benefits, showing how these factors dynamically influence the firm's capital structure decisions. Goldstein, Ju, and Leland (2001) further emphasized the role of earnings and cash flow volatility, demonstrating how profitability and economic conditions drive leverage adjustments.

This theory explains why firms may not immediately rebalance after shocks as the costs of rebalancing—such as refinancing fees or operational disruptions—delay adjustments until leverage falls outside the optimal range. Dynamic trade-off theory thus provides a more realistic understanding of capital structure behavior in practice.

Agency Theory

The **agency theory** of capital structure, developed by Jensen and Meckling (1976), focuses on conflicts of interest between managers, shareholders, and debt holders. Managers may act in their own interests rather than maximizing shareholder value, such as by pursuing empire-building projects or avoiding risky but profitable opportunities. Similarly, shareholders may take actions that benefit themselves at the expense of debt holders, such as paying excessive dividends or investing in high-risk projects.

Debt can serve as a disciplining mechanism by reducing free cash flow available for managerial discretion and requiring regular interest payments. However, excessive debt increases the risk of shareholder-debt holder conflicts, particularly in firms with volatile cash flows. Agency theory predicts that the optimal capital structure minimizes these conflicts by aligning the incentives of managers, shareholders, and creditors.

Summing Up Capital Structure Theories

In this section, we provided an overview of prominent capital structure theories, highlighting their evolution and key insights. These theories, while developed before the emergence of climate-related risks, remain foundational. They serve as a basis for analyzing how financing policies may adapt in response to new economic and environmental realities. Table 9.1 summarizes the key theories, organizing them by theory, issue addressed, and implications for leverage.

9.2   Climate Change and Capital Structure

The intersection of capital structure decisions and climate change represents a fundamental shift in how firms approach financing in a rapidly evolving economic and environmental landscape. As we have seen, capital structure theories, such as the trade-off and pecking order theories, have provided frameworks for understanding the optimal mix of debt and equity financing. However, these theories largely operate under assumptions that do not account for the growing influence of climate-related risks and opportunities. As climate change reshapes global markets, businesses must reconsider their **target capital structure** to navigate new financial and operational challenges. Regulatory pressures, such as carbon pricing and emissions standards, impose significant transition costs on firms, particularly those in carbon-intensive industries. Physical risks, including extreme weather events and resource scarcity, disrupt operations and heighten financial uncertainty, prompting firms to re-evaluate their leverage levels.

Capital rationing has emerged as a critical issue, as financial institutions and bond markets increasingly limit funding to high-risk sectors, forcing firms to rely more heavily on internal financing or equity issuance. These dynamics are further

Table 9.1 Major theories of capital structure

Theory	Issue addressed	Implications for capital structure
Modigliani & Miller (MM)	Perfect capital markets: taxes, transaction costs, and bankruptcy costs assumed absent	Leverage irrelevant under ideal conditions; firm value depends entirely on its assets and investment decisions, not financing
Trade-off Theory	Balance between tax benefits of debt and financial distress costs	Firms with stable cash flows and tangible assets use more leverage to maximize tax benefits; high bankruptcy risks reduce leverage
Pecking Order Theory	Asymmetric information between managers and investors	Internal funds are preferred; debt is the second option, and equity is used as a last resort. High leverage signals managerial confidence
Signaling Theory	Market perception of financing decisions as signals to investors	Debt issuance is seen as a positive signal of firm confidence in future cash flows, while equity issuance can indicate overvaluation or uncertainty
Dynamic Trade-off Theory	Costs of adjusting capital structure over time	Leverage fluctuates as firms gradually adjust toward an optimal capital structure, balancing adjustment costs with market opportunities
Agency Theory	Conflicts between managers, shareholders, and debt holders	Moderate leverage helps align interests by increasing monitoring (discipline), but excessive leverage may lead to conflicts or risky behavior by shareholders

compounded by shifts in investor preferences toward environmental, social, and governance (ESG) criteria.

The **relative cost of debt and equity** has undergone transformation in the climate-conscious economy. Creditors demand higher interest rates to compensate for increased default risks in climate-vulnerable sectors, while equity markets often penalize firms perceived as lagging in their sustainability transitions. In contrast, firms with robust climate strategies and ESG credentials benefit from lower capital costs and greater access to green financing instruments, such as green bonds and sustainability-linked loans. These trends suggest a profound reconfiguration of financing preferences, challenging the assumptions of traditional capital structure theories and highlighting the need for adaptive strategies.

In this section we explore how climate change influences capital structure decisions, focusing on three critical areas:

1. Changes in target capital structures
2. Capital rationing
3. Relative costs of debt and equity.

Changes in Target Capital Structure

Climate change is fundamentally altering how firms define and achieve their target capital structures. Climate-related risks, both physical and transitional, are introducing new variables that challenge conventional approaches to capital structure.

Physical climate risks, such as extreme weather events, droughts, and rising sea levels, directly affect firms' operations, cash flows, and asset values. These disruptions increase financial volatility and make it more difficult for firms to sustain high levels of leverage. Studies have shown that firms in climate-vulnerable sectors, such as agriculture, energy, and real estate, have begun to reduce their reliance on debt to mitigate the risks of financial distress caused by operational instability (Nguyen et al., 2022). For example, the impact of hurricanes and floods has led to significant asset write-downs in real estate and infrastructure, forcing firms to reassess their financing strategies to prioritize resilience and flexibility.

Transitional risks also play a critical role in reshaping target capital structures. Regulatory policies, such as carbon pricing mechanisms, emissions caps, and renewable energy mandates, impose substantial costs on carbon-intensive industries. Firms in these sectors face rising costs related to compliance, retrofitting, and transitioning to low-carbon operations, which strain their financial resources and increase the perceived risks for creditors. As a result, debt becomes more expensive or less available, leading firms to rely more on equity or internal financing. Research indicates that energy and industrial firms with high greenhouse gas emissions have significantly lowered their leverage ratios in response to these pressures (Cumming et al., 2024).

Investor preferences for environmental, social, and governance (ESG) criteria further amplify these dynamics. Firms that align with ESG standards experience better access to capital at favorable terms, while those lagging in sustainability face higher borrowing costs and limited equity market participation (Tran et al., 2024). This disparity creates a polarization in target capital structures, with sustainable firms maintaining or even increasing leverage due to their ability to attract green financing, such as green bonds and sustainability-linked loans. In contrast, carbon-intensive firms often deleverage to avoid financial distress and align their strategies with market expectations.

Capital Rationing

Capital rationing has become crucial due to climate change, affecting firms' access to external financing. This term describes limitations on capital availability imposed by financial intermediaries, bond markets, and investors. Climate risks, especially for carbon-intensive industries, heighten these constraints as lenders and investors view them as higher risk.

One of the key drivers of capital rationing is the regulatory environment shaped by central banks and financial supervisors. Central banks, recognizing the systemic risks posed by climate change, have integrated climate-related stress testing into their regulatory frameworks. The Bank of England and the European Central Bank have pioneered climate stress tests to evaluate the resilience of financial institutions under various climate scenarios (Bolton et al., 2020). These tests assess banks' exposure to climate-sensitive sectors, such as fossil fuels and heavy industry, and encourage them to limit their lending to firms with significant climate risks. Similarly, the US Federal Reserve has incorporated climate risks into its supervisory guidance, emphasizing resilience among systemically important banks. Meanwhile, the People's Bank of China has been at the forefront of incentivizing green credit by extending its low-carbon lending program, which provides financial institutions with low-cost loans to support companies engaged in carbon reduction projects (Reuters, 2024). Originally launched in 2021, the program allows banks to access up to 60% of the principal for qualifying loans at a preferential one-year lending rate of 1.75%. By extending this initiative until the end of 2027, China aims to accelerate its transition to a green economy by promoting investments in energy-efficient technologies, electric vehicles, and sustainable infrastructure. These measures, while essential for financial stability, have contributed to tighter credit conditions for carbon-intensive firms, raising borrowing costs and limiting access to debt markets (Reuters, 2024).

Bond markets further contribute to capital rationing through higher yields and limited market access for carbon-intensive firms. Investors are increasingly incorporating environmental, social, and governance (ESG) factors into their decision-making, leading to a reallocation of capital away from firms with high-carbon footprints. This trend is reflected in the growing popularity of green bonds, which offer favorable terms to firms that meet strict environmental criteria. Empirical research shows that green bonds are often issued at a "greenium"—a lower yield compared to traditional bonds—providing a financial advantage to firms with strong sustainability credentials (Flammer, 2021). However, carbon-intensive firms frequently struggle to meet these standards, effectively excluding them from this growing segment of the capital market. The higher borrowing costs and limited access to capital markets compel these firms to rely more heavily on equity financing or retained earnings.

Debt rating agencies also play a significant role in capital rationing by integrating climate risks into their assessments. Firms with poor environmental performance or insufficient climate strategies are increasingly penalized with lower credit ratings, which further raises their cost of borrowing or restricts their access to debt markets. Studies have shown that firms with low ESG scores may experience an increase in their cost of debt by up to fifty-nine basis points compared to their higher-rated peers, reflecting the market's heightened perception of climate risks (Stout, 2023). Moreover, research indicates that climate-related credit downgrades may extend beyond financing costs, potentially leading to exclusion from institutional investment portfolios, including pension funds. Fitch Ratings projects that nearly 20% of global corporations could face credit rating down-grades by 2035 due to climate vulnerabilities, which could translate into higher credit spreads and reduced investor confidence (Fitch Ratings, 2023). Similarly, the European Systemic Risk Board warns that firms with climate-exposed assets may experience constrained access to capital markets as financial institutions reassess portfolio risks (ESRB, 2024).

Institutional investors, including pension funds, are responding to these risks by adjusting their investment strategies. A coalition of twenty-six financial insti-tutions managing over $1.5 trillion in assets has publicly urged asset managers to strengthen climate action or risk exclusion from investment portfolios (Financial Times, 2024). This trend is reinforced by specific pension fund initiatives such as Nest, a UK pension fund, which has announced that it may freeze investments in companies failing to meet net-zero commitments (The Times, 2024). These shifts highlight the broader financial consequences of climate-related credit downgrades, where firms with weak ESG performance not only face higher borrowing costs but also risk being excluded from key investment pools. On the other hand, firms with strong ESG credentials benefit from more favorable credit terms and access to innovative financing instruments such as green bonds and sustainability-linked loans. These instruments often carry lower interest rates or "greeniums," incen-tivizing sustainable practices while reducing the cost of debt for qualifying firms (Flammer, 2021).

Relative Cost of Debt and Equity

The relative costs of debt and equity are significantly influenced by climate risks, affecting firms' target capital structures. As creditors and investors incorporate ESG criteria, firms with high-carbon footprints face higher borrowing costs and equity risk premiums. Studies show that carbon-intensive sectors, such as utilities and energy, encounter increased debt costs due to anticipated transition risks and regulatory compliance (Nguyen et al., 2022). Meanwhile, sustainable firms benefit from lower financing costs, as their alignment with ESG standards attracts favor-able borrowing terms and investor confidence (Bolton & Kacperczyk, 2021; Tran et al., 2024). This divergence prompts firms to adjust their financing strategies, shifting away from debt reliance toward equity or internal financing to mitigate financial risks and maintain capital access (Amiraslani et al., 2021).

Strategic Adjustments and Implications

The financial landscape shaped by climate change has compelled firms to make significant strategic adjustments to their capital structures and investment approaches. These adjustments are driven by the increasing costs of debt and equity, capital rationing, and the growing importance of aligning corporate strate-gies with environmental, social, and governance (ESG) principles. Among these strategies, green investments and green bonds have emerged as pivotal tools for firms seeking to navigate the transition to a low-carbon economy while maintaining financial stability and access to capital.

Firms are increasingly reallocating resources toward green investments, focusing on projects and technologies that reduce carbon emissions, enhance energy efficiency, and support climate adaptation. These investments are not only environmentally beneficial but also financially strategic, as they signal to investors and creditors a firm's commitment to sustainability. Research indicates that firms engaging in substantial green investments experience improved market valuation and reduced costs of equity, as these actions align with investor preferences for ESG-oriented firms (Bolton & Kacperczyk, 2021). Moreover, green investments often provide long-term cost savings, such as reduced energy expenditures and lower compliance costs associated with regulatory frameworks like carbon pricing or emissions caps. For example, utility companies transitioning to renewable energy sources have reported enhanced operational efficiency and better access to financing, illustrating the dual financial and environmental benefits of green investments (Flammer, 2021).

Green bonds have become a cornerstone of financing for these investments, offering firms a way to access capital while signaling their environmental commitment. These bonds, which earmark proceeds for environmentally sustainable projects, have witnessed exponential growth in issuance over the past decade, with firms across industries leveraging this instrument to finance renewable energy, energy efficiency, and other green initiatives. Green bonds are attractive to issuers because they often carry lower interest rates than traditional bonds, a phenomenon known as the "greenium." This cost advantage reflects strong investor demand for sustainable financial products, as well as reduced perceived risk for green investments compared to traditional projects (Flammer, 2021). Furthermore, green bond frameworks often impose transparency and reporting requirements, which can improve a firm's reputation and accountability, further enhancing investor confidence.

Empirical studies have demonstrated the positive financial implications of green bond issuance. For example, research by Amiraslani et al. (2021) found that firms issuing green bonds experienced a reduction in their cost of capital and an increase in equity valuations, particularly in industries transitioning to more sustainable practices. This effect is especially pronounced for firms in emerging markets, where green bonds serve as a bridge to attract international capital and build investor trust. However, the ability to issue green bonds is contingent on a firm's ability to meet stringent eligibility criteria, which often include robust environmental impact assessments and adherence to international green finance standards. As a result, firms in high-carbon sectors may find it challenging to qualify for green bond issuance without substantial changes to their operations and strategic priorities.

The rise of green bonds has also had broader implications for the capital markets. Their popularity reflects a reallocation of capital toward sustainable projects, creating incentives for firms to align their strategies with global climate goals.

For instance, companies that successfully leverage green bonds to fund their sustainability initiatives often find it easier to access other forms of green financing, such as sustainability-linked loans or green equity offerings (Tran et al., 2024). These financial instruments further embed sustainability into corporate financing, encouraging firms to adopt long-term strategies that prioritize environmental performance.

Despite the advantages of green investments and green bonds, challenges remain, particularly for firms in carbon-intensive industries. Transitioning to a green portfolio often requires substantial upfront capital and long-term commitment, which can strain financial resources. Additionally, firms that fail to meet investor expectations for ESG performance may find themselves excluded from green financing opportunities, facing higher costs of capital and limited market access. These dynamics highlight the importance of strategic foresight and transparency in navigating the evolving financial environment.

The implications of these strategic adjustments extend beyond individual firms to the broader capital markets and financial ecosystem. As the global economy shifts toward decarbonization, firms that adopt green investments and leverage green bonds are better positioned to maintain competitive advantage and financial resilience. These strategies not only enhance access to capital but also mitigate climate-related risks, ensuring long-term sustainability in a rapidly changing world.

Broader Implications

The evolving relationship between climate change and capital structure decisions has significant theoretical and practical implications. Climate-related risks and opportunities challenge traditional capital structure theories by introducing new variables, such as regulatory pressures, investor preferences for sustainability, and the financial consequences of physical and transition risks. These factors compel firms to reconsider long-standing assumptions about leverage, cost of capital, and financing priorities.

Theoretical Implications

From a theoretical perspective, climate change necessitates a reassessment of foundational capital structure models. In Modigliani and Miller's (MM) framework, firm value is determined solely by assets and investment decisions under perfect market conditions. However, climate-related risks introduce market imperfections that disrupt these assumptions, including asymmetric information about climate strategies, the costs of regulatory compliance, and stranded asset risks. For instance, firms in carbon-intensive industries face increased borrowing costs and constrained access to debt markets due to heightened perceptions of financial distress, directly contradicting the MM assumption of debt irrelevance.

The trade-off theory, which balances tax benefits of debt against financial distress costs, must also account for climate risks. These risks heighten the likelihood of financial distress, particularly in industries vulnerable to regulatory transition costs or physical climate impacts. This suggests a shift in the optimal debt-equity mix, with firms opting for lower leverage to maintain financial flexibility in the face of uncertainty.

Similarly, the pecking order theory's emphasis on information asymmetry and internal financing must incorporate the signaling effects of climate strategy. Firms adopting green financing instruments, such as green bonds, send positive signals about their environmental commitment, potentially reducing the costs of external financing. Conversely, firms that delay sustainability transitions risk signaling a lack of confidence, exacerbating their financing challenges.

Agency theory, which highlights conflicts between stakeholders, takes on new dimensions in the climate context. Agency theory identifies two potential conflicts: one between owners and managers; and the other between creditors and shareholders. The theory predicts that shareholders will use leverage to constrain self-dealing by managers by constraining free cashflow. However, there are substantive reasons to suggest that climate change could exacerbate conflicts between creditors and shareholders through risk-shifting. Consider a firm in the oil and gas sector that decides not to invest in green energy but rather to continue drilling. In the longer term, a significant portion of assets in the oil and gas sector can become stranded. If the same company simultaneously increases payouts to shareholders via dividends and share repurchases, the firm has effectively shifted risk to creditors! This is another reason why firms that continue with a "business as usual" philosophy could find potential creditors less willing to extend loans. These dynamics underscore the need for capital structure theories to integrate climate-related variables into their analytical frameworks.

Table 9.2 summarizes how climate change interacts with major capital structure theories:

Practical Implications

The applicability of capital structure theory to real-world situations is evident in Figs. 9.1 and 9.2, derived from the IEA, (IEA, 2021). These figures illustrate the regional differences in capital structure for clean energy investments. Advanced economies predominantly rely on debt financing, supported by mature financial systems and lower risk premiums, while emerging markets exhibit a higher reliance on equity and concessional financing due to limited access to affordable debt (International Energy Agency, 2021). Emerging economies have less institutional support as well as greater uncertainty due not only to a more fragile socioeconomic context, but also to the greater vulnerability to physical climate risk borne by these regions.

Our analysis of capital structure implies that sectors with high exposure to climate risk will exhibit lower debt levels. Drawing on data from the Damodaran

Table 9.2 Major theories of capital structure and climate change implications

Theory	Issue addressed	Implications for capital structure	Climate change integration
Modigliani & Miller	Perfect capital markets: taxes, transaction costs, and bankruptcy costs assumed absent	Leverage is irrelevant under ideal conditions; firm value depends on assets and investment decisions, not financing	Climate risks disrupt perfect market assumptions by introducing costs like regulatory compliance, stranded assets, and reputational risks, making leverage decisions critical
Trade-off Theory	Balance between tax benefits of debt and financial distress costs	Firms with stable cash flows and tangible assets use more leverage to maximize tax benefits; high bankruptcy risks reduce leverage	Heightened financial distress risks in climate-vulnerable industries shift the optimal leverage level downward, favoring equity or green financing options
Pecking Order Theory	Asymmetric information between managers and investors	Internal funds are preferred; debt is the second option, and equity is used as a last resort. High leverage signals managerial confidence	Climate strategy serves as a signal to investors; green bonds and ESG transparency reduce financing costs, while poor climate performance exacerbates financing constraints
Signaling Theory	Market perception of financing decisions as signals	Debt issuance is a positive signal of confidence; equity issuance signals overvaluation or uncertainty	Green financing signals environmental responsibility; delay in adopting sustainability strategies signals risk, raising borrowing costs
Dynamic Trade-off	Costs of adjusting capital structure over time	Leverage fluctuates as firms adjust gradually toward an optimal structure	Climate risks increase the frequency of adjustments, as firms must account for evolving regulatory and market pressures while maintaining financial resilience
Agency Theory	Conflicts between managers, shareholders, and debt holders	Moderate leverage aligns interests, but excessive leverage exacerbates conflicts	ESG performance aligns managerial and investor priorities; climate inaction creates conflicts between shareholders and creditors

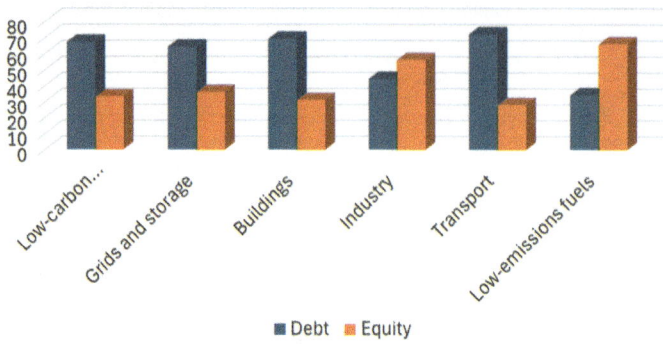

Fig. 9.1 Capital structure advanced economies (*Source* IEA [2021])

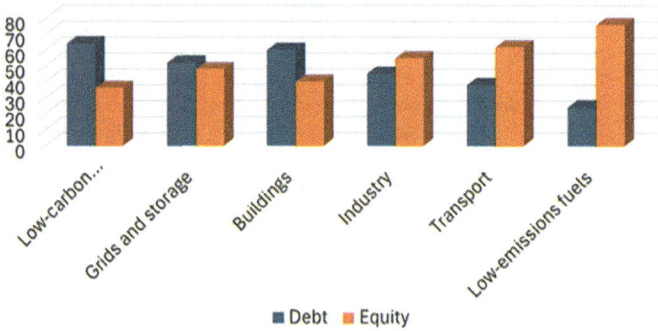

Fig. 9.2 Capital structure in emerging and developing economies (*Source* IEA [2021])

website[1] we selected several US sectors with significant exposure to climate risk due to their extensive emissions (those in hard-to-abate sectors) and examined their debt ratios over the past decade in comparison to debt ratios for firms in the Green and Renewable Energy Sector. Our results are displayed in Fig. 9.3 and compare market value debt ratios over the 2013–2023 period.

Aviation,Air transport, for example, has succeeded in taking on greater debt which may be reflective of a high degree of regulatory oversight the industry faces. With the implementation of frameworks such as CORSIA, the aviation sector enjoys a level of regulatory certainty that mitigates some of its transition climate risks. These clear guidelines for achieving net-zero emissions have provided the sector with a predictable path to align with decarbonization objectives while maintaining access to debt financing.

---

[1] The market debt ratio is defined as the ratio of the market value of debt to the market capitalization + market value of debt. Damodaran follows the common practice in finance and approximates the market value of debt by its book value.

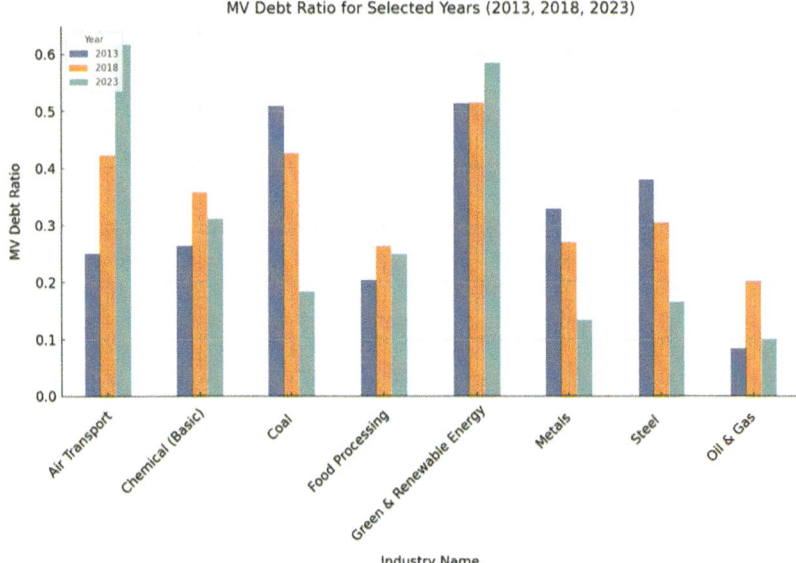

Fig. 9.3 Debt ratios for selected years across sectors (*Source* Authors' calculations based on data from Professor Damodaran's website at: https://pages.stern.nyu.edu/~adamodar/)

In contrast, the oil and gas sector exhibits highly variable debt ratios, influenced undoubtedly by the small sample size of firms represented and their vastly different net-zero strategies. ExxonMobil and Shell, as will be seen a bit later in the case study chapter, display divergent approaches to addressing climate challenges. ExxonMobil has a debt ratio around 8%, with a strategy focused on carbon capture and storage (CCS) to reduce emissions while continuing its fossil fuel extraction activities. This approach contrasts sharply with Shell, which has a debt ratio closer to 25% and is actively transforming into an integrated "energy" company, emphasizing planned investments in renewable energy and sustainable technologies.

The graphic in Fig. 9.3 also highlights the increasing leverage in the green and renewable energy sector, reflecting growing investor confidence and enhanced access to financing. This trend underscores the financial market's recognition of renewable energy as a central pillar of the global energy transition, supported by favorable regulatory environments and innovations like green bonds. Meanwhile, industries such as coal, steel, and metals have seen dramatic declines in debt ratios over the period, reflecting rising borrowing costs, reduced access to debt markets, and heightened financial risks tied to regulatory and market pressures. These trends reveal the challenges facing high-carbon industries in sustaining traditional financing models amid a rapidly decarbonizing global economy.

This visualization reinforces the evolving role of regulatory certainty, market dynamics, and firm-level strategies in shaping capital structures across industries.

The aviation sector's more certain regulatory context, the variability within oil and gas, and the rise of renewable energy highlight the differing trajectories industries are taking in response to climate and transition risks.

Summing Up

In this section, we examined how climate change affects the firm's capital structure decision. Our discussion concludes that climate change, particularly for firms vulnerable to substantial transition costs, will face significant constraints on the way they finance future investments. Many firms in hard-to-abate sectors will face credit rationing and higher borrowing costs. The Trade-off theory of capital structure will predict a natural tilt in capital structure toward internal financing. Financial flexibility in climate crisis will be a priority for firms as many will have to increasingly rely on internal sources of financing, consistent with the Pecking Order theory of capital structure. These capital structure realities, not necessarily decisions, will impact the firm's cash holdings as well as how the firm will return cash to shareholders via dividends and stock buybacks. In the next section, we will address how climate change affects these key financing decisions of the firm.

9.3   Climate Change: Impacts on Dividend Policies, Share Repurchases, and Cash Holdings

As climate change reshapes the economic landscape, firms are rethinking their financial strategies to balance shareholder returns with the need for resilience. Extreme weather events, regulatory shifts, and evolving investor preferences require companies to maintain financial flexibility while adapting to sustainability imperatives. The need for such flexibility influences corporate payout decisions (dividends and share repurchases), and cash holdings—allowing firms to navigate uncertainty while ensuring long-term value creation.

Cash Holdings: A Precautionary Approach

Maintaining liquidity has become essential for firms in climate-vulnerable sectors. Companies exposed to climate risks often increase cash reserves as a precautionary measure, ensuring they can absorb financial shocks and fund sustainability transitions. Javadi et al. (2023) found that firms in regions facing significant climate-related disruptions hold higher cash balances to mitigate operational risks and maintain financial stability. This precautionary strategy aligns with the precautionary motive for holding cash balances, which underscores liquidity as a buffer against uncertainty. Originally introduced by Keynes (1936), this motive suggests that firms retain cash to safeguard against unforeseen financial shocks, including economic downturns, regulatory changes, and supply chain disruptions. In corporate finance, this principle has been further developed to explain why firms in volatile industries, including those exposed to climate-related risks accumulate higher cash reserves (Bates et al., 2009; Javadi et al., 2023). For firms facing transition risks such as regulatory shifts in carbon pricing or sudden capital rationing, maintaining liquidity ensures financial stability and preserves strategic flexibility in a rapidly evolving economic landscape.

Payout Strategies in Climate Change: Dividends and Share Repurchases

In addition to cash holdings, firms adjust their payout strategies to enhance financial flexibility. While dividend payments are often viewed as a commitment to shareholders, share repurchases offer firms greater adaptability in managing cash flows.

Dividends

Firms in carbon-intensive sectors like oil and gas often face earnings volatility and rising compliance costs, leading them to reconsider dividend distributions. Zhu and Hou (2022) found that such firms typically reduce dividends to allocate funds for regulatory compliance and environmental investments. However, companies like BP and Shell have maintained strong shareholder payouts while investing in renewables. According to The Guardian (2024), BP balances dividends and share buybacks with decarbonization efforts, and Shell continues its dividend policy despite scrutiny over funding both distributions and green initiatives. These examples show varying approaches—some firms focus on financial flexibility due to climate risks, while others aim to sustain investor returns alongside energy transition strategies.

In contrast, firms with strong sustainability credentials—such as those in the renewable energy sector—often maintain or even increase dividends due to favorable financing conditions and investor demand for ESG-aligned assets (Balachandran, 2018). This divergence underscores how climate change is driving sector-specific shifts in dividend policies, with high-emission industries struggling to balance investor returns and decarbonization efforts.

Share Repurchase

Payout strategies are evolving. As firms navigate climate-related uncertainty, share repurchases are emerging as an alternative to dividends due to their flexibility. Unlike dividends, which create long-term payout expectations, share buybacks allow firms to return capital selectively while retaining the ability to adjust financial policies based on changing market conditions. Thus, while some firms are reducing dividends to reinvest in green initiatives, others favor share repurchases, which provide greater flexibility in managing capital allocation. Repurchases can be scaled up or paused depending on the firm's financial health and investment needs, making them an attractive option in uncertain climates. Chang et al. (2024) note that in countries with significant climate risk, firms have substituted dividends with share repurchases to retain the ability to adjust payouts based on changing conditions. For example, ExxonMobil has utilized share repurchases as part of its climate strategy, enabling the firm to manage capital allocation while funding carbon capture technologies. However, as noted above, some oil and gas firms have continued paying high dividends, despite facing financial pressure to invest in clean energy transitions. These variations illustrate the strategic choices firms must make when allocating capital under climate uncertainty.

Conclusion

This chapter highlights how climate change is reshaping corporate finance by influencing capital structure decisions and payout policies. Traditional theories of capital structure, while still foundational, must now accommodate the complexities introduced by climate risks and the transition to a decarbonized economy. Firms in carbon-intensive sectors are seeing reduced leverage and rising financing costs as regulatory and market pressures intensify, while green firms often benefit from greater access to capital at favorable terms. These dynamics underscore the evolving landscape of financing and the growing divergence between sustainable and traditional business models.

At the same time, payout policies are adapting to reflect the need for financial flexibility. Decisions on dividends, share repurchases, and cash reserves increasingly account for the uncertainties of physical risks, regulatory changes, and investor preferences for sustainability. Firms that successfully align their capital structure and payout strategies with the demands of a climate-conscious financial environment are better positioned to sustain long-term value and resilience.

The discussion in this chapter builds on the financial risks presented in Chap. 7 and the WACC analysis of Chap. 8, illustrating how these factors translate into actionable corporate finance decisions. The insights here lay the groundwork for understanding the broader implications of climate change on corporate strategy, investment, and shareholder engagement.

Key Takeaways

1. **Climate Change and Capital Structure**: Climate risks are reshaping traditional capital structure decisions. Firms in carbon-intensive sectors are reducing leverage due to increased borrowing costs and regulatory pressures, while green firms often benefit from enhanced access to financing through instruments like green bonds.
2. **The Role of Financing Costs**: The shifts in debt and equity costs described in earlier chapters are central to how firms redefine their optimal mix of financing. Higher costs for carbon-intensive firms and favorable terms for sustainable businesses create a growing divide in financing strategies across industries.
3. **Financial Flexibility and Climate Uncertainty**: Maintaining financial flexibility has become crucial in managing climate risks. Firms are adapting payout policies, balancing dividends, share buybacks, and cash holdings to ensure they can meet both immediate shareholder expectations and long-term investment needs.
4. **Relevance of Theoretical Foundations**: While traditional capital structure theories remain important, they must be adapted to incorporate the financial implications of climate risks, including physical disruptions, transition costs, and investor preferences for sustainability.
5. **Sustainability as Strategic Advantage**: Firms aligning their strategies with a decarbonized economy are better positioned to access favorable financing and maintain competitiveness. The financial market's recognition of sustainability is reshaping the landscape of corporate decision-making.

Questions

1. How does climate change influence a firm's capital structure decisions?
2. Explain how the Weighted Average Cost of Capital (WACC) is impacted by climate-related risks.
3. What role does financial flexibility play in adapting to climate uncertainty?
4. How do green bonds influence the capital structure of sustainable firms?
5. Discuss how traditional capital structure theories need to adapt to the realities of climate change.

References

Amiraslani, H., Lins, K. V., Servaes, H., & Tamayo, A. (2021). A new dawn for capital structure decisions in emerging markets. *Emerging Markets Review, 47*, 101724. https://doi.org/10.1016/j.ememar.2021.101724

Balachandran, B. (2018). The impact of carbon risk on dividend policy in Australia. *Journal of Financial Economics, 142*(2), 499–516.

Bates, T. W., Kahle, K. M., & Stulz, R. M. (2009). Why do U.S. firms hold so much more cash than they used to? *Journal of Finance, 64*(5), 1985–2021. https://doi.org/10.1111/j.1540-6261.2009.01492.x

Bolton, P., & Kacperczyk, M. (2021). Do investors care about carbon risk? *Journal of Financial Economics, 142*(2), 517–549. https://doi.org/10.1016/j.jfineco.2021.01.010

Bolton, P., Despres, M., Pereira da Silva, L. A., Samama, F., & Saltzman, R. (2020). The green swan: Central banking and financial stability in the age of climate change. Bank for International Settlements. https://www.bis.org/publ/othp31.pdf

Chang, Y., He, W., & Mi, L. (2024). Climate risk and payout flexibility around the world. *Journal of Banking & Finance.* Retrieved November 4, 2024, from ScienceDirect.

Cumming, D., Duppati, G., Fernando, R., Singh, S. P., & Tiwari, A. K. (2024). Dynamics of carbon risk, cost of debt, and leverage adjustments. *The British Accounting Review.* https://doi.org/10.1016/j.bar.2024.101353

Durand, D. (1952). Costs of debt and equity funds for business: Trends and problems of measurement. Conference on Research in Business Finance. National Bureau of Economic Research.

Fischer, E. O., Heinkel, R., & Zechner, J. (1989). Dynamic capital structure choice: Theory and tests. *Journal of Finance, 44*(1), 19–40. https://doi.org/10.1111/j.1540-6261.1989.tb02402.x

European Systemic Risk Board. (2024). *Climate-related risks and financial stability: A systemic perspective.* ESRB. Retrieved February 20, 2024, from https://www.esrb.europa.eu/pub/pdf/reports/esrb.report202404_climaterelatedrisks~2311dfaee2.en.pdf

Financial Times. (2024, January 5). Long-term investors split with asset managers over climate risk. *Financial Times.* Retrieved February 20, 2024, from https://www.ft.com/content/0a703624-37ba-4d87-af67-3d7d15caf306

Fitch Ratings. (2023, August 3). Climate risk-related downgrade may affect 20% of global corporates by 2035. *Fitch Ratings.* Retrieved February 20, 2024 from https://www.fitchratings.com/research/corporate-finance/climate-risk-related-downgrade-may-affect-20-of-global-corporates-by-2035-08-03-2023

Flammer, C. (2021). Corporate green bonds. *Journal of Financial Economics, 142*(2), 499–516. https://doi.org/10.1016/j.jfineco.2021.05.009

Goldstein, R., Ju, N., & Leland, H. E. (2001). An EBIT-based model of dynamic capital structure. *Journal of Business, 74*(4), 483–512. https://doi.org/10.1086/322893

IEA (International Energy Agency). (2021). *Financing clean energy transitions in emerging and developing economies.*

Javadi, S., Masum, A. A., Aram, M., & Rao, R. P. (2023). Climate change and corporate cash holdings: Global evidence. *Financial Management, 52,* 253–295.

Jensen, M. C., & Meckling, W. H. (1976). Theory of the firm: Managerial behavior, agency costs and ownership structure. *Journal of Financial Economics, 3*(4), 305–360. https://doi.org/10.1016/0304-405X(76)90026-X

Keynes, J. M. (1936). *The general theory of employment, interest, and money.* Macmillan.

Kraus, A., & Litzenberger, R. H. (1973). A state-preference model of optimal financial leverage. *Journal of Finance, 28*(4), 911–922.

Leland, H. E. (1994). Corporate debt value, bond covenants, and optimal capital structure. *Journal of Finance, 49*(4), 1213–1252. https://doi.org/10.1111/j.1540-6261.1994.tb02452.x

Leland, H. E., & Pyle, D. H. (1977). Informational asymmetries, financial structure, and financial intermediation. *Journal of Finance, 32*(2), 371–387.

Miller, M. H. (1977). Debt and taxes. *Journal of Finance, 32*(2), 261–275.

Modigliani, F., & Miller, M. H. (1958). The cost of capital, corporation finance, and the theory of investment. *American Economic Review, 48*(3), 261–297.

Modigliani, F., & Miller, M. H. (1963). Corporate income taxes and the cost of capital: A correction. *The American Economic Review, 53*(3), 433–443.

Myers, S. C., & Majluf, N. S. (1984). Corporate financing and investment decisions when firms have information that investors do not have. *Journal of Financial Economics, 13*(2), 187–221.

Nguyen, T., Bai, M., Hou, G., & Truong, C. (2022). Drought risk and capital structure dynamics. *Accounting & Finance, 62*(4), 5645–5671.

Reuters. (2024, August 11). China's central bank to extend low-carbon lending tool until 2027. *Reuters.* https://www.reuters.com/sustainability/climate-energy/chinas-central-bank-extend-low-carbon-lending-tool-end-2027-2024-08-11/

Robichek, A. A., & Myers, S. C. (1965). *Optimal financing decisions.* Prentice-Hall.

Ross, S. A. (1977). The determination of financial structure: The incentive-signaling approach. *Bell Journal of Economics, 8*(1), 23–40.

Stout. (2023). Integrating ESG risks into the corporate valuation process. *Stout.* https://www.stout.com/en/insights/article/integrating-esg-risks-corporate-valuation-process

The Guardian. (2024). *BP profits defy expectations amid climate transition pressures.* Retrieved November 4, 2024, from https://www.theguardian.com/business/article/2024/jul/30/bp-profits-forecasts-oil-gas-dividend-buybacks

Tran, L. T. H., Ho, T., Ho, H. T., & Phung, N. D. (2024). Climate vulnerability and capital structure: Moderating effect of financial development, financial constraints, and 2015 Paris Agreement. *International Review of Economics & Finance, 96*, (PC). https://ideas.repec.org/a/eee/reveco/v96y2024ipcs1059056024007032.html

Zhu, B., & Hou, R. (2022). Carbon risk and dividend policy: Evidence from China. *International Review of Financial Analysis, 84*, 102360. https://doi.org/10.1016/j.irfa.2022.102360

Climate Change and Green Technology: Beyond Emission Mitigations

10

Introduction

This chapter explores the climate crisis and the critical role of green technologies in addressing it, focusing on emissions, emerging innovations, and the financing and corporate strategies for a sustainable future. At the heart of this challenge lie greenhouse gases (GHGs), silent but potent drivers of global warming, setting the stage for an exploration into the world of green technologies. We delve into the technological specifics and the nuanced stages of their development and the critical role of financing in bringing these innovations from ideation to impactful reality. As we traverse the terrain of climate financing, we confront the dual imperatives of transition and adaptation—financing not just the technologies of tomorrow but also fortifying our societies against the already-unfolding impacts of climate change. Amid this narrative is the slow-burning nature of the climate threat, a deceptive temporality that demands a balancing act between curbing emissions and advancing negative emission technologies. This chapter further illuminates how corporations, in response, are reshaping strategies and financial models, measuring and managing their impact with an eye toward a sustainable future. Here, we lay bare the complexities, the challenges, and the innovative pathways emerging at the nexus of environmental crisis, technological advancement, financial strategy, and corporate responsibility.

10.1 GHGs and the Big Problem

Chapter 2 introduces the climate science that attributes the phenomenal problem to excess greenhouse gas emissions (GHGs) and their accumulation in the Earth's atmosphere. Chapter 5 further summarizes the analysis of GHGs by sectors and highlights the chief culprit, the fossil fuels and industry. Over 36 billion tons of

Table 10.1 Greenhouse gas emissions: Types, contributions, and sources[1]

Types of greenhouse gases	% of Total GHG emissions	CO2-equivalent	Sources (% by Source)
Carbon Dioxide (CO2)	~76%	About 36 billion tons	Energy (73%) Deforestation and Other Land Use (18–24%)
Methane (CH4)	~16%	Varies (21–25 times more potent than CO2)	Agriculture (especially livestock, ~40%) Energy Production (33%) Waste Management (27%)
Nitrous Oxide (N2O)	~6%	Varies (298 times more potent than CO2)	Agricultural Soil Management Animal Manure Management Fossil Fuel Combustion
Fluorinated Gases	~2%	Varies (thousands of times more potent than CO2)	Industrial Processes Refrigeration Air Conditioning Systems

CO2 are emitted annually from fossil fuels and industry. Methane, a much more potent GHG, sees annual emissions of approximately 600 million tons. This is caused chiefly by fossil fuel dependency in energy production. The global economy relies heavily on fossil fuels for energy, contributing significantly to GHG emissions. About 84% of the world's energy comes from fossil fuels, with daily consumption including billions of tons of coal and over 90 million barrels of oil.

But singling out fossil fuels leads to under-examined complexity of energy and the entire economy. Deforestation, agriculture, and industrial processes are major drivers of GHG emissions, significantly contributing to climate change. Deforestation reduces carbon absorption capacity, releasing about 10% of human-induced CO2 emissions annually. Agriculture, especially livestock farming, emits methane and nitrous oxide, which are far more potent than CO2, accounting for 24% of global GHG emissions. Industrial sectors like cement and steel production add substantially to CO2 output, with the cement industry alone responsible for 7–8% of emissions (Table 10.1).

While natural carbon sinks like forests, soil, and oceans absorb atmospheric CO2, human activities above emit over 50 billion tons of CO2-equivalent, more than the capacity of these carbon sinks. This carbon cycle imbalance amounts to about half of the CO2 emitted by human activities. The cumulative effect of

---

[1] The estimates and percentages provided here are based on global averages and are subject to variation. For the most accurate and recent data, refer to sources like the Global Carbon Project, the Intergovernmental Panel on Climate Change (IPCC), and the World Resources Institute.

increased GHG concentrations in the Earth's atmosphere is a rise in the Earth's average temperature. It has increased by approximately 1.1 °C since pre-industrial times and is accelerating. This is the Big Problem.

The enormity of the problem we face extends far beyond its sheer scale; its roots are deeply entwined in the complex web of human activity and progress. This challenge is not simply a matter of atmospheric science but a reflection of centuries of industrial and economic development that have shaped our very civilization. The carbon cycle imbalance is a direct consequence of the foundational practices that have driven prosperity and progress, embedded in the institutions that govern our societies. This complexity adds layers of difficulty to addressing the issue, as it requires a fundamental re-evaluation and transformation of the systems that underpin modern life. We stand at a pivotal juncture in human evolution, facing a paradox where the very advancements that signified our progress now loom as harbingers of potential peril. This moment in history presents a unique and daunting challenge to contemporary humanity, as it grapples with the dawning realization that our continued prosperity, perhaps even our survival, hinges on our ability to reimagine and reshape our relationship with the planet. The problem, therefore, is not just "big" in magnitude; it is profoundly intricate, woven into the fabric of our collective existence and demanding a transformative approach to how we live, work, and sustain our world.

And yet in the face of such existential threats, history has shown that humanity often turns to its most potent tool: innovation. It is within this crucible of crisis that the seeds of green technologies are sown and nurtured, offering a beacon of hope. These emerging technologies represent our collective ingenuity and determination to confront and mitigate the dire consequences of our past actions, marking a pivotal shift toward a more sustainable and resilient future.

10.2 Green Technologies

In the realm of green technology, the ancient and the future interweave in a tapestry of innovations. The primal use of heat from the Earth's crust, once a mere whisper of potential in the hot springs of antiquity, now resonates through advanced geothermal systems. The sails that harnessed the winds of the Nile have evolved into the majestic blades of modern wind turbines. Ancient rudimentary lenses, precursors to our understanding of solar energy, have paved the way for today's sprawling solar farms. Each of these technologies, rooted in the wisdom of the past, blooms into the futuristic aspirations of sustainable living, bridging epochs in their quest to harness the Earth's natural forces.

There is more to green technology than solar panels and wind turbines. An everything-you-need-to-know website lists fifteen.[2] Our analysis below includes

---

[2] Source: Greenly Earth blog, https://greenly.earth/en-us/blog/ecology-news/everything-you-need-to-know-about-green-technology-in-2022, accessed on December 3, 2024.

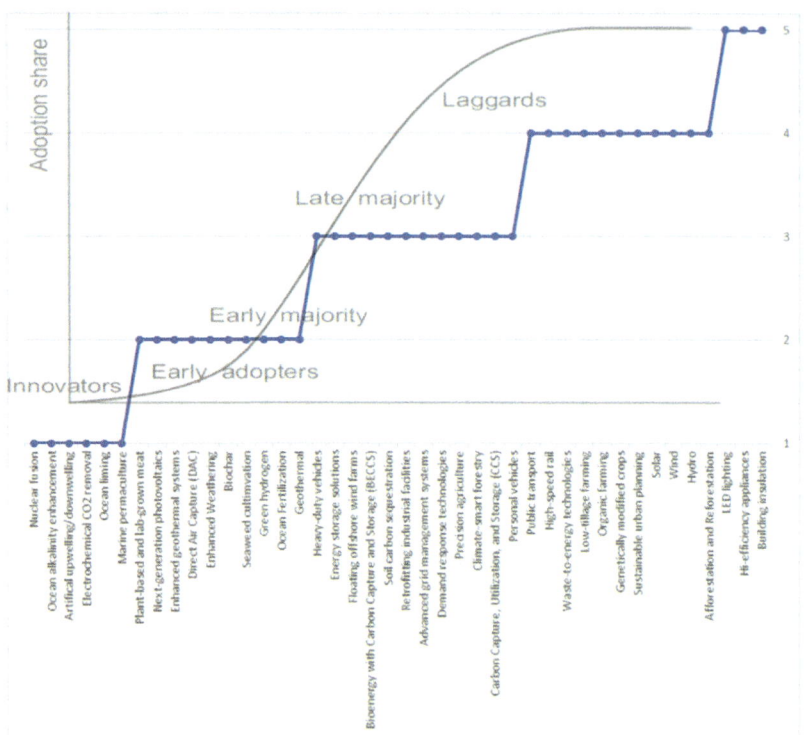

Fig. 10.1 Stage of green technology development, S-curve

40 or so green technologies (see Figs. 10.1 and 10.2) (Kumar & Eswari, 2023; Takalo et al, 2021). We classify them simply but profoundly in two categories, Emission Mitigation Technologies (EMTs) and Negative Emission Technologies (NETs). The literature supports the conceptual distinction between EMTs and NETs, acknowledging their distinct roles and implications in climate change mitigation strategies. This straightforward classification helps in understanding the multifaceted approaches required to tackle climate change, encompassing both immediate reduction strategies and longer-term carbon removal solutions.

Emission Mitigation Technologies (EMTs)

EMTs encompass a broad range of approaches and innovations designed to decrease the emissions of greenhouse gases, particularly carbon dioxide, into the atmosphere. These can include renewable energy technologies (solar, wind power, geothermal power), energy efficiency improvements (LED lighting, efficient appliances, heat pumps, smart grids, industrial energy optimization), electric vehicles

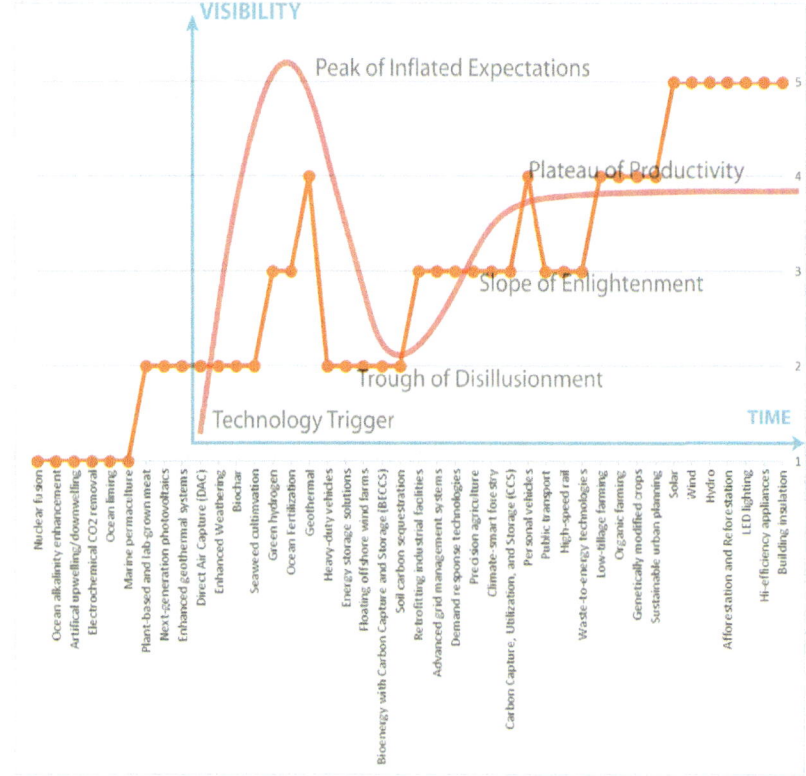

Fig. 10.2   Stage of green technology development, hype cycle

(EVs) and battery technology, biofuels (ethanol, advanced biofuels), carbon capture and storage (CCS), combinatory technologies (bioenergy with CCS), and changes in land-use practices, among others.

Negative Emissions Technologies (NETs)

NETs are emerging or relatively novel approaches and innovations designed to absorb atmospheric carbon dioxide. These include, direct air capture, bioenergy with carbon capture and storage, enhanced weathering, and marine-based carbon dioxide removals. Their brief descriptions are listed below.

- Direct Air Capture (DAC) is a technology designed to capture carbon dioxide (CO2) directly from the atmosphere. This process involves using chemicals or other materials to absorb CO2 from the air, effectively reducing the concentration of this greenhouse gas in the atmosphere. Once captured, the CO2 can be stored underground or used in various applications, such as in the production of synthetic fuels or other materials.

- Bioenergy with Carbon Capture and Storage (BECCS) involves burning biological material (biomass) like wood, crops, or waste materials for energy production, then capturing the CO2 emissions produced during this process and storing them underground or in other stable forms.
- Enhanced Weathering is a geoengineering technique to accelerate the natural process of weathering to remove CO2 from the atmosphere. It involves spreading finely ground silicate rocks, like basalt or olivine, over large land areas. These minerals react with CO2 in the atmosphere, forming stable carbonates that effectively lock away the carbon. This method not only helps reduce atmospheric CO2 levels but can also improve soil quality and reduce ocean acidification, contributing to environmental sustainability.
- Marine-based NETs (mNETs) are a group of largely experimental technologies focused on reducing atmospheric carbon dioxide levels by enhancing the ocean's natural carbon storage processes and leverage its vast capacity to store carbon. These include ocean fertilization, which involves adding nutrients to stimulate phytoplankton growth, thereby increasing CO2 absorption through photosynthesis. Another approach is artificial upwelling or downwelling, which facilitates the transfer of CO2 from surface waters to deeper ocean layers.
- There are a few more. Appendix A provides more detailed descriptions of the various NETs.

The importance of NETs has emerged recently due to the increasing urgency to address climate change (Intergovernmental Panel on Climate Change, 2018). The realization that current efforts to reduce emissions may not be sufficient to meet the goals of the Paris Agreement has pushed forward additional solutions. NETs offer a way to actively remove CO2 from the atmosphere, complementing traditional emission reduction strategies. This shift in focus is driven by scientific assessments indicating that simply reducing emissions will not be enough to avoid severe climate impacts, necessitating technologies that can reverse some of the existing atmospheric CO2 accumulation.

10.3   Stage of Green Technology Development

Green technology development typically progresses through research, pilot testing, and market adoption, a process that reflects human progress and societal choice. Understanding the process is akin to deciphering a roadmap of our collective journey—a journey marked by milestones of creativity, technical breakthrough, productivity, adaptation, and at times, profound challenges. Each stage of the process, from initial conceptualization to widespread adoption, offers unique insights into how technologies evolve, interact with societal needs, and ultimately reshape the web of our lives. Understanding the phases of technology development is essential. It helps us not only appreciate the detailed progression of inventions and their evolving forms and practical impacts but also anticipate future trends. This understanding is key to guiding investments and shaping policies that match our goals for a sustainable and technologically advanced society.

In the domain of technology management research, two models stand out for their insightful portrayal of innovation dynamics: the S-curve and the Hype Cycle. The S-curve, with its elegant, sigmoidal shape, captures the lifecycle of a technology from its nascent beginnings, through slow growth, rapid growth, and eventually to maturity and stabilization. Contrastingly, the Hype Cycle provides a more socio-psychological perspective, tracing the public's expectations and emotions through a journey of inflated anticipation, disillusionment, and eventual understanding and adoption. While the S-curve emphasizes the actual development and adoption rates of technology, the Hype Cycle focuses on perception and expectations. Together, they offer a comprehensive lens: the S-curve grounds us in the realistic progression of technological capabilities, while the Hype Cycle navigates the turbulent seas of public sentiment and market expectations. Their interplay is crucial, painting a fuller picture of not just how technologies grow and mature, but also how they are perceived and accepted by society at large (Shi & Herniman, 2023).

Figures 10.1 and 10.2 list 42 green technologies and plot their stage of development based on the S-curve and the technology hype cycle. Determination of the development stage is based on both the understanding of the two distinct models and descriptions found in the various references listed in Appendix B—References on Green Technology Development.

A vast majority of the green technologies are between the early adopter stage and maturity (solar, wind, GMO crops in late majority and geothermal, floating offshore wind farms, precision agriculture in early majority adoption stages). Widely popular LED lighting and high-efficiency appliances have probably passed their maturity, entering the stagnation phase, in which further innovation may promise a renewal. On the other end of the S-curve, nuclear fusion and the marine-based carbon dioxide removal innovations are so new that only a few know or are paying close attention to.

The "hottest" ones are plant-based and lab-grown meat, next-gen photovoltaics, DACs, and several others. In the hype cycle, they are near or at the peak of inflated expectations. Green hydrogen, ocean fertilization, climate-smart forestry, and many others have already dipped from the peak and are near or at the trough of disillusionment. This is not necessarily all bad, because emerging from the trough is the chance to climb the slope of enlightenment for productivity gain, which geothermal, personal EVs, low-tillage farming, organic farming, and GMO crops have experienced.

The transition from innovative ideas to market-ready solutions is not just a journey of scientific and technological advancement but also a formidable financial challenge. The strategies for financing green technologies are as varied as the technologies themselves, involving a blend of public funding, private investment, and innovative financial instruments. This phase is pivotal, as securing adequate and appropriate funding can dramatically accelerate the deployment of green technologies, enabling them to make a tangible impact in our quest for a sustainable future. Thus, understanding the nuances of financing is as crucial as the technology development itself, forming the backbone that supports the leap from potential to reality in the world of green technology.

10.4   Financing Green Technology Development and Deployment

Financing the further development of green technologies at various stages of their lifecycle requires a multifaceted approach, tailored to the specific characteristics and risks associated with each stage. Financing approaches for technologies are related to their stages of development.

Government grants and subsidies are crucial for initial research and development, especially for technologies with long-term potential but high upfront costs and risks. University and research institution funding supports fundamental research and early development. Venture and angel investing are ideal for high-risk, high-reward early-stage innovations like emerging carbon capture technologies.

For growth-stage technologies, public–private partnerships (PPPs) are effective for technologies like smart grid systems and advanced renewable energy, where there is a clear path to commercialization but still significant capital requirements. Corporate investments and strategic partnerships are suitable for technologies like green hydrogen and advanced energy storage, which can benefit from corporate expertise and networks. Bank loans and bonds are appropriate for more established technologies with demonstrable revenue potential and lower risk profiles.

Mature technologies in wide adoption stage can be financed through the stock market. For successful technologies ready for large-scale deployment, such as established renewable energy sources, the public market is often the most efficient financing mechanism, which can also include green bonds and climate bonds, which are useful for large-scale, capital-intensive projects like renewable energy farms or energy efficiency retrofitting. For specific large projects with predictable revenue streams, such as wind farms or solar parks, project finance is often used.

Innovative approaches to financing, such as crowdfunding and community financing can be used at various stages, especially for technologies with strong community or environmental appeal. Carbon credits and trading can support project or venture financing when revenue from carbon credits can support technologies that demonstrably reduce greenhouse gas emissions. Other innovative financing mechanisms, such as climate innovation funds or blended finance combining public, philanthropic and commercial capital, are continually structured to support technology developments that are of experimental nature and long-term potential. Policy-driven financing like feed-in tariffs, tax credits, or renewable energy certificates can incentivize investment across various stages of technology development.

Table 10.2 provides a quick summary of the above discussion with illustrations of various green technologies. It is important to note that the choice of financing mechanism should consider not only the development stage of the technology but also other factors like market readiness, regulatory environment, potential for scale, and alignment with broader environmental and sustainability goals. Effective financing strategies often involve a combination of different mechanisms, adjusted over time as the technology matures and market conditions evolve.

The financing mechanisms in Table 10.2 summarizes the approaches suited to different stages of technology development for green technologies. While each mechanism serves a unique purpose based on the technology's maturity and risk profile, they also have certain strengths and weaknesses. Below is a critique of each financing approach by stage.

University and Research Institution Funding

Strengths: Research grants from entities like the U.S. Department of Energy or state governments are crucial at this stage, as they provide early-stage funding for highly speculative technologies. These funds help push the boundaries of innovation in nascent fields like nuclear fusion, enhanced weathering, and ocean-based solutions.

Weaknesses: These funding mechanisms are often limited in scope and duration, may be highly competitive, and are tied to specific political or research agendas. This can result in a lack of flexibility for pursuing disruptive or high-risk innovations. Additionally, reliance on government funding can delay progress due to bureaucratic processes or changing political priorities.

Innovative Financing Mechanisms (e.g., Climate Innovation Funds, Blended Finance).

Strengths: Blended finance, which combines public funds with private capital, is a powerful tool for leveraging private investment into high-risk sectors like experimental technologies. It reduces the perceived risk for private investors while still advancing innovation.

Weaknesses: This approach can be complex to structure and manage, with the need to align the interests of both public and private stakeholders. Furthermore, it may require significant transaction costs and administrative overhead. There's also a risk that blended finance might still be too cautious in supporting truly transformative technologies.

Table 10.2 Financing green technology development and deployment

Technology stage	Financing approaches and references
Experimental and Emerging Stage Nuclear Fusion Enhanced Weathering Ocean-based Solutions	University and Research Institution Funding (e.g., U.S. Department of Energy, State of California, and other entities offer funding to support climate research programs), Innovative Financing Mechanisms (e.g., climate innovation funds,[3,4] blended finance[5])
Early-Stage Direct Air Capture (DAC) Bioenergy with Carbon Capture and Storage (BECCS)	Venture Capital (Morrison, 2023). Angel Investing, Government Grants and Subsidies[6,7]
Growth Stage Advanced Renewable Energy Technologies Green Hydrogen Smart Cities and IoT Carbon Capture and Storage	Public–Private Partnerships (Mortensen, 2024) Corporate Investments and Strategic Partnerships Bank Loans and Bonds

(continued)

---

[3] EDF Climate Innovation Funding Tracker, https://innovationtracker.edf.org/, accessed December 3, 2024.

[4] European Commission Climate Action Innovation Funding, https://climate.ec.europa.eu/eu-act ion/eu-funding-climate-action/innovation-fund_en, accessed December 3, 2024.

[5] Source: LSE Explainers - How can 'blended finance' help fund climate action and development goals?
https://www.lse.ac.uk/granthaminstitute/explainers/how-can-blended-finance-help-fund-cli mate-action-and-development-goals/, accessed December 3, 2024.

[6] U.S. Department of Commerce has an Evergreen Climate Innovations investment scheme, https://www.eda.gov/funding/programs/build-to-scale/past-grantees/2023-capital-challenge/Evergreen-Climate-Innovations, access December 3, 2024.

[7] Source: IEA report: How governments support clean energy startups, https://www.iea.org/reports/how-governments-support-clean-energy-start-ups/financing, accessed December 3, 2024.

Table 10.2 (continued)

Technology stage	Financing approaches and references
Mature Technologies (Wide Adoption) Renewable Energy Sources (Solar, Wind, Hydro, Geothermal) Energy Efficiency Improvements Electrification of Transportation Smart Grid Technologies Sustainable Agriculture Practices	Stock Market Financing (IPOs)[8] Green Bonds and Climate Bonds[9] Project Finance[10]

Angel Investing, Government Grants, and Subsidies

Strengths: Angel investing can provide the initial capital that is often needed to bridge the gap between basic research and commercialization. Government grants and subsidies also play a critical role in de-risking early-stage technologies by providing non-dilutive capital.

Weaknesses: Angel investors typically invest with a high-risk tolerance, but the amount of capital they provide may not be sufficient to scale up technologies. Government grants can sometimes be insufficient or poorly aligned with market needs. Moreover, grants and subsidies might not incentivize private sector innovation to the same extent as more market-driven approaches.

Growth Stage

Public–Private Partnerships (PPPs)

Strengths: PPPs can provide a blend of public sector incentives with private sector efficiency, which is key for technologies that are moving toward commercial viability but still need significant investment. They allow for shared risk and pooling

---

[8] An example of an Initial Public Offering (IPO) in the green technology sector is Enphase Energy's IPO. Enphase Energy, a global energy technology company and a leading supplier of solar microinverters, went public in 2012. The company's IPO was a significant step in raising capital to expand its innovative solar technology solutions. Enphase's success in the stock market reflects the growing investor interest in sustainable and renewable energy technologies.

[9] In 2020, NextEra Energy, one of the world's leading clean energy companies, issued a green bond of $1.5 billion. This bond was specifically aimed at financing or refinancing, in part or in full, new, or existing eligible green projects like renewable energy generation, battery storage, and electric vehicle charging stations. The issuance of this green bond underscored NextEra Energy's commitment to sustainability and its role in the transition to a low-carbon economy.

[10] A notable example of project finance in mature-stage green technology is the financing of the Hornsea Project One offshore wind farm in the UK. This project, led by Ørsted (formerly DONG Energy), achieved financial close in 2016. The project finance involved multiple financial institutions and was one of the largest financing operations for an offshore wind farm. The financing enabled the construction of what was, at the time, the world's largest offshore wind farm, demonstrating the significant scale and potential for renewable energy projects.

of expertise, making them ideal for large-scale infrastructure projects like smart cities, green hydrogen, and carbon capture technologies.

Weaknesses: While PPPs offer substantial benefits, they can also suffer from bureaucratic delays and inefficient management. These partnerships often require extensive negotiation, and the allocation of risk and reward between public and private entities can be contentious. There is also the risk that these partnerships are overly influenced by political agendas rather than long-term sustainability goals.

Venture Capital (VC)

Strengths: Venture capital is well-suited to fund early-stage technologies like Direct Air Capture (DAC) and Bioenergy with Carbon Capture and Storage (BECCS). VC can provide not only funding but also strategic guidance, industry connections, and the drive needed to scale these technologies.

Weaknesses: VC funding typically comes with high expectations for rapid scaling and financial returns, which may not align with the long-term, uncertain timelines associated with green tech. Furthermore, the high level of competition for VC funding may leave some promising technologies without the resources they need, especially if they don't promise immediate returns.

Corporate Investments and Strategic Partnerships

Strengths: Corporate partnerships bring substantial financial resources and industry expertise. They can help drive the adoption of growth-stage technologies like green hydrogen or smart cities solutions by providing access to established market channels and customer bases.

Weaknesses: Corporate investments may come with strings attached, including pressure to meet short-term financial targets or to pivot technologies in ways that might not align with the original sustainability objectives. There's also the risk of "greenwashing," where companies invest in ESG initiatives for reputation rather than actual environmental impact.

Bank Loans and Bonds

Strengths: Bank loans and green bonds provide a significant amount of capital for scaling up technologies in the growth stage. Green bonds, in particular, offer an attractive option for institutional investors seeking to fulfill ESG mandates, and they provide companies with a reliable funding source.

Weaknesses: The downside of debt financing is the obligation to repay loans, which can strain a company's cash flow, especially if the technology isn't yet generating revenue. Moreover, bond issuance may be challenging for companies in nascent industries that lack a proven track record.

Stock Market Financing (IPOs)

Strengths: IPOs are a strong financing tool for mature green technologies, providing access to large-scale capital. This route can help companies in renewable energy and other sectors with wide adoption to finance expansion or research into further innovations.

Weaknesses: The IPO process can be expensive and time-consuming, requiring significant upfront costs and regulatory compliance. Additionally, public companies are subject to market volatility, which can impact their stock price and access to capital. Companies in emerging sectors may also struggle to attract investor interest, as they may be seen as riskier than more established industries.

Green Bonds and Climate Bonds

Strengths: Green and climate bonds are growing in popularity due to their environmental appeal and the increasing demand from institutional investors for ESG-compliant assets. These bonds provide a relatively low-cost and scalable financing option for companies in mature green tech sectors.

Weaknesses: The market for green bonds is still evolving, and not all green bonds live up to their environmental promises, leading to concerns over "greenwashing." Furthermore, bond issuance still carries risks related to interest rates, and the financial performance of the underlying technology can still impact the attractiveness of the bonds to investors.

Project Finance

Strengths: Project finance is ideal for funding large-scale infrastructure projects like renewable energy plants or smart grid technologies, where the project itself can generate revenue. It is structured to allocate risk in a way that protects the investors, making it an attractive option for funders.

Weaknesses: Project finance can be complex, requiring extensive due diligence and legal arrangements. If the project fails to deliver expected returns, investors may face significant losses. Additionally, project financing is often limited by the size and scale of the project, which may not always be flexible enough to accommodate rapid technological changes.

The financing approaches at each technology stage play critical roles in enabling green technology development and deployment. From the early-stage risk tolerance required for venture capital and angel investments to the more structured and stable financing of green bonds and stock market financing in mature technologies, each mechanism must align with the unique needs of the technology and its stage of development. While these financing tools offer significant opportunities for scaling green solutions, they also come with challenges such as high risk, complex regulatory environments, and the potential for greenwashing or mis-alignment of interests. By carefully considering these factors, industries can better match their green technology initiatives with the most appropriate and effective financing strategies.

Adaptation and Financing

Green technology development and climate adaptation are two critical facets of our collective response to climate change, each requiring robust financing to be effective. Investing in green technologies paves the way for innovative solutions that reduce emissions and promote sustainability. Meanwhile, financing climate adaptation focuses on strengthening resilience against the impacts of climate change, especially in vulnerable communities. By aligning the financing mechanisms for both these approaches, we can ensure a balanced and comprehensive strategy toward climate resilience. This harmonized financial support not only fosters the development and deployment of cutting-edge green technologies but also bolsters adaptive capacities, ensuring that communities are equipped to manage the changing climate. Thus, a unified approach in financing green technology and climate adaptation is crucial for leveling the playing field, making sure that advancements in sustainability are coupled with necessary adaptations, ultimately leading to a more resilient and sustainable future for all.

Adaptation to climate change involves strategies and actions to reduce the vulnerability of natural and human systems against actual or expected climate change effects. Financing these adaptation approaches varies depending on the scale, scope, and region. Table 10.3 outlines common adaptation approaches, their financing mechanisms, and estimated financing needs, along with key references:

These estimates are indicative and vary significantly based on the specific context, scale of implementation, and regional needs. They underscore the vast financial resources required for effective adaptation to climate change.

10.5   The Nature of the Threats and Technology Development as Social Choice

The kaleidoscopic parade of green technologies demonstrates the raw, unbridled power of human ingenuity when it collides with the pressing, desperate need for a greener world. It is also a vivid reflection of our societal choices, our long-held dreams, and our deep-seated anxieties. As we delve into the profound nature of the climate crisis, it becomes increasingly clear that our response hinges on critical choices between Emission Mitigation Technologies (EMTs) and Negative Emission Technologies (NETs). These represent distinct pathways in our societal approach to the environmental challenge. EMTs focus on reducing greenhouse gas emissions at the source, addressing the root cause of climate change. On the other hand, NETs aim to rectify the existing imbalance in our atmosphere by actively

Table 10.3 Climate change adaptation approaches and financing

Adaptation approach	Financing mechanisms	Estimated financing needed	Key reference
Coastal Protection (e.g., sea walls, restored mangroves)	Public funding, international grants (e.g., Green Climate Fund), bonds	Tens to hundreds of billions USD annually	IPCC reports[11]
Water Resource Management (e.g., efficient irrigation, rainwater harvesting)	Government budgets, development aid, public–private partnerships	Tens of billions USD annually	World Bank[12,13]
Disaster Risk Reduction (e.g., early warning systems, resilient infrastructure)	Government funding, international assistance, insurance schemes	Billions to tens of billions USD annually	World Bank[14]
Agricultural Adaptation (e.g., drought-resistant crops, climate-smart agriculture)	National budgets, international development aid, climate funds	Tens of billions USD annually	FAO[15]

(continued)

---

[11] Source: IPCC Sixth Assessment Report (AR6), Chapter 3: Oceans and Coastal Ecosystems and Their Services; Cross-Chapter Paper 2: Cities and Settlements by the Sea; IPCC Tar Climate Change 2001: Impact, Adaptation, and Vulnerability.

[12] Source: The World Bank Feature Story: Floods and Droughts: An EPIC Response to These Hazards in the Era of Climate Change, June 17, 2021, https://www.worldbank.org/en/news/feature/2021/06/17/floods-and-droughts-an-epic-response-to-these-hazards-in-the-era-of-climate-change, accessed December 3, 2024.

[13] World Bank Water Global Practice; Global Water Security and Sanitation Partnership. 2022. Water Resources Management (WRM): Strengthening Climate-Informed Project Design. © Washington, DC. http://hdl.handle.net/10986/37123, accessed December 3, 2024.

[14] Source: The World Bank Fact Sheet, World Bank Extends New Lifeline for Countries Hit by Natural Disasters, December 1, 2023. https://www.worldbank.org/en/news/factsheet/2023/12/01/world-bank-extends-new-lifeline-for-countries-hit-by-natural-disasters; Also, World Bank. 2013. Building Resilience: Integrating Climate and Disaster Risk into Development. © Washington, DC. http://hdl.handle.net/10986/16639, both accessed December 3, 2024.

[15] FAO Reports: The State of Food and Agriculture 2023, https://doi.org/10.4060/cc7724en; FAO Strategy on Climate Change 2022–2031, https://www.fao.org/3/cc2274en/cc2274en.pdf, both accessed December 3, 2024.

Table 10.3 (continued)

Adaptation approach	Financing mechanisms	Estimated financing needed	Key reference
Health System Strengthening (e.g., disease surveillance, emergency preparedness)	Domestic health budgets, global health funds, bilateral aid	Billions USD annually	WHO[16]
Urban Planning and Green Infrastructure (e.g., flood-resistant buildings, green spaces)	Municipal budgets, green bonds, private investments	Tens of billions to hundreds of billions USD annually	C40 Cities[17] UN-Habitat[18]
Ecosystem Restoration and Conservation (e.g., reforestation, wetlands restoration)	Environmental funds, NGO initiatives, carbon offset programs	Billions to tens of billions USD annually	UNEP[19] Strassburg et al. (2020)

removing carbon dioxide. This section invites a deeper exploration of these options not as competing solutions, but as complementary strategies that require a delicate balancing act. We must weigh their respective impacts, costs, and feasibility to orchestrate a coordinated response to the climate crisis.

Anthropogenic climate change occurs over decades and centuries. It represents a chronic, long-term threat to humanity, rather than an acute one. Climate change's threats, unlike sudden, acute dangers, are a slow-burning crisis. They have been scientifically forecasted for decades—rising sea levels, more frequent and severe weather events, disrupted ecosystems—yet are only now piercing the public consciousness. This looming, gradual menace is finally shaping an awakening awareness, a collective realization of the profound and lasting impacts these changes herald for our planet.

---

[16] Source: WHO Climate Change and Health, https://www.who.int/teams/environment-climate-change-and-health/climate-change-and-health/evidence-monitoring; Also, WHO Report: Operational framework for building climate resilient and low carbon health systems, November 9, 2023, https://www.who.int/publications/i/item/9789240081888, both accessed December 3, 2024.

[17] Founded in 2005, C40 connects more than 90 of the world's greatest cities, representing over 700 million people and one-quarter of the global economy. The network emphasizes that the large cities of the world have a significant role to play in combating climate change, as cities are both major sources of carbon emissions and highly vulnerable to climate impacts. For more details, go to www.c40.org

[18] UN-Habitat Climate Change Highlighted Publications, https://unhabitat.org/topic/climate-change, accessed December 3, 2024.

[19] UNEP Policy and Strategy: Seven Lessons on Using Ecosystem Restoration for Climate Change Adaptation, May 4, 2023, https://www.unep.org/resources/policy-and-strategy/seven-lessons-using-ecosystem-restoration-climate-change-adaptation-0, accessed December 3, 2024.

The impacts of climate change are the result of cumulative greenhouse gas emissions. These emissions have been building up in the atmosphere since the industrial revolution, leading to a gradual increase in global temperatures and associated climate effects. They are widespread and will persist over lengthy periods. The recovery or reversal of these effects, if possible, will also take considerable time. This delay means that even if emissions were drastically reduced today, some level of further warming and climate impact would still be inevitable due to past emissions.

Addressing climate change effectively requires long-term strategies, policies, and changes in human behavior. Mitigation and adaptation efforts need to be sustained over long periods to be effective. The effects of climate change are not limited to the current generation. The actions taken today, or the lack thereof, will have consequences for future generations, further underscoring its chronic nature.

Climate change, while fundamentally a chronic issue, can manifest as an acute threat under certain conditions or in specific contexts. It is challenging to predict a precise timeline for when climate change might transition into an acute threat globally, as this depends on a complex interplay of environmental, social, and economic factors. However, there are signs and scenarios that can indicate such a transition:

Exceeding Tipping Points. Climate science has concluded (Chapter 2) that the Earth's climate system has a number of tipping points, which, if crossed, could lead to rapid and irreversible changes. These include the melting of major ice sheets, the dieback of rainforests, or the release of copious amounts of methane from thawing permafrost. Crossing these thresholds could abruptly worsen the impacts of climate change.

Technologies that play a role under these challenges include **Early Warning Systems (EWS)** such as satellite-based monitoring technologies, remote sensing, and advanced data analytics are essential for detecting early signs of tipping points. These systems can monitor changes in ice sheet volume, forest health, methane levels in permafrost, and other critical indicators, providing early warnings that could trigger timely global or regional interventions. **Geoengineering Technologies** such as solar radiation management might be considered as potential, though highly debated, measures to counteract the effects of crossing critical tipping points.

Rapid Increase in Extreme Weather Events. A significant uptick in the frequency and severity of extreme weather events, such as hurricanes, heatwaves, droughts, and floods, could signal an acute phase. These events can lead to immediate and severe impacts on human lives, infrastructure, and ecosystems.

In addition to EWS, **Advanced Climate Models** that use machine learning and supercomputing power can help predict the frequency and intensity of extreme weather events. These models are essential for preparing for and mitigating the effects of hurricanes, heatwaves, floods, and other natural disasters. **Disaster Resilience Technologies** aimed at increasing infrastructure resilience to extreme weather events, such as flood barriers, storm-resistant buildings, and infrastructure retrofitting, are key in reducing the damage caused by these events. Additionally, early warning systems, mobile apps for disaster response, and drones for search and rescue operations are crucial in managing acute events. **Smart Infrastructure**, Sensors and Internet of Things (IoT)-enabled systems can monitor infrastructure health and provide real-time data to mitigate damage during extreme weather events. Smart grids, water management systems, and predictive maintenance technologies can help urban areas better withstand climate impacts.

Critical Resource Shortages. Acute threats may arise if climate change leads to severe shortages of essential resources like water, food, or habitable land. This could happen due to prolonged droughts, loss of agricultural productivity, or sea-level rise.

Water Technologies—Advanced water filtration, desalination technologies, and water recycling systems can play a critical role in ensuring access to fresh water during droughts or after contamination from extreme weather events. These technologies can also support water conservation efforts. **Food Technologies**— Innovations in crop genetics, precision farming, and climate-resilient seeds are key to maintaining food production in changing climates. Technologies that optimize water use, reduce pesticide dependency, and ensure nutrient management in changing climates will be essential in preventing food shortages. Vertical farming, aquaponics, and lab-grown meats can help diversify food sources and reduce reliance on traditional agriculture, which is vulnerable to climate change. These technologies could prevent or mitigate food crises brought on by extreme weather and resource shortages. **Resource Efficiency Technologies**—Advances in materials science, energy efficiency, and circular economy models (e.g., upcycling, zero-waste manufacturing) can help reduce overall consumption of essential resources and make more sustainable use of what's available.

Mass Displacement and Conflict. Large-scale displacement of populations due to climate impacts (like sea-level rise or desertification) could precipitate acute crises, potentially leading to increased conflicts and humanitarian emergencies.

Climate-Resilient Urban Planning involves technologies that support the development of climate-resilient cities, such as sustainable architecture, green infrastructure, and urban cooling technologies, which can help prevent mass displacement by making cities more adaptable to climate change. **Migration Management** utilizes tools like geospatial information systems (GIS) and big data that help predict migration patterns and inform government responses to displacement caused by climate events (e.g., sea-level rise). Drones and satellite imagery can also assist in disaster recovery and relocation efforts. **Conflict Resolution Technologies** include advanced communication platforms, mediation tools, and peace-building technologies (e.g., AI for conflict prediction, virtual negotiations) can help prevent or mitigate conflicts exacerbated by climate-induced migration and resource scarcity.

Rapid Economic or Social Collapse. If climate change severely disrupts economic systems, for example, through the destruction of key infrastructure or the loss of entire industries (such as fisheries or agriculture), it could trigger acute economic or social crises.

Blockchain technology can offer transparency and traceability in the distribution of humanitarian aid or in tracking the use of funds for climate adaptation and mitigation projects. It can also be used for efficient resource allocation and rebuilding efforts post-disaster. **FinTech** refers to financial technologies that can facilitate rapid economic recovery by providing quick access to capital, insurance payouts, and social safety nets for affected populations. Digital currencies and mobile payment systems can ensure that funds reach those in need, even during crises. **Supply Chain Automation** leverages smart logistics, AI, and robotics that can help rebuild disrupted supply chains more efficiently after climate disasters. These technologies can also ensure that essential goods and services are delivered to affected areas, reducing the risk of economic collapse. **Social Impact Metrics** involve technologies that track and measure social resilience and the socioeconomic impact of climate disruptions (e.g., AI-driven data analysis) can help mitigate economic collapse by providing targeted interventions to the most vulnerable populations.

Health Crises. The emergence of widespread health issues directly attributable to climate change, such as heat-related illnesses, the spread of vector-borne diseases, or respiratory problems due to poor air quality, could represent an acute phase.

Telemedicine—As health systems are stretched during climate-induced crises, telemedicine technologies can provide critical healthcare services remotely, especially in underserved or disaster-stricken areas. AI-powered diagnostics and remote monitoring can reduce the strain on physical healthcare infrastructure. **Disease Surveillance** uses advanced technologies, such as sensor networks and AI for predictive health analytics, which can help track the emergence of diseases (e.g., vector-borne diseases due to changing climate conditions) and provide early alerts to public health authorities. **Climate-Smart Healthcare** involves technologies that make healthcare facilities more energy-efficient and resilient to extreme weather (e.g., solar-powered clinics, energy-efficient HVAC systems), which can prevent breakdowns in the healthcare system during times of crisis. **Air Quality Monitoring** uses real-time air pollution monitoring systems, combined with AI-powered predictive analytics, which can provide vital data on the health risks posed by poor air quality, helping mitigate respiratory and other climate-related health issues.

If global mitigation, absorption, and adaptation efforts fail to keep pace with the rate of climate change, the accumulated impacts could reach a point where they cause acute crises. The technological role in these climate scenarios is multifaceted, ranging from preventative measures like early warning systems and carbon capture, to adaptation strategies such as climate-smart infrastructure and telemedicine. Each technology must be leveraged at the appropriate stage of development and in the right context to effectively mitigate or adapt to climate-induced crises. Technology not only helps manage current impacts but also provides critical pathways to preventing or alleviating acute threats in the future.

It is important to recognize that in some regions or for certain populations, climate change has already taken on acute characteristics, particularly in vulnerable areas like small island nations, coastal regions, and communities dependent on climate-sensitive resources. Monitoring the signs and responding proactively with effective mitigation and adaptation strategies is crucial to prevent climate change from becoming an acute global threat.

Balancing EMT and NET Development as Social Choice

The question of whether it is timely to advocate for a balancing act between Emission Mitigation Technologies (EMTs) and Negative Emission Technologies (NETs) in the context of net-zero commitments is complex and multifaceted. It is true that current net-zero commitments largely focus on EMTs. This is primarily because reducing emissions is the most direct and immediate way to mitigate climate change. EMTs are generally better understood, more developed, and often more cost-effective than NETs.

While EMTs are crucial, they may not be sufficient to achieve net-zero targets, especially considering the historical and ongoing emissions. The Intergovernmental Panel on Climate Change (IPCC) has indicated that meeting the goals of the Paris Agreement will likely require the use of NETs to remove CO_2 from the atmosphere. Currently, most NETs are in their infancy in terms of development and deployment. However, their role is expected to become more prominent over time. Investing in and developing these technologies now can be seen as preparing for the future, where they might be needed on a larger scale.

There is a risk that heavy reliance on undeveloped NETs in the future could lead to a situation where necessary technologies are not ready or scalable when they are critically needed. The economic models and policy frameworks for NETs are not as mature as those for EMTs. Focusing solely on NETs might lead to moral hazard, where the promise of future solutions or perceived assurance provided by NET progresses could reduce the urgency to reduce emissions now. Policymakers, businesses, and individuals might believe that future technologies will resolve the issue, undermining current mitigation efforts. Focusing on NETs might divert financial and intellectual resources away from proven emission reduction strategies. This could slow the progress in areas like renewable energy, energy efficiency, and sustainable transportation. There is also a concern about equity and justice. Relying on NETs might disproportionately affect vulnerable populations, who could be most affected by potential side effects of NET deployment while being least responsible for the elevated concentration of CO_2 in the atmosphere.

Thus, a balanced approach prioritizes emission reduction while integrating the role of NETs. A balanced approach would involve continuing to aggressively pursue EMTs while also investing in the development of NETs. Focusing on balancing EMTs and NETs now can stimulate the development of these necessary frameworks. It is arguably timely to start incorporating NETs into climate strategies to ensure a comprehensive approach to achieving net-zero goals, even though the primary focus will likely remain on emission reductions for the near future.

Corporations face limitations in resources, including capital, human resources, and technology. Prioritizing investments between EMTs, which may offer more immediate reductions in emissions, and NETs, which are crucial for long-term climate stabilization, requires a strategic balancing act of responding to market dynamics and heeding to corporate risk management. EMTs generally have more established markets, while NETs are still emerging and may require significant

R&D investment. Balancing these technologies involves navigating these different market dynamics. EMTs often involve mature technologies with known risks, whereas NETs can be more speculative with uncertain technological feasibility and scalability. Balancing investments between these categories helps manage overall risk. The market readiness and scalability of these technologies vary.

Policies and regulations may favor one type of technology over the other, and these frameworks are continually evolving. Corporations must navigate these changing landscapes, balancing their investments to align with current and anticipated regulatory environments. While EMTs are critical for immediate impact, the long-term goals of climate change mitigation will likely require the deployment of NETs. Balancing these technologies aligns with a long-term sustainability vision, recognizing the need for immediate action and future-proofing strategies.

There is a growing expectation for businesses to act responsibly toward the environment. Balancing EMTs and NETs allows corporations to address these expectations comprehensively, demonstrating commitment both to reducing future emissions and addressing past and current environmental impacts. The concept of "balancing" in this context reflects the strategic decision-making required in allocating resources, managing risks, navigating market and regulatory environments, and fulfilling corporate social responsibilities toward effective climate change mitigation.

The implications of a balancing act for corporate finance are significant and multifaceted. Here are several key considerations at corporate, business, and functional levels:

Long-Term Planning and Scenario Analysis. Corporations will need to engage in long-term planning and scenario analysis to understand how the mix of EMTs and NETs will evolve over time and impact their business models and financial health.

Aligning with Net-Zero Commitments. For corporations committed to net-zero targets, balancing EMTs and NETs is not just a financial decision but also a strategic imperative. This balance will be critical in demonstrating progress toward these commitments.

Corporate Strategy and Reputation. Incorporating NETs into a corporation's strategy can enhance its reputation as a leader in sustainability. This could have positive implications for brand value, customer loyalty, and investor attractiveness.

Diversification of Investment Portfolio. Corporations will need to diversify their investment portfolios to include both EMTs and NETs. This diversification can spread risk, as these technologies are at various stages of maturity and market readiness. However, it also requires a nuanced understanding of the unique risks and returns associated with each technology type.

Capital Allocation Challenges. Allocating capital between proven EMTs and emerging NETs will be a major challenge. EMTs might offer more predictable returns in the short to medium term, while NETs, though riskier and with longer-term payoff horizons, may be essential for long-term sustainability and regulatory compliance.

Risk Management. Investing in NETs, many of which are in experimental or pilot stages, involves higher risks, including technological, market, and regulatory risks. Corporations will need sophisticated risk assessment and management strategies to navigate these uncertainties.

Innovation and R&D Investment. Corporations will need to invest more in research and development, particularly for NETs. This might involve direct investment in R&D or strategic partnerships with startups and research institutions.

Regulatory and Policy Impacts. As governments and international bodies implement policies to combat climate change, these could significantly affect the viability and attractiveness of various green technologies. Corporations need to be proactive in understanding and responding to these policy shifts.

Financing Instruments and Structures. There might be a need to develop or use innovative financing instruments and structures suited to the unique characteristics of NETs. This could include green bonds, sustainability-linked loans, or public–private partnerships.

Reporting and Disclosure. There will be increased pressure for transparent reporting and disclosure on how corporations are balancing their investments in EMTs and NETs, especially from investors who are increasingly focusing on environmental, social, and governance (ESG) criteria.

The balancing act between EMTs and NETs presents a complex array of challenges and opportunities for corporate finance. It requires a strategic approach to investment, risk management, innovation, compliance, and sustainability reporting.

An effective balancing act requires assessing whether the current pace of green technology development and deployment will be sufficient to avoid a climate change catastrophe. It is a challenging task involving several complex factors. Here is an overview of the current situation and challenges:

Emissions Trajectory. Current global greenhouse gas emissions are still on a trajectory that could lead to significant warming by the end of the century. Despite progress in green technologies, emissions need to be reduced more rapidly to meet the goals of the Paris Agreement, namely keeping global warming well below 2 °C above pre-industrial levels and pursuing efforts to limit it to 1.5 °C.[20]

Technology Development Pace. While green technologies are advancing, some crucial technologies, especially in the field of carbon capture and storage (CCS) and negative emissions, are not developing at the pace required to significantly impact global emissions in the short term.

Implementation and Scaling. Deployment of existing technologies like renewable energy and energy efficiency solutions is growing but needs to accelerate.

---

[20] Source: UN Climate Change: New Analysis of National Climate Plans: Insufficient Progress Made, COP28 Must Set Stage for Immediate Action, November 14, 2023. https://unfccc.int/news/new-analysis-of-national-climate-plans-insufficient-progress-made-cop28-must-set-stage-for-immediate, accessed December 3, 2024.

The main challenge is scaling these solutions globally, especially in developing countries where financial and infrastructural barriers exist.[21]

Policy and Investment. Achieving the necessary scale and speed of technology deployment depends heavily on policy support and investment. While there is significant investment in green technologies, it is still below the levels needed for a global transition to a low-carbon economy.[22]

Socioeconomic Factors. The transition to green technologies must also consider socioeconomic factors, including job transitions, energy access, and affordability, to ensure it is just and equitable.

Natural Feedback Loops. Climate change includes potential tipping points and feedback loops (like thawing permafrost releasing methane). If these are triggered, they could accelerate warming, making it harder to control despite technological advances (Tollefson, 2023).

Mitigation vs. Adaptation. While mitigation technologies are crucial, adaptation strategies are equally important to deal with the impacts of climate change that are already occurring and will continue regardless of mitigation efforts.

Special Considerations for NETs

While considerable progress is being made in developing and deploying green technologies, the pace and scale of these efforts may still be insufficient to avoid significant impacts of climate change. It is a race against time, and avoiding a climate catastrophe will likely require an unprecedented global effort in terms of rapid technology deployment, robust policy frameworks, substantial investment, and coordinated action across all sectors of society. Negative Emission Technologies (NETs) could significantly impact global carbon markets in several ways:

Increasing Supply of Carbon Credits. The deployment of NETs, such as carbon capture and storage (CCS) or direct air capture (DAC), can potentially generate a large volume of carbon credits by removing CO_2 from the atmosphere. This increased supply could influence the pricing and dynamics of carbon markets.

Enhancing Market Confidence. Successful implementation of NETs could enhance confidence in carbon markets by providing tangible solutions for carbon removal, which is essential for achieving net-zero targets. This could attract more participants and investment into the carbon markets.

Market Diversification. With the advent of NETs, carbon markets may diversify beyond traditional emission reduction projects (like renewable energy) to include

---

[21] Source: UNEP Report. Renewable energy and energy efficiency in development countries: Contributions to reducing global emissions, November 15, 2016. https://www.unep.org/resources/report/renewable-energy-and-energy-efficiency-developing-countries-contributions-reducing, accessed December 3, 2024.

[22] Source: BloombergNEF, Global low-carbon energy technology investment suggest past $1 trillion for the first time, January 26, 2023, https://about.bnef.com/blog/global-low-carbon-energy-technology-investment-surges-past-1-trillion-for-the-first-time/, accessed December 3, 2024.

a variety of carbon removal projects. This diversification could appeal to a broader range of investors and emitters with different risk appetites and sustainability goals.

Regulatory Adjustments. The integration of NETs into carbon markets may require adjustments in regulatory frameworks. This could include setting standards for measuring, reporting, and verifying (MRV) the amount of CO_2 removed, ensuring the environmental integrity of carbon credits from NETs.

Price Effects. If NETs can remove CO_2 efficiently and at scale, they could potentially lower the price of carbon credits by increasing the supply. Conversely, if NETs are expensive and their implementation is slow, this could maintain or even increase carbon prices.

Compliance Strategies. For companies with carbon neutrality goals or those under regulatory obligations to reduce emissions, NETs offer an additional strategy to meet these targets. This could shift some focus from emissions reduction to emissions removal in compliance strategies.

Innovation and Investment. The potential for NETs in carbon markets may spur innovation and investment in this field. Companies and governments might invest more in developing and scaling up NETs to capitalize on the emerging market opportunities.

Global Equity Considerations. The development of NETs and their integration into carbon markets could have implications for global equity. There could be concerns about who has access to these technologies, how they are financed, and how the benefits and burdens are distributed globally.

Long-term Market Stability. As NETs become more established, they could contribute to the long-term stability and sustainability of carbon markets by providing a reliable method of offsetting emissions that are difficult to eliminate.

Potential for Market Saturation. In the long term, if NETs are phenomenally successful and widespread, there could be a risk of market saturation where the supply of carbon credits exceeds demand, leading to significant market adjustments.

NETs have the potential to reshape carbon markets by increasing the supply of carbon credits, diversifying market options, influencing prices, and necessitating regulatory changes. However, their impact will depend on technological advancements, cost-effectiveness, scalability, and integration into existing market and regulatory frameworks.

10.6   Impact Measurement and Management

Indeed, the journey to 2050 and the goal of achieving net-zero greenhouse gas emissions is laden with critical milestones. These milestones serve as checkpoints to gauge progress and recalibrate actions as needed. Table 10.4 illustrates some key milestones to watch for:

Cross-referencing climate milestones with green technology development and deployment involves a strategic approach to ensure that investments in green technologies are accountable and aligned with the broader goals of climate change

Table 10.4 Milestones of human efforts achievements toward climate change

Milestone	Description	Indicators	When to watch
2030 Emissions Targets	The year 2030 is a crucial short-term milestone with ambitious emission reduction targets under the Paris Agreement	Reduction in global GHG emissions, progress in renewable energy capacity, and energy efficiency improvements	Ongoing assessments leading up to 2030, with significant attention in the late 2020s
Global Stocktake in 2023 and Every Five Years Thereafter	A five-yearly global stocktake to assess collective progress toward the Paris Agreement's long-term goals	Collective assessment of emission reductions, adaptation measures, and financial flows	Starting in 2023 and every five years following
Phase-out of Coal and Fossil Fuel Subsidies	Significant reduction or total phase-out of coal usage and fossil fuel subsidies	Closure of coal-fired power plants, changes in energy policies, and investment shifts	Progressively through the 2020s and 2030s
Major Technological Milestones	Developments in carbon capture, battery storage, and renewable energy technologies	Commercial viability and scalability of innovative technologies, breakthroughs in energy storage and efficiency	Continuous, with advancements expected throughout the 2020s and 2030s
Adaptation Progress	Implementation of strategies to minimize climate impact on vulnerable communities	Development and implementation of national and regional adaptation plans, resilience of infrastructure	Ongoing, with increasing focus in each successive decade
Sustainable Development Goals (SDGs) Deadline in 2030	SDGs related to climate action, energy, and sustainable cities, with a deadline in 2030	Progress in SDGs related to environmental and energy sustainability	By 2030
Green Financing and Investments	Shift in financial flows toward green investments	Levels of green bonds issuance, sustainable investments, and divestment from fossil fuels	Increasingly important throughout the 2020s and 2030s

(continued)

Table 10.4 (continued)

Milestone	Description	Indicators	When to watch
Net-Zero Commitments by Corporations	Tracking private sector progress toward net-zero commitments	Corporate emission reduction achievements, sustainability reports	Ongoing, with specific targets set by companies
Political Commitments and International Cooperation	Continuous commitment and collaboration among nations on climate issues	Policy developments, international agreements, and cooperation on climate issues	Ongoing, especially during international climate summits

mitigation. This is one important way to identify greenwashing. For example, it is noticeable how many heavy emitters say they will solve their emission problems with carbon capture, yet there is dearth of progress in the technology development to make carbon capture technically feasible, let alone economically viable.

Setting Clear Targets and KPIs. Develop specific, measurable targets and key performance indicators (KPIs) for green technology investments. These should be aligned with the broader climate milestones such as 2030 emissions targets or the SDG deadlines. For example, if a company commits to 75% renewable energy by 2030, investment in renewables should have clear KPIs like gigawatts of capacity installed annually.

Integrating Milestones into Investment Plans. Incorporate climate milestones directly into green technology investment plans. This ensures that funding is channeled into projects that contribute to these milestones. Prioritize funding for technologies that directly reduce emissions in sectors where the fastest and most substantial cuts can be made by 2030.

Regular Monitoring and Reporting. Establish a robust system for monitoring and reporting the progress of green technology projects. This should involve regular assessments against the set KPIs and climate milestones. Biannual or annual sustainability reports detailing progress in terms of technology deployment and its impact on emissions.

Leveraging Financial Instruments. Use financial instruments that inherently promote accountability, such as green bonds or sustainability-linked loans, where the financial returns are tied to the achievement of environmental targets. Issuing green bonds where the funds are exclusively used for projects that contribute to specific climate milestones.

Engaging Stakeholders. Actively engage all stakeholders, including investors, governments, and communities, in the planning and assessment process. This promotes transparency and collective responsibility. For example, establishing multi-stakeholder committees to oversee and evaluate green technology projects.

Data-Driven Decision-Making. Use data and analytics to guide investment decisions and evaluate their impact. This includes using climate and environmental data to identify the most effective technologies and areas for investment. For example, using emission reduction potential as a key criterion for selecting green technology investments.

Collaboration and Benchmarking. Collaborate domestically and internationally to benchmark and learn from best practices in green technology investment. Sharing knowledge and experiences can help in refining approaches to ensure alignment with global climate milestones. Participating in international forums to share progress and strategies on green technology deployment.

Conclusion

As the climate crisis intensifies, the dual imperatives of mitigating emissions and adapting to inevitable changes have propelled green technology to the forefront of corporate and societal priorities. The technological advancements and financial strategies discussed in this chapter demonstrate humanity's capacity to innovate and adapt, yet they also underscore the scale and complexity of the challenge ahead. Achieving a sustainable future requires more than a portfolio of promising technologies—it demands meticulous planning, strategic investment, and judicious resource allocation.

This pivotal intersection is where capital budgeting becomes essential. The transition from ideation to implementation hinges on the ability to evaluate competing projects, weigh risks and returns, and align financial decisions with long-term sustainability goals. As we delve into the tools of project valuation in the next chapter, we will explore how corporations can quantify the potential of green technologies, prioritize investments, and ensure that each dollar spent accelerates progress toward a net-zero future. These methodologies will provide the financial scaffolding necessary to navigate the complexities of climate-focused initiatives, bringing us one step closer to a resilient and sustainable global economy.

Key Takeaways

1. **The Urgency of Climate Action**

   Climate change is driven by cumulative greenhouse gas emissions, creating both chronic and acute threats that demand immediate and long-term solutions. The dual strategies of mitigation and adaptation are essential for a sustainable future.
2. Emission Mitigation vs. Negative Emission Technologies (EMTs and NETs)

   EMTs focus on reducing emissions at the source through renewable energy, energy efficiency, and sustainable practices.

   NETs aim to actively remove CO_2 from the atmosphere, such as through carbon capture or reforestation, providing a complementary approach to EMTs.

   A balanced strategy combining both is crucial for achieving net-zero goals.
3. Stages of Technology Development

Green technologies progress through distinct stages—from research and pilot testing to market adoption. The S-curve and Hype Cycle provide valuable frameworks for understanding these transitions and guiding strategic investments.

4. Financing Green Technology

Different technologies require tailored financing approaches at each stage of development, from government grants and venture capital in early stages to green bonds and public markets for mature technologies.

5. Corporate Decision-Making

Companies face strategic choices in balancing investments between EMTs and NETs, considering factors such as risk, scalability, regulatory changes, and alignment with sustainability goals.

6. Climate Change Milestones

Achieving key climate milestones, such as 2030 emissions targets and net-zero commitments, depends on accelerating technology deployment, robust policies, and increased investments in both mitigation and adaptation strategies.

7. Measuring Impact

Transparent reporting, clear KPIs, and data-driven decision-making are essential for aligning green technology investments with broader climate milestones and ensuring accountability.

8. Economic and Social Considerations

The deployment of green technologies must balance financial viability with equitable transitions, minimizing disruptions to jobs and ensuring affordability and accessibility.

9. Carbon Markets and NETs

The integration of NETs into carbon markets has the potential to reshape pricing dynamics, enhance market confidence, and diversify compliance strategies, but requires robust regulatory frameworks and technological scalability.

10. Role of Innovation

Technological innovation, supported by strategic financing, policy alignment, and corporate action, is humanity's most powerful tool to address the existential threat of climate change.

Questions

1. What are the primary differences between Emission Mitigation Technologies (EMTs) and Negative Emission Technologies (NETs)?
2. Explain the significance of the S-curve and Hype Cycle in understanding green technology development.
3. Identify and describe at least three financing mechanisms suitable for early-stage green technologies.
4. How do EMTs and NETs complement each other in achieving net-zero targets?

5. Discuss the role of capital allocation in balancing investments between EMTs and NETs.
6. What are the key challenges in scaling green technologies globally?
7. How can corporations integrate climate milestones into their green technology investment strategies?
8. Provide examples of how green technologies can align with climate adaptation approaches.
9. Explain the potential impact of NETs on carbon markets and pricing dynamics.
10. Discuss how transparent reporting and data-driven decision-making support green technology investments.

Appendix A: Detailed Description of Negative Emission Technologies

This table lists various NETs and provides a summary description and relative strengths and weaknesses in terms of potential scaled deployment for each of them.

Technology	Description	Strengths	Weaknesses
Carbon Capture and Storage (CCS)	CCS involves capturing CO_2 emissions from sources like power plants before they are released into the atmosphere, then storing them underground	Effective at reducing emissions from large point sources	High cost, energy-intensive, requires significant infrastructure
Direct Air Capture (DAC)	DAC uses chemical processes to extract CO_2 directly from ambient air	Can be located anywhere, independent of emission sources	Currently expensive and energy-intensive
Bioenergy with Carbon Capture and Storage (BECCS)	BECCS involves growing biomass, burning it for energy, and capturing and storing the CO_2 emissions	Provides energy while removing CO_2 from the atmosphere	Land-use concerns, potential competition with food production
Enhanced Weathering	Speeds up natural weathering processes to remove CO_2 by spreading finely ground minerals that react with CO_2 in the air	Uses natural processes, potentially large-scale	Uncertainties about side effects, enormous amounts of minerals needed

(continued)

(continued)

Technology	Description	Strengths	Weaknesses
Afforestation and Reforestation	Planting new forests or restoring lost ones to absorb CO_2 through photosynthesis	Enhances biodiversity, provides other ecosystem services	Requires large areas of land, potential competition with agriculture
Soil Carbon Sequestration	Agricultural practices increase soil carbon content, e.g., no-till farming, cover cropping	Improves soil health, synergistic with sustainable agriculture	Variable effectiveness depending on soil and climate conditions
Biochar	Organic material is pyrolyzed to create a stable form of carbon for soil addition	Enhances soil fertility, long-term carbon storage	Requires biomass and can be energy intensive
Ocean Fertilization	Adding nutrients to the ocean to boost phytoplankton growth, which absorbs CO_2	Potential for significant carbon sequestration	Ecological risks, potential for unintended consequences
Ocean Alkalinity Enhancement	Adding alkaline substances to the ocean to increase its capacity to absorb CO_2 from the atmosphere	Potentially large-scale CO_2 removal, can counteract ocean acidification	Requires massive quantities of materials, potential ecological impacts
Artificial Upwelling and Downwelling	Pumping nutrient-rich deep water to the surface to stimulate phytoplankton growth or transporting surface carbon to deeper waters	Enhances natural biological carbon sequestration	Energy-intensive, potential impacts on ocean ecosystems
Seaweed Cultivation	Growing seaweed farms for CO_2 absorption, which can be used for bioenergy or sunk for long-term storage	Additional benefits include biofuel production, habitat creation	Requires large areas, potential ecological impacts
Electrochemical CO_2 Removal	Renewable energy-driven electrochemical reactions in seawater convert CO_2 into stable minerals or compounds	Could be highly efficient, producing valuable byproducts	Technologically complex, high demand for energy

(continued)

(continued)

Technology	Description	Strengths	Weaknesses
Marine Permaculture	Installing floating platforms to facilitate the growth of marine ecosystems that sequester carbon	Enhances marine biodiversity, and can support fisheries	Needs significant area, potential navigational hazards

Appendix B: Key Sources of References on Green Technology Development

Renewable Energy Sources
Reports by the International Renewable Energy Agency (IRENA)

International Energy Agency (IEA).

Energy Efficiency Improvements

International Energy Agency (IEA) Energy Efficiency reports

U.S. Department of Energy (DOE) resources.

Electrification of Transportation
Electric Vehicles Initiative (EVI) reports

IEA's Global EV Outlook.

Smart Grid Technologies
Smart Electric Power Alliance (SEPA) publications

DOE resources.

Sustainable Agriculture Practices
Food and Agriculture Organization (FAO) reports

scientific journals on agricultural science.

Advanced Renewable Energy Technologies
Research articles on photovoltaics, wind energy, and geothermal systems.

Carbon Capture and Storage (CCS)
Intergovernmental Panel on Climate Change (IPCC) Special Reports

IEA CCS Technology Roadmap.

Green Hydrogen

Hydrogen Council reports

IEA's Future of Hydrogen.

Smart Cities and IoT
IEEE publications

Smart Cities Council resources.

Nuclear Fusion
International Atomic Energy Agency (IAEA) publications

Nuclear physics journals.

Direct Air Capture (DAC)
Environmental science journals

Carbon Capture & Storage Association (CCSA) reports.

Bioenergy with Carbon Capture and Storage (BECCS)
IPCC reports

Bioenergy journals.

Enhanced Weathering
Geoscience journals

IPCC assessment reports.

Afforestation and Reforestation
World Wildlife Fund (WWF) research

United Nations Environment Programme (UNEP) reports.

Ocean-based Carbon Dioxide Removal Solutions
CDR Primer https://cdrprimer.org

Ocean CDR Roadmap, https://oceanvisions.org/roadmaps/

Navigating Potential Hype and Opportunity in Governing Marine Carbon

References

Kumar, T., & Satya Eswari, J. (2023). Review and perspectives of emerging green technology for the sequestration of carbon dioxide into value-added products: An intensifying development. *Energy & Fuels, 37–5*, 3570–3589. https://doi.org/10.1021/acs.energyfuels.2c04122

Morrison, C. J. (2023) Founders alert: Climate tech funding outlook for 2023, Goodwin Law Insights. https://www.goodwinlaw.com/en/insights/publications/2023/03/03_09-climate-tech-funding-outlook-for-2023

Mortensen, L., Kørnøv, L., Gjerding, A. N., Rattigan, E., & Schlüter, L. (2024). Middle-out evolution of greenfield eco-industrial parks: The journey of GreenLab, Denmark. *Journal of Industrial Ecology, 28*, 1816–1829. https://doi.org/10.1111/jiec.13569

Shi, Y., & Herniman, J. (2023). The role of expectation in innovation evolution: Exploring hype cycles. *Technovation, 119*. https://doi.org/10.1016/j.technovation.2022.102459

Strassburg, B. B. N., Iribarrem, A., Beyer, H. L., et al. (2020). Global priority areas for ecosystem restoration. *Nature, 586*, 724–729. https://doi.org/10.1038/s41586-020-2784-9

Takalo, S. K., Tooranloo, H. S., & Parizi, Z. S. (2021). Green innovation: A systematic literature review. *Journal of Cleaner Production, 279*. https://doi.org/10.1016/j.jclepro

Tollefson, J. (2023). Catastrophic change looms as Earth nears climate 'tipping points', reports say. *Nature, 624*, 233–234. https://doi.org/10.1038/d41586-023-03849-y

Part III

Practical Tools and Applications

Project Valuation in Climate Crisis

Introduction

The financial landscape of corporate investment is undergoing a profound transformation as businesses grapple with the implications of climate change. Meeting the Net-Zero goal by 2050 requires an unprecedented scale of investment, with estimates suggesting $9.2 trillion in global capital investment annually for the next 30 years to achieve decarbonization (Cutler & Linder, 2023). Beyond mitigation, adaptation investments in developing countries alone are projected to range between $140 billion to $300 billion annually by 2030, climbing to $280 billion to $500 billion annually by 2050. The Global Commission on Adaptation estimates that a $1.8 trillion investment in resilience from 2020 to 2030 could yield $7.1 trillion in benefits (Fayolle, 2022).

Firms in high-carbon sectors face mounting risks, including regulatory pressures, stranded assets, and declining market demand as renewable energy gains traction. Physical risks—ranging from droughts impacting agricultural yields to water shortages disrupting beer production—further underline the pervasive nature of climate challenges across industries and geographies. These dynamics make it clear that every investment decision will be influenced by climate considerations to varying degrees.

This chapter focuses on the essential tools and frameworks needed to navigate today's complex and dynamic business environment shaped by climate challenges. It introduces advanced capital budgeting techniques specifically designed to account for climate-related risks and opportunities.

Supplementary Information The online version contains supplementary material available at https://doi.org/10.1007/978-3-031-83487-5_11.

The chapter begins with an analysis of the Levelized Cost of Electricity (LCOE), a fundamental metric for comparing the lifecycle costs of renewable energy projects against those of fossil fuels. It then explores methods for enhancing financial evaluations, including:

- **Net Present Value (NPV) Adjustments**: Incorporating climate uncertainties into investment decisions.
- **Shadow Pricing**: Accounting for the financial impact of carbon emissions.
- **Real Options Analysis (ROA)**: Valuing flexibility in decision-making under uncertainty.

Using detailed case studies such as the Evergreen Wind Farm and Granite Peak Mine, the chapter illustrates practical applications of these advanced methodologies.◄

11.1   Levelized Cost of Electricity (LCOE)

The Levelized Cost of Electricity (LCOE) is a widely used metric for comparing the cost-effectiveness of different power generation technologies by calculating the per-unit cost of electricity (e.g., \$/MWh) over the lifetime of a project, accounting for both capital and operational expenses. As the world transitions to a low-carbon economy, LCOE becomes increasingly relevant for comparing renewable energy projects (e.g., wind, solar, hydro) to traditional fossil fuel-based power generation.

LCOE allows financial stakeholders to assess whether a renewable energy project can compete with conventional power sources, taking into account capital costs, maintenance costs, fuel costs, and the expected energy output over time. Climate change can impact the availability of certain renewable resources. For instance, wind patterns or solar radiation levels may be affected by long-term climate shifts. A robust LCOE analysis could incorporate climate scenario modeling to estimate future resource variability or intermittency, helping to better assess the true cost of renewables under changing environmental conditions.

LCOE can reflect the impact of subsidies, tax credits, and other government incentives (such as renewable energy targets or carbon pricing) on the competitiveness of low-carbon energy technologies. These incentives can help make renewable projects more attractive in the face of rising fossil fuel costs, carbon taxes, and emissions regulations. As countries work toward achieving net-zero emissions, the cost-effectiveness of renewable energy technologies will become an increasingly important metric for investors. LCOE helps determine whether clean energy projects will become more competitive relative to fossil fuel-based projects as carbon pricing and environmental regulations increase.

By starting with LCOE, we can effectively benchmark various energy projects, such as wind, solar, coal, or Carbon Capture, Utilization, and Storage (CCUS).

▶ There are slight differences in the calculation of the LCOE depending on the source and level of detail. For example, some metrics include carbon costs as part of the equations and others do not. The elements typically included in the LCOE calculation are given in the following expression:

$$LCOE = \frac{\sum_{t=1}^{n} \frac{I_t + O_t + F_t + C_t + D_t}{(1+r)^t}}{\sum_{t=1}^{n} \frac{E_t}{(1+r)^t}}$$

where:

- $I_t =$ Capital expenditures in year t (including construction, land acquisition, and equipment costs). These upfront costs are distributed over the plant's lifetime in the formula.◀
- $O_t =$ Operating and maintenance costs in year t. This includes the annual costs of running the plant, maintaining equipment, labor, and other fixed or variable costs. For renewable energy, these costs are relatively low compared to fossil fuel plants.
- $F_t =$ Fuel costs in year t (if applicable, e.g., for coal, natural gas, or nuclear). For fossil fuel plants (e.g., coal, natural gas), fuel costs are an important component. Renewables like wind and solar have negligible fuel costs.
- $C_t =$ Carbon costs (such as emission pricing) in year t.[1]
- $D_t =$ Decommissioning costs in year t represent the expense of safely retiring the plant at the end of its useful life.
- $E_t =$ Electricity generated in year t. The total amount of electricity generated by the plant over its lifetime is considered. Technologies with higher capacity factors (like nuclear or natural gas) will produce more electricity, lowering their LCOE compared to intermittent renewables like wind or solar.
- $r =$ Discount rate (reflects the cost of capital). The discount rate is crucial in the calculation as it reflects the time value of money, representing the opportunity cost of capital. A higher discount rate will increase the LCOE by giving more weight to upfront capital costs.

---

[1] Carbon Costs: The **emission factor** represents the amount of carbon dioxide (CO_2) emitted per unit of electricity generated. It varies depending on the source of emissions and the efficiency of the power plant. Generally, the emission factor for coal is higher than for other fossil fuels like natural gas. The IEA and IPCC, among others provide region-specific and fuel-specific emission factors based on empirical data from energy producers. The carbon cost for the firm depends upon its emissions factor and the cost of carbon. If the emissions factor is 0.85/MWh and the cost of carbon is \$30 /ton, then the carbon cost is: \$30 × 0.85 = \$25.50/MWh.

- **n** = Economic lifetime of the plant (typically 20–40 years).

Application of LCOE

The IEA has an LCOE calculator available (International Energy Agency, 2020) which allows the user to view the impact of varying key assumptions (e.g., discount rate). The website also provides access to global LCOE for different technologies and sources that you can download. To illustrate we show results for Coal Generation in Table 11.1.

Table 11.1 illustrates how the Levelized Cost of Electricity (LCOE) varies significantly across different countries and technologies. Notably, China and India have the lowest LCOE for coal generation, primarily due to lower capital and operational expenses. Conversely, in the US, deployment of Carbon Capture, Utilization, and Storage (CCUS) technology makes coal generation markedly more expensive compared to other regions, globally and within the US. Notably, for comparison purposes, the IEA default assumptions assume a discount rate of 7% for all locations and technologies. However, in the LCOE calculator this, and other assumptions, can be modified.

LCOE in Practice[2]
Traditionally tied to power generation, LCOE now guides decisions in various other sectors. In the industrial sector, it identifies cost-effective solutions for reducing greenhouse gas emissions, like integrating renewable energy or using Carbon Capture. In transportation, LCOE evaluates the shift from fossil fuels to electric vehicles and hydrogen fuel cells. In agriculture, it assesses renewable energy options such as biogas and solar-powered irrigation. For residential and commercial buildings, LCOE compares costs of energy systems like solar panels, energy-efficient materials, and smart grid technologies.

The LCOE metric offers a standardized way to compare energy generation costs across technologies, but it assumes stability in inputs like fuel costs, carbon pricing, and operational conditions. In reality, climate change introduces significant uncertainties—scientific, modeling, and socioeconomic—that can disrupt these assumptions and impact long-term project outcomes. Section 11.2 explores how these uncertainties complicate capital budgeting and require adjustments to traditional evaluation techniques to account for risks and variability in a rapidly evolving climate landscape.

---

[2] While LCOE is the most referenced levelized cost metric due to its direct relevance in comparing power generation technologies, other levelized costs such as Levelized Cost of Storage (LCOS), Levelized Cost of Carbon Abatement (LCCA), Levelized Avoided Cost of Energy (LACE), and Levelized Cost of Hydrogen (LCOH) are also gaining traction. These metrics, though used less frequently than LCOE, are becoming increasingly important as the energy landscape evolves and new technologies emerge. Their application is particularly prevalent in niche sectors such as energy storage, carbon reduction initiatives, and hydrogen production, where they provide critical insights into the cost-effectiveness and long-term viability of various projects.

Table 11.1 LCOE Coal Generation

Country	Plant type	Capital costs	O&M	Fuel (Thermal)	Fuel (Electricity)	Carbon	CHP heat revenues	LCOE
India	Ultra-supercritical (pithead) (400 MW)	12.87	8.53	11.89	26.43	22.7	0	70.54
China	Ultra-supercritical (347 MW)	8.97	14.97	12.61	28.02	22.7	0	74.67
Korea, Republic of	Ultra-supercritical (954 MW)	12.9	11.62	11.89	27.47	23.6	0	75.59
United States of America	Supercritical pulverized (650 MW)	28.95	17.2	7.31	17.51	24.47	0	88.13
United States of America	Pulverized (650 MW)	27.78	17.01	7.31	18.23	25.48	0	88.5
Australia	Supercritical pulverized (722 MW)	27.28	8.69	11.89	29.66	25.47	0	91.1
Japan	Ultra-supercritical (749 MW)	27.12	19.31	11.89	28.73	24.68	0	99.84
India	Ultra-supercritical (load centered) (400 MW)	12.46	38.65	11.89	26.43	22.7	0	100.25
United States of America	Pulverized (140 MW)	38.65	22.25	7.31	20.36	28.46	0	109.72
Australia	Supercritical pulverized (CCUS) (633 MW)	50.34	19.34	11.89	39.58	3.4	0	112.66
United States of America	Supercritical pulverized (CCUS) (650 MW)	52.18	30.15	7.31 ara>	22.38	12.51	0	117.23
United States of America	Coal (641 MW)	46.61	29.5	7.31	17.16	23.99	0	117.27
United States of America	Pulverized (CCUS) (650 MW)	51.62	30.45	7.31	23.5	13.14	0	118.72

(continued)

Table 11.1 (continued)

Country	Plant type	Capital costs	O&M	Fuel (Thermal)	Fuel (Electricity)	Carbon	CHP heat revenues	LCOE
United States of America	Pulverized (138 MW)	49.13	30.47	7.31	20.19	28.22	0	128.02
United States of America	Coal (CCUS) (499 MW)	67.17	42.96	7.31	23.32	13.04	0	146.49

Source International Energy Agency (2020)

11.2   Climate Change Uncertainties and Capital Budgeting

Using a capital budgeting framework to evaluate projects like green initiatives, oil sector expansion, or climate adaptation (e.g., heat-resistant runways) is increasingly complex as the climate crisis progresses. To many, "science" means hard facts. However, we may forget that climate change science and its global impact analysis are continually evolving.

In an earlier chapter when we explored various climate futures, we consistently emphasized that these futures were not "predictions" but rather possible futures. In fact, there is significant ambiguity surrounding how global warming (but not its anthropogenic origins) will evolve which affects both the severity of physical climate risk and stringency of transition policies aimed at containing temperature rise. The Intergovernmental Panel on Climate Change (IPCC) underscores "uncertainties" in its assessments to highlight the complexities and variabilities inherent in climate science, predictions, and impacts. We can think about these uncertainties in three ways:

- **Scientific Uncertainty:** Despite advancements in climate science, there remain gaps in our understanding of the climate system (Intergovernmental Panel on Climate Change Chapter 7, 2021). For instance, feedback mechanisms like cloud formation[3] and ice-albedo[4] are complex and not fully understood, leading to variations in climate model projections.

- **Modeling Uncertainty:** Climate models are essential tools for predicting future climate conditions. However, they incorporate assumptions and simplifications of the real world, leading to uncertainties in their projections. Different models can produce a range of outcomes for the same scenario, depending on the inputs and climate sensitivity assumed (Intergovernmental Panel on Climate Change, Chapter 4, 2021).
- **Socioeconomic Uncertainty:** Future levels of greenhouse gas emissions depend significantly on socioeconomic factors, including economic growth, technological advances, policy decisions, and societal behavior changes. The uncertainty in these factors makes it challenging to foresee future emissions and climate impacts accurately (Intergovernmental Panel on Climate Change Chapter 3, 2022).

Carbon Sinks

The role of natural sinks to absorb carbon exemplifies these uncertainties. To reach net zero there is an expansive menu of available tools, (Sustainable Aviation Fuel, carbon capture techniques, green hydrogen, regenerative agriculture, electric cars, etc.) but we also rely extensively on natural carbon sinks. Natural carbon sinks, such as forests, soils, and oceans, have historically absorbed approximately 55% of anthropogenic carbon dioxide (CO_2) emissions, effectively mitigating the rate

---

[3] Cloud formation feedback is one of the feedback mechanisms that affect the climate sensitivity, which is the degree of global warming in response to a given change in greenhouse gas concentrations. Clouds can have both cooling and warming effects on the climate, depending on their type, altitude, and location. Low-level clouds tend to reflect incoming solar radiation back to space, thus having a cooling effect. High-level clouds tend to trap outgoing infrared radiation from the earth, thus having a warming effect. The net effect of clouds on the climate is uncertain, as it depends on how cloud cover, thickness, and distribution change with rising temperatures and atmospheric moisture. Some studies suggest that cloud feedback is positive, meaning that clouds amplify global warming by reducing the amount of heat that escapes the earth. Other studies suggest that cloud feedback is negative, meaning that clouds dampen global warming by increasing the amount of heat that reflects into space. The uncertainty in cloud feedback is one of the main sources of uncertainty in climate sensitivity and model projections.

[4] Ice-albedo is a feedback mechanism that affects the Earth's climate. Albedo is the measure of how much solar radiation is reflected by a surface. Ice and snow have a high albedo, meaning they reflect most of the incoming sunlight, while water and land have a low albedo, meaning they absorb more of the incoming sunlight. This means that ice and snow help to cool the planet by reducing the amount of heat absorbed by the surface. However, as the climate warms, ice and snow melt, exposing more water and land, which have lower albedo. This leads to more heat absorption, which in turn causes more melting, creating a positive feedback loop that amplifies the warming effect. This is one of the reasons why the polar regions are warming faster than the rest of the world. Ice-albedo is an example of a complex and uncertain process that affects the climate system and its projections.

of global warming. Specifically, the ocean absorbs about 31% of these emissions, while terrestrial ecosystems account for the remaining 24% (National Centers for Environmental Information [NCEI], n.d.).

These sinks are under threat due to rising global temperatures, droughts, wildfires, and other extreme weather events. An article appearing in The Guardian in 2024 discusses how in 2023, land carbon absorption temporarily collapsed with forests and soil absorbing almost no carbon (The Guardian, 2024)! Similar concerns were raised about the oceans, which are experiencing disruptions in processes that store carbon due to warming waters and melting ice. According to The Guardian reporting, researchers are concerned that rapid deterioration of natural carbon sinks calls into question the reliability of climate models, and by extension the necessity for more stringent policies aimed at curbing GHG emissions. The bottom line is that consideration of risk in capital budgeting takes on a whole new meaning. Unlike the past, where conventional risks and returns were the focus, today's capital budgeting tools must also intricately weave in climate resilience and adaptation strategies to ensure sustainability and viability.

11.3 Capital Budgeting and Decision Rules

Before getting to complexities in capital budgeting in climate crisis, we will situate ourselves in a bit of a corporate finance "comfort zone" by reviewing some basic techniques of project valuation.[5]

When reviewing project profitability we can apply several analytic approaches. There is an accompanying spreadsheet showing calculations for techniques described: *"Chapter 11: Redfern and Green Skies."*

Payback Period: The payback period measures how long it takes for an investment to recoup its initial cost from the cash inflows it generates. Shorter payback periods are preferred as they indicate quicker recovery of the investment.

Suppose Redfern Inc. is considering a $100,000 investment in new machinery, expecting to generate Free Cash Flow (FCF) of $30,000 annually for 5 years. The firm's cost of capital is 10%. The payback period is 3.33 years ($100,000 / $30,000). The firm has an arbitrary rule of accepting all projects that pay for themselves in at least four years, so this project is attractive to the firm. Of course, we have completely ignored the time value of money by using the Payback Period criterion. It is simple to understand though.

---

[5] We assume familiarity with capital budgeting basics. If you are a bit forgetful you can go back and have a look at any basic corporate finance textbook.

Now consider an aviation firm, Green Skies, that wants to investigate replacing conventional jet fuel with green hydrogen. The company will begin by identifying initial costs of $1 billion which include initial infrastructure, prototype development, and R&D. Modest cash flows of $50 million per year are projected for the early operation phase (years 6 through 15). In years 16 through 24 the firm thinks that a combination of reduced operating costs and greater demand will provide FCFs of $100 million annually and rise to $200 million annually in years 25 through 40. The Payback period for this project is 20 years (the sum of FCFs years 1 through 20 = 1000). The cost of capital is 10% but we do not need that number to determine the payback period. Actually, the company does not have any kind of Payback threshold that even approaches 20 years. Too much could go wrong! The reality is that many mitigation and adaptation projects require major capital expenditure with long useful lives but often modest cash flows when the project launches.

Discounted Payback Period: This method is like the payback period but considers the time value of money by discounting future cash inflows and therefore provides a more accurate picture of the time required to recover the investment.

For Redfern Inc. future FCFs are discounted 10%. The discounted payback period would be longer than the simple payback period, reflecting the reduced present value of distant cash flows. The discounted payback period in our example is approximately 4.26 years. This means the investment's initial cost is recovered in just over 4 years when accounting for the time value of money. In this instance, the firm would gauge the discounted payback of the project against an arbitrary threshold established by the firm.

Green Skies is a disaster. The accompanying spreadsheet shows how each future FCF was discounted back to the present value at the firm's cost of capital. We then determined when the initial investment would be recovered by these discounted cash flows—NEVER!

Internal Rate of Return (IRR): The IRR is the discount rate that makes the Net Present Value of all cash flows from a particular project equal to zero. In other words, the IRR is the discount rate that makes the present value of future cash flows equal to the initial investment. Projects with an IRR above the firm's cost of capital are generally considered attractive.

Table 11.2 Project Metrics
for Redfern and Green Skies

Decision Tool	Redfern	Green Skies
Payback	3.33 Years	20 Years
Discounted Payback (10%)	4.26 Years	Never
IRR	15.24%	6.27%
NPV (10%)	$13,723	−$512.51

Redfern Inc.'s investment produces an IRR of just over 15% (see the accompanying spreadsheet). Since this exceeds the firm's required rate of return of 10%, the project should be accepted.

The IRR of the Green Skies investment is 6.27%. Since the IRR is less than the firm's cost of capital the project will be rejected.◄

Net Present Value (NPV): NPV calculates the difference between the present value of cash inflows and outflows over a project's lifetime. A positive NPV indicates the project is expected to generate value beyond its cost.

In our Redfern investment example, cash flows are discounted at 10% resulting in an NPV of $13,723. The present value of future cash flows exceeds the initial capital expenditure of $100,000. The project earns its keep and should be undertaken by the firm.

The NPV of the Green Skies investment is −$512.51. Since NPV is negative the project should be rejected.◄

Table 11.2 summarizes the project evaluation results for Redfern and Green Skies.

Decision Rules for Climate Change

Not all the techniques we just illustrated are effective when analyzing capital projects in the context of the climate crisis. Capital budgeting for such projects must consider the long-term nature of physical and transition risks. While these techniques incorporate the useful life of a project, many analyses for projects with significant climate consequences span a lengthy duration. The timeline (useful life) for mitigation and adaptation projects extends to a "foreseeable future," often beyond 2050, which is established as the target for achieving net-zero emissions. Within this 25-year time frame (2025 to 2050) to address climate change impacts, numerous variables can go astray, including projected physical climate changes, policy responses, and the technology underpinning climate change solutions, potentially affecting project viability.

Payback Period: It is impractical to consider capital budgeting techniques that rely on arbitrarily designated "Paybacks." For many large-scale mitigation projects, such as the transition from conventional jet fuel to Sustainable Aviation Fuel (SAF), initial cash flows are negative, rendering "Payback" less relevant. Moreover, in the aviation sector, this switch, whether profitable or not, is mandated by policy.[6] Of course, climate policy is the most effective lever to encourage investment in green opportunities to mitigate GHG emissions. The greater the price on carbon, the more attractive are mitigation investments.

Internal Rate of Return: The IRR technique may present unique challenges, particularly in the context of the climate crisis. A well-known issue in capital budgeting analysis is that the IRR may not yield a unique solution due to the nature of cash flows associated with certain projects, especially those related to climate crisis investments. Non-conventional cash flows, those alternating between positive and negative values, can result in multiple rates satisfying the IRR equation, leading to multiple IRRs.

This phenomenon arises because the IRR is defined as the discount rate that sets the net present value (NPV) of all cash flows from a project to zero. Climate crisis projects often involve substantial initial outlays followed by complex patterns of returns and additional expenditures over an extended period. For instance, deploying new technologies or infrastructure for carbon mitigation might incur high initial costs, followed by periods of savings or revenue, and subsequent reinvestment requirements. Each shift between negative and positive cash flows can introduce another solution to the IRR equation.

Given the unsuitability of Payback Rules and IRR criteria, we will focus the remainder of our discussion on capital budgeting to NPV and address uncertainty inherent in capital budgeting in climate change using Real Options Analysis.

11.4   Net Present Value (NPV)

In traditional NPV models, climate risks—such as the likelihood of natural disasters, supply chain disruptions due to extreme weather, or the impact of carbon regulations—may not be fully accounted for. Under climate change, NPV calculations should factor in the potential uncertainty and variability introduced by both physical and transition risks. For instance, the future cash flows of a project could be affected by the increasing costs of complying with carbon pricing, regulatory changes, or climate-induced disruptions in operations.

The discount rate used in NPV calculations might need to be adjusted to reflect the additional risks posed by climate change. Projects exposed to higher

---

[6] For most large-scale mitigation projects, favorable tax treatment, advantageous financing, and subsidies are required to turn a negative NPV to positive. Further, higher carbon prices would inflate savings associated with undertaking such projects, again favorably impacting NPV.

transition or physical risks (e.g., fossil fuel-based projects, projects in vulnerable locations) may require a higher discount rate to account for the increased uncertainty and long-term impacts of climate change. Conversely, projects with low-carbon attributes (e.g., renewable energy) or those with strong adaptation plans might have lower discount rates due to perceived lower risks.

Projects in sectors sensitive to climate change, such as agriculture or coastal infrastructure, may require additional investment to mitigate climate risks (e.g., flood defenses, drought-resistant crops). NPV models can help assess the financial feasibility of these investments by evaluating the future cost of adaptation and incorporating the resulting savings or cost avoidance into the cash flow model. For projects aligned with the energy transition (e.g., renewable energy, energy storage, electric vehicle infrastructure), NPV can help investors evaluate whether the future benefits (e.g., lower operating costs, green energy incentives, long-term resilience) outweigh the upfront capital expenditure. NPV can also be used to assess projects' alignment with ESG (Environmental, Social, and Governance) goals, an increasingly important factor for investors.

Due to the comprehensive consideration of all cash flows and the ability to provide a clear measure of value creation, most of our project analyses will begin with NPV. In this section, we expand upon the simplified calculation of NPV in the previous section and illustrate the application of NPV analysis to Evergreen Wind Farm. We start with the formal presentation of the formula for NPV and then illustrate the application for Evergreen Wind Farm.

▶

$$NPV = \sum_{n=0}^{N} \frac{FCF_n}{(1+r)^n}$$

where:
$FCF_n = Free\ Cash\ Flow\ from\ the\ project\ for\ period\ n.$
$r = the\ discount\ rate\ to\ apply\ to\ future\ cashflows$

Evergreen Wind Farm

Evergreen Wind Farm, a renewable energy company, is considering expanding its existing onshore wind farm in California by adding ten new wind turbines. The existing wind farm has twenty turbines, and each turbine produces 2.5 MW of power. The new turbines will also have a capacity of 2.5 MW each. The company has a Power Purchase Agreement (PPA) to sell electricity at $50 per MWh. In energy-related projects, like wind farms, it is crucial to estimate how much electricity a turbine can produce over the course of a year. To do this, we need to calculate the total number of hours available in a year during which energy could potentially be generated.

▶ **Capacity Factor:** The Evergreen Wind Farm has an average capacity factor of 35%, meaning each turbine operates at 35% of its maximum capacity on average throughout the year. This same factor is expected for the new turbines. A turbine at full capacity runs 8,760 hours annually (24×365), but wind conditions prevent constant full capacity operation. The capacity factor indicates the average performance relative to maximum capacity. To estimate revenue from the Evergreen expansion, use this calculation per turbine:

Power Output per Turbine $= 2.5 \times 8760$ x $0.35 = 7665$ MWh per year

The annual revenue per turbine is calculated by multiplying the total power output by the electricity price (PPA):

Annual Revenue per Turbine $= 7665 \times \$50 = \$383,250$

Capital Expenditure: The installation cost for each new turbine is $40 million, covering the turbine, foundation, installation, and internal electrical infrastructure. In addition, the wind farm will also incur a $2 million grid connection upgrade cost to accommodate the additional output from the ten new turbines. The turbines are depreciated on a straight-line basis over a 30-year useful life. The salvage value at the end of 30 years is assumed to be 10% of the initial capital investment and basically represents the value of scrap materials. Further project analysis reveals that after the first 10 years of operation the turbine blades must be replaced at a cost of $300,000 per turbine. The useful life of the replacement blades is 10 years, and we assume there is no salvage value. In the following 10 years, the blades must again be replaced at the same cost and with the same useful life.

Operating and Maintenance (O&M) Costs: O & M expenses are significant for wind farms and include monitoring the turbines and the overall facility, managing the day-to-day operations to ensure optimal performance, staffing for control centers, remote monitoring, and local operations teams. The turbines also require regular inspections, repairs, and servicing, Annual O & M expenses are estimated at $100,000 per turbine.

Additional Information: Evergreen Wind Farm faces a corporate tax rate of 25%. The cost of capital for this project is 6%. Relevant project information is summarized in Table 11.3. Project analysis is detailed in the accompanying spreadsheet: *"Chapter 11: Evergreen Wind Farm."*

▶ FCF Evergreen Wind Farm.

We will show the free cashflow of the project for the first year. You can refer to the spreadsheet to see the evolution of the cashflows over the life of the project.

Begin by calculating EBITDA (Earnings before interest, taxes, depreciation & amortization). We did this on a per turbine basis.

EBITDA $=$ Revenue $-$ Operating Expenses $= \$383,250–\$100,00$
$= \$283,250.$

Table 11.3 Evergreen wind farm

Category	Details
Initial CapEx	Installation cost per turbine: $4 million, Grid connection upgrade: $2 million, Total Initial CapEx: $42 million
Depreciation (Initial CapEx)	Total Initial CapEx: $42 million, Depreciation period: 30 years, Annual Depreciation: $1.26 million
Blade Replacement Costs	Replacement frequency: Year 10 and Year 20, Cost per turbine: $300,000, Total cost per cycle: $3 million
Depreciation of Blade Replacements	Depreciation period: 10 years, Annual Depreciation: $300,000 starting from Year 10 and Year 20
Operating and Maintenance (O&M) Costs	Incremental O&M cost per turbine: $100,000, Total annual O&M cost: $1,000,000
Salvage Value	Expected salvage value: 10% of initial turbine cost, Total: $4.2 million
Tax Treatment	Corporate tax rate: 25%, Depreciation provides tax savings
Cost of Capital (Discount Rate)	6% based on industry benchmarks
Depreciation Schedule (Years 1 to 10)	Annual depreciation from initial CapEx: $1.26 million
Depreciation Schedule (Years 11 to 20)	Annual depreciation from initial CapEx: $1.26 million, Additional depreciation from first blade replacement: $300,000, Total: $1.56 million
Depreciation Schedule (Years 21 to 30)	Annual depreciation from initial CapEx: $1.26 million, Additional depreciation from second blade replacements: $300,000, Total: $1.56 million

EBITDA for 10 turbines = $283,250 × 10 = $2,832,500.

Calculate EBIT (Earnings before interest & taxes) = EBIT − Depreciation.

EBIT = $2,832,500–$1.26 million = $1,572,500.

Calculate Tax = Tax rate × EBIT.

Tax = 0.25 × $1,572,500 = $393,125.

Calculate CFAT (Cash Flow after Tax) = EBITDA − Tax + Depreciation.

CFAT = $2,832,500–$393,125 + $1,260,000 = $3,699,375.

TIP: Do not forget to add back depreciation in the determination of CFAT. Depreciation is a non-cash expense that generates a tax saving. Remember to account for the Salvage Value at the end of the project. This is a cash inflow that must be discounted back to the present value at the firm's cost of capital. For Evergreen

Wind Farm the present Value of Salvage is: $4,200,000/ (1.06)^{30}$
$= \$731,262.50$.

Summary of Evergreen Wind Farm.

The NPV of the Evergreen Wind Farm expansion is positive for discount rates of 4%, 5%, and 6%. At the upper end (6%), NPV is \$12,054,313.83. Therefore, the project is **financially viable** under our current assumptions. The project benefits from a steady revenue stream, with an electricity price of \$50 per MWh, which is reasonable for wind projects in California. This is consistent with industry benchmarks, as California's competitive energy market and renewable energy goals support higher Power Purchase Agreement (PPA) prices, typically ranging from \$40 to \$60 per MWh. However, it is essential to acknowledge that PPA prices are influenced by market conditions, renewable energy policies, and demand fluctuations. As the renewable energy sector continues to evolve, actual electricity prices may differ from the assumed \$50 per MWh. Regular monitoring of market trends is crucial to ensure the long-term profitability of the project.

Additionally, the assumed O&M costs of \$100,000 per turbine annually are aligned with industry standards but represent a key factor in the financial outcome. Maintenance costs can vary based on turbine age, site conditions, and operational efficiency. Effective maintenance strategies will be critical to minimizing downtime and ensuring the turbines operate efficiently over their 30-year lifespan.

Our analysis of the Evergreen project did not consider the effect of additional working capital that might be required for this expansion project. However, we can safely ignore this issue since working capital requirements for wind farm projects are likely to be insignificant: the PPA agreements protect quite well against changes in receivables and payables and there is no inventory attached to the wind farm.

11.5   Capital Budgeting and Shadow Prices

Shadow pricing is a technique used to assign a value to externalities or unpriced environmental and social impacts, such as the cost of carbon emissions, water usage, or pollution, that are not reflected in market prices. With rising carbon prices and regulatory pressure, incorporating a shadow carbon price into project evaluations can help projects align with future policy trends. For example, many countries are introducing or planning carbon taxes or cap-and-trade systems to drive the reduction of greenhouse gas emissions. By assigning a shadow price to carbon emissions, financial models can reflect the future cost of emissions and help determine whether a project will be financially viable in a world with higher carbon costs.

Beyond carbon pricing, shadow pricing can also be used to quantify the external costs of climate change—such as damages from extreme weather events, impacts on biodiversity, or water scarcity—on the project's long-term viability. For example, the shadow price of water could influence the economic feasibility of projects in drought-prone areas, while the shadow cost of flood risk could impact infrastructure investment decisions. Shadow pricing can help ensure that environmentally and socially sustainable projects are financially viable, even when the costs of environmental damage are not directly captured in market prices. By incorporating the true social cost of carbon or other environmental impacts, investors and financiers can incentivize projects that have positive environmental outcomes and avoid projects that generate negative externalities. As countries and companies move toward net-zero emissions, shadow pricing carbon and other pollutants can help companies assess the potential impact of future regulations and align projects with long-term sustainability goals. For investors, shadow pricing provides a more comprehensive understanding of a project's true costs and potential liabilities.

Incorporating Shadow Carbon Prices into Projected Cash Flows

What is an appropriate carbon price? At a most basic level, we can think of carbon prices as an estimate of the present and future cost of carbon emissions. The World Bank offers guidance on carbon pricing in general and makes the distinction between shadow prices and the social cost of carbon (World Bank, 2024).

The shadow price of carbon represents the price of carbon consistent with a specific climate goal such as achieving Net zero, while the social cost of carbon estimates the damages caused by carbon emissions. Firms are free to select what value (if any) for a shadow carbon price, although it is reasonable to expect that candidate shadow prices would reflect best-estimates of future carbon prices as mandated by climate policy.

There are a number of avenues available to select an appropriate shadow price. For example, we could use the price of carbon posted in the carbon market.[7] Bloomberg reports an average carbon price in the EU market at €65[8] in the first quarter of 2024 but predicts €194 in 2026 (BloombergNEF, May 2024).

The various scenario brands we discussed in an earlier chapter also include projections of future carbon prices. Bloomberg compares implied carbon prices of net-zero scenarios sourced from the NGFS and the IEA. In the NGFS scenarios, carbon prices are the main tool for reducing emissions in both the smooth and abrupt net-zero scenarios and reflect the marginal cost of abatement. A rising price for CO_2 triggers rapid emission reduction measures and lowers fossil fuel demand. Prices soar to very high levels across different regions in the scenarios to reflect different reduction costs, with levels for Europe perhaps exceeding $1,000 by 2050, followed by China at over $640 (BloombergNEF, May 2024).

---

[7] We discuss in a subsequent chapter how carbon markets work. The carbon market is a system that allows countries, companies, and individuals to trade emissions allowances or credits that represent a certain amount of greenhouse gas emissions. The carbon market aims to create an economic incentive for reducing emissions and to achieve emission reduction targets in a cost-effective way. Carbon markets include mandatory cap-and-trade systems and voluntary schemes.

[8] Carbon prices refer to price per metric ton of CO_2 equivalent (CO_2e).

The IEA Net Zero Emissions (NZE) scenario emphasizes energy and technology policy solutions, such as a ban on new internal combustion engines in the auto industry by 2035. As a result, the prices of carbon in IEA scenarios are significantly lower than those anticipated by the NGFS scenarios. The IEA projections set carbon prices in advanced economies of $250 by 2050 and in developing economies carbon price projects are forecast at just over $180 p (BloombergNEF, May 2024).

There is a wide range of values assigned to the social cost of carbon due to the challenges of capturing all the risks and costs associated with climate change. Valuing the social cost of carbon is a contentious issue with values ranging from $40 to $600 per ton according to one study that also suggested that by 2100 the social cost of carbon could reach $1000 (Moore & Diaz, 2015). Very recent modeling (Bilal & Känzig, 2024) places a $1000.00 price on carbon as well, under only a moderate warming scenario. Currently, a "conventional" benchmark for the social cost of carbon is around $150.

Firm Use of Shadow Carbon Prices

Currently, there is no standard "carbon price." Not only is there a wide range of prices used at the firm level with significant regional differences, but even regulators are also not in agreement on what the carbon price should be. In recent US legislation, for example, a carbon price of $200.00 was set as a reflection of the social cost of carbon (itself not having unanimity of opinion) while the IMF suggests that a carbon price of $85 by 2030 is an appropriate price.

Using too low a carbon price, or none at all, implies that those businesses are significantly over-valuing some projects while others are forgoing profitable investment opportunities because the cost savings from avoided carbon emissions are not fully considered. You might read that companies use shadow carbon pricing to guide decision-making—how to allocate capital expenditure. However, this viewpoint frankly underestimates the significance of the choice of shadow price. As the climate crisis unfolds, there will be redoubled efforts to require firms to pay for their emissions and such projections are considered by the various scenario brands we referenced in an earlier chapter. Thus, using a correct shadow carbon price might at the moment be a choice but it could well become a requirement in the future.

The use of shadow carbon prices varies across firms depending on the sector, region, and level of ambition. According to a 2021 report by CDP, more than 1300 companies reported using an internal carbon price or were planning to do so within two years (CDP, 2021). Average carbon price used by these companies was $28 per ton of CO_2e, ranging from $1 to $800, although a more recent Reuter's study reported a carbon price of $1600 that was set by California drug maker Amgen (Reuters, 2023). However, the CDP report noted that only 15% of the companies disclosed how they applied their carbon prices and what impact they had on their business decisions.

In the next section we show how a hypothetical coal firm, Western Mining Corporation, could use shadow carbon prices to decide whether to continue operations at Granite Peak Mine.

11.6   Granite Peak Mine Investment

Granite Peak Mine Investment

The Granite Peak Coal Mine, located in the southwestern United States, has been a key supplier of thermal coal to regional utilities for decades. However, the mine is facing aging infrastructure and rising operational costs, leading to a crucial decision point for the company that owns it, Western Mining Corporation (WMC).

To continue operations for another 20 years, WMC must decide whether to make a $500 million investment to upgrade the mine and extend its lifespan. The company turned to their team of in-house analysts to evaluate this investment decision. The investment would allow the mine to continue producing 2.14 million tons of coal annually. At the end of its useful life, the mine will have a negative salvage value (-$50 million) associated with environmental remediation expenses. To support the mine an initial investment in working capital of $30 million is required. This amount is recovered at the end of the mine's useful life. Table 11.4 outlines the key financial and market assumptions. To see complete calculations for Granite Peak Mine Investment refer to the spreadsheet "*Chapter 11 Granite Peak Mine.*"

Base Case

WMC needs to evaluate whether this $500 million investment is worthwhile using NPV calculations. The company is not yet considering carbon shadow pricing in this part of the analysis but wants to know whether continuing operations will

Table 11.4 Granite peak coal mine

Key financial and market assumptions	Values
Initial investment	$500 million
Annual production	2.14 million tons of coal
Annual revenues	$150 million
Annual operating costs	$50 million
Depreciation (Straight line)	$25 million per year
Corporate tax rate	21%
Discount rate	8%
Salvage value	−$50 million
Working capital (recovered in year 20)	$30 million
Carbon emissions	6.12 million tons of CO_2 annually

Table 11.5 NPV & IRR for Base Case Granite Peak

Discount Rate	PV FCF Base Case	NPV Base Case	IRR
6%	$1,246.85	$716.85	20%
8%	$1,068.34	$538.34	
10%	$927.13	$397.13	

generate enough returns given the expected revenues, costs, and market conditions. Specifically, Is the investment justified based on projected cash flows and an 8% discount rate; and is the project still profitable for discount rates of 6% and 10%? To answer these questions, we have to determine the NPV of the project:

For each year (1–19), the **FCF** is calculated as: EBITDA – Taxes + Depreciation. In years 1 through 19, we project an annual FCF of $109.25; and in year 20 accounting for negative salvage value and recovery of working capital, we think FCF will be $89.25.

Table 11.5 illustrates the Present Value of Free Cash Flows (FCF) and the NPV of the Granite Peak Mine Investment for different discount rates.

The initial calculations for the NPV of the Granite Peak mine project look very promising. The IRR at 20% exceeds a range of discount rates between 6 and 10%. In fact, although WMC prefers to use NPV criterion for project analysis, the team is encouraged by the IRR, which at the least suggests that the company could withstand a significant increase in their cost of capital and have the project remain profitable. However, the overall analysis seems somewhat simplistic. While the US does not currently have a unified approach to pricing emissions from coal mining, various federal and state-level initiatives are moving in that direction. To address potential shifts in the regulatory climate, the WMC team decided to redo the analysis by incorporating a shadow price for carbon.

Shadow Prices and Granite Peak Mine Investment

Not all team members agreed on how to handle the issue of shadow carbon prices. Some thought that a shadow price of $50.00 was about right while others wanted to use a higher carbon price more reflective of how carbon might be priced in the future. The team decided to consider potential mandatory carbon costs in a separate section. The NPV analysis incorporating the shadow price is available in the Granite Peak Mine spreadsheet under the Baseline Cash Flows Tab and accompanying NPV. The results of the NPV analysis are presented in Table 11.6.

The new analysis reveals that imposing carbon pricing adds an additional $306 million per year to the project's costs (calculated based on 6.12 million tons of CO_2 emissions per year).[9] After factoring in carbon pricing, the adjusted FCF

---

[9] Note that in this scenario, the analysts did not consider any tax implications arising from the imposition of a shadow carbon price of $50.00.

Table 11.6 NPV of Granite Peak Mine with Shadow Carbon Price

Discount Rate	PV FCF Shadow Carbon Price of $50	NPV Shadow Carbon Price of$50
6%	($2,262.94)	($2,792.94)
8%	($1,936.01)	($2,466.01)
10%	($1,678.02)	($2,208.02)

turns negative at –$196.75 million per year (for years 1–19). Because of the additional carbon costs, the NPV of the project is negative across all discount rates and therefore the mine project upgrade should be rejected. Since the carbon costs are hypothetical, there was no real impact on the investment. In many ways, the results are informative but not binding on whether the project might be profitable. Moreover, it remains an open question of if and when carbon pricing policy might result in mandatory fees for the Granite Peak Mine. However, there is still work to be done. The team of analysts worried about the accuracy of projected cash flows of the investment—are the estimates of free cash flow too optimistic?

Granite Peak Mine Scenario Analysis

The Granite Peak mine upgrade is a significant investment for WMC. The team began by consulting an extensive study produced by Carbon Tracker in 2017 and was alarmed by the jarring headline: **"Death spiral goes global – 42% of global operating fleet unprofitable in 2018 and 72% by 2040 independent of additional climate or air pollution policy"** (Carbon Tracker Initiative, 2017). While a lot of angst had been associated with a potentially impending regulatory environment that might see carbon prices go through the roof, the team had ignored some fundamentals: What about the competition? In the case of coal, the competition originated with renewables! The team quickly realized that their analysis of Granite Peak would need to consider changing market dynamics in the energy sector including a very real possibility of stranded assets. To better assess the risks, WMC decided to review how the coal market might evolve over the next 20 years. Using projections gathered from the International Energy Agency (IEA) and the Network for Greening the Financial System (NGFS), along with in-house analysis WMC evaluated several possible future scenarios. They decided to focus on market dynamics initially and then if the analysis warranted further exploration, costs of emissions could be accounted for.

11.7   Real Options Analysis in Climate Change Valuation

NPV is an "all or nothing" approach to project valuation. If NPV is negative, then we reject the project. Of course, we can create various scenarios where market conditions or the regulatory environment shifts, as we did with Granite Peak mine, but in the end, it comes down to Yes or No today! In reality, the manager does not commit to a project then forgets about how things are going and moves on to the next project on her checklist. Rather, there is continuous oversight: did the cash flows not live up to expectations—should we cut our losses; technological advances significantly reduced operating expenses—should we expand production; concern over continued global warming led to an increase in carbon prices globally—perhaps we should consider how to mitigate emissions; our competitors are investing in mitigation technologies that are not yet at scale but it's something we need to consider as well or we might miss out on the opportunity to do so!

Real Options Analysis (ROA) is a method that values flexibility in decision-making, considering the option to make future decisions as market conditions evolve. It is often applied to projects with significant uncertainty and where the value of waiting or adjusting decisions over time can be substantial. Climate change introduces significant uncertainty about future environmental conditions, policy shifts, and technological advancements. ROA is particularly useful in this context because it allows investors to value the flexibility to adapt a project over time. For instance, the option to scale up a renewable energy project or switch to a cleaner energy source in the future could be valuable in an environment of increasing regulatory pressure or shifting consumer demand for clean energy.

For projects that could be affected by the energy transition, such as fossil fuel-based energy or carbon-intensive industries, ROA can help assess the value of keeping future options open. For example, an oil or gas company might evaluate the option to invest in renewable energy or carbon capture technologies as future policy and market conditions evolve. ROA allows project financiers to value such options and weigh them against the risks of maintaining the status quo. ROA can help investors assess how the flexibility to adapt to future climate-related events (e.g., rising sea levels, increased regulation, new technologies) can add value to a project. For example, renewable energy projects may benefit from the option to expand, add storage capacity, or partner with other clean energy initiatives as the market for renewables grows and government policies evolve.

Real Options Analysis can be applied across a portfolio of projects to optimize the allocation of resources in uncertain, changing climates. This can help project financiers prioritize investments that offer the most flexibility to respond to evolving energy market dynamics, such as changes in consumer preferences or regulatory regimes.

Ideally, we want to remain flexible, and that flexibility can have value. Climate change presents unique challenges and opportunities for investment in long-term projects, particularly those involving green energy and decarbonization efforts. Traditional capital budgeting techniques often fall short in capturing the full spectrum of value and uncertainty inherent in these projects. Real Options Analysis (ROA) offers a sophisticated alternative by incorporating the value of managerial flexibility and strategic decision-making under uncertainty.

In this section we explore how ROA is suited for valuing projects in the context of climate change, discuss various kinds of real options, explain valuation using the Black–Scholes model, and provide relevant examples.

Types of Real Options

Real Options Analysis (ROA) is a powerful tool for project valuation, particularly in the context of climate change and other highly uncertain environments. Traditional methods like Net Present Value (NPV) and Internal Rate of Return (IRR) often struggle to capture the full scope of value and uncertainty in long-term projects. There are two steps in real option analysis. First of all, we need to identify if in fact there are real options embedded in a project and if so, what kind of options? Once we determine real options exist, we can value them. Notably, we can identify more than one embedded real option, so the key is to determine which are most important—we need to prioritize the real options. We will consider three general categories of real options:

Option to Expand: Allows a company to increase investment based on favorable outcomes from an initial phase.

- For example, an automaker considering the transition from internal combustion engines (ICE) to electric vehicles (EV) would likely begin with a preliminary Phase 1 investment to mitigate risks and gather critical data. This initial step allows the company to assess market reception, refine production processes, and adapt to emerging technological advancements. By starting small, the automaker can make informed decisions and avoid potential pitfalls, ultimately paving the way for a more substantial commitment to EV manufacturing in the future. The primary focus here is on scaling up an already established project based on positive initial outcomes.

Abandonment Option: Provides the flexibility to halt a project if it becomes economically unviable.

- An example of an abandonment option in the coal sector might involve a mining company that has invested in a coal extraction project but faces increasing regulatory pressure and declining coal prices due to the transition to renewable energy sources. The company might initially invest in the necessary infrastructure and begin operations. However, as global carbon pricing becomes more stringent and technological advancements in renewable energy make coal increasingly uncompetitive, the company could exercise its abandonment option. This would involve halting coal extraction, decommissioning the mine, and potentially selling off equipment and land for other uses, such as renewable energy projects or carbon offset initiatives. By incorporating the abandonment option, the company retains the flexibility to exit the project if it becomes economically unviable, thereby mitigating potential losses.

Timing Option (Option to Defer): Allows a company to delay investment until conditions are more favorable.

- Consider an energy company facing high initial costs and uncertain future regulatory frameworks. The company might choose to wait until there is more clarity on carbon pricing mechanisms and advancements in CCUS technology before making a large investment. This delay allows the company to gather more information and reduce the risks associated with an immediate commitment. By exercising the option to defer, the company retains the flexibility to invest when conditions are more favorable, thereby optimizing the timing of their investment to capitalize on future opportunities or mitigate potential losses.

Black–Scholes Model: Valuing Flexibility

Although there are several methods available for conducting Real Options Analysis (ROA), we will utilize a standard corporate finance approach, which models the option value using the Black–Scholes framework. This approach, originally developed for pricing financial options, is highly effective for valuing the flexibility inherent in investment projects. By applying the Black–Scholes model, we can capture the value of managerial flexibility to make strategic decisions (such as delaying, expanding, or abandoning projects) in response to changes in market conditions or new information. This method enhances the robustness of our investment evaluations by providing a more comprehensive assessment of the potential value and risks associated with our projects. The Black–Scholes model might appear daunting at first, but it is easy to understand. Let's start! The value of an option to expand or an option to wait are both examples of real call options. The formula below shows how to determine the value of a Call, assuming continuous compounding of future cash flows.

▶   Black–Scholes Call Option

$$C = S_0 \cdot N(d_1) - Xe^{-r_f \cdot T} \cdot N(d_2)$$

$$d_1 = \frac{ln\left(\frac{S_0}{X}\right) + \left(r_f + \frac{\sigma^2}{2}\right)T}{\sigma\sqrt{T}}$$

$$d_2 = d_1 - \sigma\sqrt{T}$$

where:
- C = value of the call option
- S_0 = current value of the underlying asset (in this case, the present value of the future cash flows of the project)
- X = exercise price (the investment cost, or what you need to pay to exercise the option and invest in the project)
- r_f = risk-free interest rate
- σ = volatility of the project's underlying asset or cash flows
- T = time to expiration (the wait time)

- $N(d_1)$ and $N(d_2)$ are the cumulative normal distribution values at d_1 and d_2

Practically speaking we will assume that future values are discounted at discrete intervals rather than continuous, so we take a short-cut on the calculation of d_1.

$$d_1 = \frac{ln\left(\frac{S_0}{X}\right)}{\sigma\sqrt{T}} + \frac{\sigma\sqrt{T}}{2}$$

$$d_2 = d_1 - \sigma\sqrt{T}$$

$$C = S_0 \cdot N(d_1) - X \cdot N(d_2)$$

Let's take a closer look at the formula[10] and think about what it means intuitively. The ratio S/X for determining d_1 can be thought of as a measure of how much the current value of the underlying asset (S) compares to the strike price (X). This ratio indicates whether exercising the option would be profitable. In a way, it is like a profitability index because if (S) is much larger than (X), the intrinsic value of the call option is high. A higher (S/X) ratio implies a greater likelihood of the option being "in the money" (profitable to exercise).

The term $\sigma\sqrt{T}$ (volatility multiplied by the square root of time) measures how much the value of the underlying asset (S) is expected to fluctuate over the life of the option. Higher volatility means the underlying asset value is more likely to make significant moves, both up and down. For a call option, higher volatility increases the chance that the value of the asset will rise above the strike price, making the option more valuable. The more time an option has until it expires, the greater the potential for the underlying asset price to change significantly. Thus, a longer time to expiration gives the option more opportunity to become valuable.

There is a catch: where do we find a measure of volatility? Sometimes the volatility of the underlying stock returns might be a useful indicator. For firms looking at projects that create avoided emissions like carbon capture, we could try to evaluate future carbon prices and their volatility. We might, for example, start with volatility of carbon credits traded in carbon markets, or we could consider

---

[10] Tongue in cheek comment: don't look all that closely. By assuming discrete compounding and then discounting future cashflows at the cost of capital we are violating an underlying assumption of the Black-Scholes approach that investors are risk neutral. At the same time, we assume that the strike price of the real option (X) which represents the investment at the time of exercise (T), is known with certainty and thus discounted at the risk-free rate. Therefore, if we discount future cash flows at WACC, we are effectively mixing risk-adjusted returns with risk-neutral valuation, which can lead to incorrect pricing of options and misinformed investment decisions.

A few other assumptions underpinning the Black-Scholes model are that underlying project returns are follow a lognormal distribution, and both volatility and the risk-free rate are constant.

volatility in fossil fuel prices as a departure point. It is all rather up in the air but volatility assumptions of 30% to 40% would be considered moderate.

The underlying asset (S) in the real options world is the present value of cash flows associated with the option to wait or expand, for example, we could wait if we think technological advancements will reduce operating expenses, or the value of avoided emissions increase due to more stringent climate policy.

The last piece that might need a bit of clarification is the role of cumulative normal distribution values. In the context of real options analysis, particularly when using the Black–Scholes model to value real options (e.g., deferring an investment, expanding a project, or abandoning an operation), $N(d_1)$ and $(N(d_2)$ have specific interpretations:

- $N(d_1)$ represents the probability that the value of the underlying asset (e.g., the project value) will exceed the exercise price (e.g., the investment cost) at the time the option expires.
- $N(d_2)$ represents the probability that the option will be exercised at expiration, which in real options corresponds to the probability that the project will be worth more than the investment cost at the expiration time.

Evaluating the Option to Defer Investment

The analysis for the option to defer investment is shown in the associated spreadsheet: *"Simple Real Options Analyses."*

A company is considering investing in a renewable energy project. The project is expected to generate equal annual cash flows of $3 million over a useful life of 20 years. The company can either invest now, requiring an initial investment of $30 million, or wait for 5 years to observe how the market evolves. The cost of capital is 11%. The company wants to evaluate whether it should wait or invest now using the Black–Scholes option pricing model. The company assumes a risk-free rate of 3%, volatility of 40%, and a waiting period of 5 years to determine the value of waiting before committing to the investment.

▶   **Analysis**

Refer to Options Example Spreadsheet, for detailed calculations. First let's just evaluate the project with a conventional NPV. The NPV is negative, so the project is rejected.

$$\text{PV FCF}_0 = \frac{\$3,000,000}{(1+0.11)^1} + \frac{\$3,000,000}{(1+0.11)^2}$$

$$+ \frac{\$3,000,000}{(1+0.11)^{20}} = \$23,889,984.35$$

$$\text{NPV} = \$23,889,984.35 - \$30,000,000 = \$-6,110,015.65$$

However, what if we were to wait 5 years before undertaking the investment? It would still require a $30,000,000 investment and produce annual cash flows of $3,000,000 per year. A lot could happen in five years to make the project more attractive. Our next step is to value the option to wait.

▷ • Waiting provides an FCF stream of $3,000,000 per year in years 5 through 25. The present value of this future cashflow represents the value of the underlying asset when applying the Black-Scoles model.

$$S_0 = \text{PV FCF} = \$23,889,984.35/(1.11)^5 = \$14,177,542.94$$

• The investment in 5 years is certain and is $30,000,000. To restate in present value terms, we discount at the risk-free rate of 3%. This will be our strike price (X) when we apply the Black–Scholes formula.

$$X = \text{PV CAPEX} = 30,000,000/(1.03)^5 = \$25,878,263.53$$

• Insert values into the Black–Scholes formula to find d_1 and d_2:

$$d_1 = \frac{\ln\left(\frac{\$14,177,542.94}{\$25,878,263.53}\right)}{0.40\sqrt{5}} + \frac{0.40\sqrt{5}}{2} = -0.23$$

$$d_2 = -0.23 - 0.40\sqrt{5} = -1.12$$

• Determine $N(d_1)$ and $N(d_2)$ in excel using NORM.S.DIST:

$$N(d_1) = 0.41$$

$$N(d_2) = 0.13$$

• Calculate the value of the option to wait to invest:

$$C = \$14,177,542.94x.41 - \$25,878,263.53x.13 = \$2,424,376.61$$

There is a significant value attached to waiting! By our reckoning, the value of the option to wait as modeled by the Black–Scholes Call option formula is almost $2.5 million dollars. If the company waits five years before investing in the renewable energy project, several factors could potentially make the project more attractive. The market for renewable energy might grow due to regulatory changes or increased consumer demand for green energy, which could lead to higher revenues. Technological advancements could also improve energy output, resulting in higher cash flows. Future government policies might offer new subsidies, tax credits, or incentives, reducing costs or increasing after-tax cash flows. Additionally, stricter carbon pricing mechanisms could increase demand for renewable energy projects. The costs of setting up the project may decline over time due to technological advancements and economies of scale, and reduced financing costs could make borrowing cheaper. Lower volatility in energy markets could make future cash flows more predictable, reducing risk and increasing attractiveness. Over time, market maturity and better information could reduce uncertainty, allowing for better-informed investment decisions. Opportunities for strategic partnerships might also arise, sharing the investment burden and potentially improving returns. In other words, the future could be very rosy! The decision to wait is like buying time for these potential upsides to materialize, making the project more valuable and reducing the risks. This is why the real option to delay investing has value—because it preserves the flexibility to take advantage of better future conditions.

Evaluating the Option to Expand

The analysis for the option to expand is shown in the associated spreadsheet: *"Simple Real Options Analyses."* Westmount Steel is considering investing in a new steel manufacturing plant. The initial investment required to build the plant is $50 million, which would allow the company to produce 500,000 tons of steel annually for a value of $5 million annually. The company has the option to expand the plant's capacity by an additional 500,000 tons if the market demand for steel increases significantly. This expansion would require an additional investment of $20 million after 5 years. If the expansion is undertaken, the additional capacity is expected to generate cash flows of $8 million per year for the next 15 years, starting from year 6. The company's cost of capital is 10% and it wants to assess whether it should include the potential expansion option in its decision-making. The company uses a risk-free rate of 3% and assumes a cash flow volatility of 35% for the expansion project due to the uncertainty of future steel market conditions.

▶ **Analysis:** Should Westmount Steel undertake the project?
Refer to Options Example Spreadsheet, for detailed calculations.
First let's just evaluate the project with a conventional NPV. The present value of the FCF is:

$$\text{PV FCF}_0 = \frac{\$5,000,000}{(1+0.10)^1} + \frac{\$5,000,000}{(1+0.10)^2} + \dots$$

$$+ \frac{\$5,000,000}{(1+0.10)^{20}} = \$42,570,000$$

$$\text{NPV} = \$42,570,000 - \$50,000,000 = \$-7,430,000.$$

With a negative NPV, the project should be rejected. The C-Suite of Westmount Steel are optimists. Even though many of their competitors are taking steps to invest in projects that would curtail GHG emissions, they were a "glass-half-full" crowd and wanted to plan for future expansion and a robust steel market, in part due to demands for climate resilient infrastructure. So, they decided to value the flexibility of expanding using the Black–Scholes Option Pricing model and valuing the real call option to expand. The value of the Call is just over $22,000,000! The complete calculations are shown in the accompanying spreadsheet. Westmount Steel decided to go ahead with the investment of $50,000,000 today, feeling quite confident that the option to expand in five years would turn a negative into a positive, although in our example it does seem like a bit of a stretch: there is a lot of negative NPV to climb over even with the valuable real option!

Evaluating the Option to Abandon

The analysis for the option to abandon is shown in the associated spreadsheet: "*Simple Real Options Analyses.*"

And now: introducing the "glass-half-empty" team at PetroPanic Inc. It is probably understandable that working in fossil fuel extraction, with what seemed like daily assaults on their business strategy, eventually things might catch up with them. So, it was company policy (although not publicly disclosed) that all new oil and gas expansion analysis would include an "exit" strategy.

PetroPanic Inc. is evaluating a new drilling project in Western Canada. The initial investment required to start the drilling operations is estimated to be $80 million, covering exploration, drilling, and initial infrastructure setup. The project is expected to generate $10 million per year in cash flows for 10 years,[11] starting one year after the initial investment. The company uses a discount rate of 12% for evaluating this project.

Given the uncertainties in the oil market, PetroPanic Inc. has built in an exit strategy: after 5 years, the company estimates that it could abandon the project and sell the drilling equipment and infrastructure for $20 million. However, if the company abandons the project, it forfeits the remaining cash flows

---

[11] Oil wells often experience a steep decline in production rates after the first few years. This means that while production can continue for many years, the economic value decreases significantly. In such cases, companies may use a shorter useful life for their economic analysis to reflect the period during which most of the profitable production occurs.

from years 6 to 10. To evaluate the option to abandon, the company assumes a volatility of 30% due to fluctuations in oil prices and market conditions, and a risk-free rate of 3%.

▶ The option to abandon is a Put option. Using the simplified Black–Scholes model that assumes discrete compounding, we can express the Put Value as:

$$\text{Put Value} = X * N(-d_2) - S * N(-d_1)$$

where S is the present value of forfeited cash flows, X is the salvage value or the price at which the project can be abandoned and the equipment sold; and all other variables are as previously defined. Intuitively, we can think of $N(-d_2)$ as the probability that the project will be abandoned when it is optimal to do so; and $N(-d_1)$ adjusts this probability by considering the drift and risk characteristics of the underlying project over time. The negative signs in $-d_2$ and $-d_1$ indicate that we are dealing with a put option, which benefits from a decrease in the value of the underlying asset. Plugging values into the formula for Put Value we have:

$$\text{Put Value} = 17.25 * 0.53 - 20.45 * 0.27 = \$3.5 \text{ million}$$

The option to abandon adds some strategic value by providing a safety net, but it is not enough to make the project financially viable. In the accompanying spreadsheet, you can see that the NPV of the project without consideration of the option to abandon is $-23.50 million. PetroPanic Inc. should consider avoiding this investment to prevent potential losses. Instead, they could look for other opportunities with a more favorable risk-reward balance or wait for better market conditions before proceeding with similar projects.

Summing Up

After examining the applications of Levelized Cost of Electricity (LCOE), Net Present Value (NPV), Real Options Analysis (ROA), and Shadow Pricing, it is important to note that these tools are interconnected. As climate change influences investment landscapes, these techniques must adapt to address the challenges and opportunities of a low-carbon economy. The evolving use of LCOE, NPV, shadow pricing, and ROA highlights the need for dynamic financial analysis in a carbon-constrained world. As firms align with net-zero goals, these tools will guide investment decisions and influence the path of sustainable growth. With this perspective, we can now turn to the chapter's conclusion to summarize the key points and implications for capital budgeting in the context of climate change.

Market Analysis
Revenues

- The US thermal coal market, especially in the southwestern region, is likely to shrink as more utilities transition to solar, wind, and natural gas. Renewable energy, particularly in regions like California and Texas, is expected to dominate power generation.
- According to the IEA's World Energy Outlook 2024, US coal demand is projected to decline significantly, with domestic consumption expected to fall by 80% by 2035 in the IEA's *Stated Policies Scenario* (STEPS) (International Energy Agency, 2024).
- Assuming the price of thermal coal remains volatile and may average around $40 to $50 per ton over the next decade, the mine's annual revenue would be approximately $85.6 million to $107 million (2.14 million tons x $40 to $50 per ton).
- This revenue stream could shrink as more coal plants shut down or convert to natural gas. Utilities in the region are investing heavily in renewable energy, leading to decreasing reliance on thermal coal. This market shrinkage might reduce projected revenues by 50% or more within the next decade. The team decided to examine a scenario where revenues decreased by 50% within the next decade.

Costs

- The team originally considered annual operating costs of $50 million but given the upgrade of mine facilities perhaps this estimate was too high. They decided to set operating costs at $30 million. They were well aware, however, that costs could be much higher if emissions were priced.

The Granite Peak Mine was re-evaluated based on an assumed decline in revenue of 50% and constant annual costs of $30 million. Table 11.7 displays the NPV of the mine upgrade at different discount rates. A higher rate, 12%, was also evaluated as WMC might well face rising financing costs given the volatile coal market and the many risks, chief among them perhaps being regulatory!

The WMC analysts launched into a heated discussion regarding the newest results. True enough, the mine expansion project remained profitable even when revenues were slashed in half but there were so many implicit assumptions underpinning the NPV calculations. Revenues declined by 50% through a combination of lower prices and reduced output. Was it reasonable to assume that operating expenses could remain constant at $30 million annually—especially when the initial estimate placed

Table 11.7 NPV of Granite Peak Mine Expansion (Revenue Decline of 50%)[12]

Discount Rate	PV FCF	NPV
6%	$899.62	$369.62
8%	$783.46	$253.46
10%	$690.47	$160.47
12%	$615.04	$85.04

Table 11.8 Revenues Decline by 50% and Operating Expenses = $50 million

Discount Rate	PV FCF	NPV
6%	$718.40	$188.40
8%	$628.34	$98.34
10%	$555.96	$25.96
12%	$497.02	($32.98)

operating expenses at $50 million? The team evaluated the project NPV incorporating many different combinations of costs and revenues. However, they decided to present management with only a few of these: the initial baseline calculations, the scenario accounting for shadow carbon prices, and one more (for now) that assumed revenues fall by 50% over the next decade and costs stay at $50 million (presumably accounting for some real costs to emissions over the next decade). Table 11.8 displays the results for this new set of NPVs.

The newest set of calculations illustrated how vulnerable mine profitability was in the context of climate crisis! Perhaps it was better to delay the investment for a few years while WMC considered other alternatives such as green energy investment. This would give WMC some wiggle room to perhaps even find a way to unload the negative salvage value of the mine investment.

Conclusions

This chapter has explored how capital budgeting in the era of climate change requires a transformation in both mindset and methodology. Climate-related risks are now central to financial decision-making, meaning that firms must not only consider traditional metrics of profitability but also account for complex variables like carbon pricing, regulatory shifts, and physical climate risks.

The techniques covered—from Levelized Cost of Electricity (LCOE) to Real Options Analysis (ROA) and shadow carbon pricing—demonstrate how to quantify and incorporate these climate impacts into financial models. For instance, LCOE allows firms to compare energy projects on a lifecycle basis,

---

[12] See the Granite Peak Mine spreadsheet.

revealing the long-term cost advantages of green energy projects relative to fossil fuels. Shadow carbon pricing, by assigning a cost to emissions, enables firms to anticipate and incorporate future carbon costs, aligning investment decisions with possible regulatory and market changes. These tools represent critical advancements for any firm grappling with emissions-intensive projects, especially those in high-carbon sectors like coal, oil, or heavy industry.

We also emphasized the importance of ROA (real options analysis) for investment under uncertainty framework particularly suited to climate-sensitive projects. Whether deciding on a wind farm expansion, a carbon capture upgrade, or a pivot to green hydrogen, ROA enables firms to assign value to flexibility, providing options to defer, expand, or abandon projects as the landscape evolves. For example, an abandonment option for a coal mine or an expansion option for a green energy project allows companies to adapt based on market and policy shifts, thus enhancing the resilience and viability of their portfolios.

As this chapter concludes, the key takeaway is clear: effective climate risk integration into capital budgeting is a cornerstone of strategic resilience. As regulatory, environmental, and market pressures intensify, firms that incorporate these frameworks will be better positioned not only to protect value but to thrive in a low-carbon economy.

Key Takeaways

1. **Advanced Capital Budgeting for Climate Crisis**
   The chapter highlights the importance of adapting traditional capital budgeting techniques to integrate climate-related risks and opportunities.
   Tools like Levelized Cost of Electricity (LCOE), Net Present Value (NPV), Shadow Pricing, and Real Options Analysis (ROA) are explored.
2. **Tools for Climate-Sensitive Investment Decisions**
   LCOE: Provides a lifecycle cost framework to compare renewable and fossil fuel projects.
   NPV Adjustments: Incorporates climate uncertainties, regulatory impacts, and cash flow variability.
   Shadow Pricing: Assigns a cost to emissions, helping align investments with carbon pricing mechanisms.
   ROA: Values flexibility in decision-making, such as delaying, expanding, or abandoning projects based on changing market or policy conditions.
3. **Case Studies**
   Evergreen Wind Farm:
   • Illustrates the application of NPV and LCOE in evaluating a renewable energy expansion.

- Emphasizes factors like depreciation, operational costs, and capacity factors in financial viability.

Granite Peak Mine:

- Evaluates the economic and environmental trade-offs of continuing coal operations versus transitioning to battery storage.
- Highlights the impact of carbon pricing on long-term project sustainability.

4. **Climate Change Uncertainties**

   Explores how scientific, modeling, and socioeconomic uncertainties impact project evaluation.

   Stresses the need for dynamic models that factor in evolving climate risks, such as disruptions in natural carbon sinks.

5. **Transition Risks and Opportunities**

   High-carbon sectors face increasing regulatory and market risks, such as stranded assets and declining demand.

   Emphasis on aligning projects with decarbonization goals to mitigate risks and capture growth opportunities.

6. **Real Options Analysis (ROA)**

   A sophisticated tool for valuing flexibility in uncertain environments.

   Provides strategic insights for adapting projects to market changes, regulatory shifts, and technological advances.

7. **Integration of Sustainability Goals**

   The chapter underscores the growing need to align corporate investments with environmental, social, and governance (ESG) objectives.

   Encourages the use of shadow pricing and LCOE to support sustainability-oriented decision-making.

8. **Limitations of Traditional Methods**

   Techniques like Payback Period and IRR may fail to capture the complexities of long-term, climate-sensitive projects.

   Focus shifts toward NPV and ROA for comprehensive financial analysis.

9. **Strategic Implications**

   Firms must embed climate resilience into financial models to remain competitive.

   Leveraging these advanced tools can help firms adapt to a low-carbon economy and align with regulatory frameworks.

Questions and Problems

1. What is the significance of the Levelized Cost of Electricity (LCOE) in comparing renewable and non-renewable energy projects?
2. Explain how shadow pricing can be used to integrate carbon costs into capital budgeting decisions.
3. Why is Real Options Analysis (ROA) particularly relevant in the context of climate-sensitive investments?
4. Using the Evergreen Wind Farm example, calculate the annual revenue per turbine and the total annual revenue for ten turbines.

5. Discuss the limitations of the Payback Period and IRR methods in the context of climate-related projects.
6. In the Granite Peak Mine case, how does incorporating a shadow carbon price impact the project's financial viability?
7. Describe how the Black–Scholes model is adapted for evaluating real options in capital budgeting.
8. What are the potential benefits of delaying an investment in a renewable energy project as illustrated in the chapter?
9. Explain how firms can determine an appropriate shadow carbon price and its implications for investment decisions.
10. Summarize the key strategic takeaways for firms integrating climate risks into their capital budgeting processes.

Problems

Problem 1: SAF

Gold Wings Airline is one of the largest passenger carriers in the world. Head-quartered in Germany, Gold Wings is subject to emissions regulations in the EU and in particular with CORSIA now poised for implementation, Gold Wings worries about the costs of aviation emissions. The EUETS allowances are currently trading around €90 while Offsets in the Voluntary Carbon Market are around €20.00. Gold Wings has a number of alternative transition pathways, one of which is switching from conventional jet fuel to SAF (Sustainable Airline Fuel). In fact, Gold Wings is about to sign a partnership agreement with SAF Finland to produce SAF for Gold Wings.

So far, the company has gathered the following information that they think is relevant for analysis of SAF investment. Annual revenue in year 1 of operation is expected to be €307,530.75. Their analysis suggests that revenues could grow at 1% annually by increasing load factor, which currently stands at 80%. This would allow Gold Wings to increase revenue without incurring additional fuel consumption. Therefore, the company expects that their fuel consumption over the life of the project will be a steady 10.6 billion of liters of fuel.

Research shows that emissions associated with conventional fuel are roughly 2.5 × fuel used. As for SAF, emissions are roughly 0.5 × fuel used. Switching from 100% conventional jet fuel to 50% SAF within 10 years is a stated objective of Gold Wings' Sustainability plan. In the first year of operation, Goldwings will use a blend of 10% SAF and 90% conventional fuel. The percentage of SAF in the fuel mix will increase linearly such that by year 10 the 50% target will be attained.

Goldwings needs to analyze whether the savings in emissions offset the cost of SAF fuel. Although initially the price of SAF is steep (€1.75) compared to price of conventional fuel (€0.65) the company decided that as SAF becomes more available (due to production at scale and technological advancements) the cost will drop: in the first 10 years they assume that the price of SAF will fall by 3% every year, followed by a decline of 5% annually throughout the remainder of the useful life of the project. Current estimates suggest that the capital expenditure necessary to get the project off the ground is €250 million with a salvage value expected at the end of the project of €25 million. Goldwings is subject to a corporate tax rate of 30%. The firm's cost of capital is 8%. Is it feasible for Goldwings to set this ambitious target for SAF adoption?

Problem 2: Coal Mine Redevelopment

Granite Peak Mine Battery Storage
The example in this chapter assumed that Granite Peak Mine faces the risk of expensive clean-up due to asset stranding. WMC, the parent company, is looking at redeveloping the mine as a battery storage facility, a strategy that has been deployed throughout the UK. To determine if this is a viable option, the team at WMC undertook an analysis of the relevant costs and revenues associated with the project.

i. Initial capital expenditure of $400 million is necessary to convert the mine into a battery storage facility. This cost includes the cost of batteries, installation, grid connection, and other infrastructure. Straight line depreciation is assumed.

ii. Battery storage generates revenue primarily by charging during periods of low electricity prices (off-peak) and discharging during periods of high prices (peak). Revenue will therefore depend on the price arbitrage between off-peak and peak times. The team at WMC assumes the battery storage system has a capacity of 100 MW with 4 hours of storage per cycle—400 MWh total capacity. If the system operates 365 cycles per year which assumes one cycle per day, the annual throughput will be 146,000 MWh. Assuming an average price differential of $50 per MWh between charging and discharging, potential annual revenues from the redevelopment project are $7.3 million (146,000 x $50/MWh).

iii. Ongoing operating and maintenance costs for a battery storage facility tend to be lower than for a coal mine but still include maintenance of the battery systems and grid connection fees. WMC assumes annual operating costs of $1 million.

iv. Useful Life and Salvage Value: The WMC team assumes a useful life of 20 years for the battery storage system. This reflects the typical lifespan of battery systems before replacement or major upgrades are required. To keep things simple, the WMC team assumes zero salvage value but acknowledges that modifications could be required to cover decommissioning costs.

v. To calculate the NPV of the Granite Peak Mine redevelopment project, the analysts decided to use a discount rate of 6% which reflects the lower cost of capital associated with green projects.

vi. For simplicity, we are assuming that an initial investment of $1 million dollars in working capital is required which is recovered at the end of the useful life of the project.

Based on the information for Granite Peak mine in this chapter, and the additional information provided in this problem, calculate the NPV of the redevelopment project to convert the mine into a battery storage facility. Assume a tax rate of 21%. Compare your answer to Table 11.7. What are the most important factors that WMC should consider in evaluating the redevelopment project?

Problem 3: Runway Adaptation to Extreme Heat

An airport is considering an investment in upgrading its runway materials to withstand extreme heat events. The cost of upgrading the runway materials is $10 million. Predictive models indicate that there is a 30% chance of experiencing heatwaves exceeding 45 °C (113 °F) in the next five years, which would necessitate immediate runway repairs costing $15 million if not upgraded. Alternatively, if the runway is upgraded now, it will prevent the need for repairs regardless of the temperature. The risk-free rate of interest is 5% and the airport's cost of capital is 7%.

Should the upgrade be carried out today or should the airport wait? What is the ROA (Real Option Value) of waiting? What qualitative factors can you bring to your decision to wait versus repairing immediately?

References

Bilal, A., & Känzig, D. R. (2024). The macroeconomic impact of climate change: Global vs. local temperature (NBER Working Paper No. 32450). National Bureau of Economic Research. Accessed November 4, 2024, from: https://www.nber.org/papers/w32450

BloombergNEF. (2024, May 1). *EU ETS market outlook 1H 2024: Prices valley before rally.* BloombergNEF. Accessed November 4, 2024, from: https://about.bnef.com/blog/eu-ets-market-outlook-1h-2024-prices-valley-before-rally/

BloombergNEF. (2022, May 1). *Carbon prices can both help and hinder a path to net-zero.* BloombergNEF. Accessed November 4, 2024, from: https://about.bnef.com/blog/carbon-prices-can-both-help-and-hinder-a-path-to-net-zero/

Carbon Tracker Initiative. (2017). *Powering down coal: Navigating the economic and financial risks in the last years of coal power.* Accessed November 4, 2024, from: https://carbontracker.org/reports/coal-portal/

CDP. (2021). *Putting a price on carbon: Integrating climate risk into business planning.* Accessed November 4, 2024, from: https://cdn.cdp.net/cdp-production/cms/reports/documents/000/005/651/original/CDP_Global_Carbon_Price_report_2021.pdf

Cutler, Z., & Linder, S. (2023, August 15). *Capital projects are critical for a green future.* McKinsey & Company. Accessed November 4, 2024, from: https://www.mckinsey.com/industries/travel-logistics-and-infrastructure/how-we-help-clients/global-infrastructure-initiative/voices/capital-projects-are-critical-for-a-green-future

Fayolle, V. (2022, March 1). *Climate adaptation: Identifying the investment opportunity.* Mott MacDonald. Accessed November 4, 2024, from: https://www.mottmac.com/en/insights/climate-adaptation-changing-the-narrative-from-cost-benefit-ratio-to-investment-opportunity/

International Energy Agency. (2020, December 9). *Levelised cost of electricity calculator.* Accessed November 4, 2024, from: https://www.iea.org/data-and-statistics/data-tools/levelised-cost-of-electricity-calculator

International Energy Agency. (2024). *World energy outlook 2024*. IEA. Accessed November 4, 2024, from https://www.iea.org/reports/world-energy-outlook-2024 (Licence: CC BY 4.0 [report]; CC BY NC SA 4.0 [Annex A]).

Intergovernmental Panel on Climate Change. (2021). Chapter 7: The Earth's energy budget, climate feedbacks, and climate sensitivity. In V. Masson-Delmotte, P. Zhai, A. Pirani et al. (Eds.), *Climate change 2021: The physical science basis*. Contribution of Working Group I to the Sixth Assessment Report of the Intergovernmental Panel on Climate Change. Cambridge University Press. https://www.ipcc.ch/report/ar6/wg1/

Intergovernmental Panel on Climate Change. (2022). Chapter 3: Mitigation pathways compatible with long-term goals. In P. R. Shukla, J. Skea, R. Slade et al. (Eds.), *Climate change 2022: Mitigation of climate change*. Contribution of Working Group III to the Sixth Assessment Report of the Intergovernmental Panel on Climate Change. Cambridge University Press. https://www.ipcc.ch/report/ar6/wg3/

Intergovernmental Panel on Climate Change. (2021). Chapter 4: Future global climate: Scenario-based projections and near-term information. In V. Masson-Delmotte, P. Zhai, A. Pirani et al. (Eds.), *Climate change 2021: The physical science basis*. Contribution of Working Group I to the Sixth Assessment Report of the Intergovernmental Panel on Climate Change. Cambridge University Press. https://www.ipcc.ch/report/ar6/wg1/

Le Quéré, C., Andrew, R. W., Canadell, J. G., Sitch, S., Korsbakken, J. I., Peters, G. P., Friedlingstein, P., et al. 'Global Carbon Budget 2016.' *Earth System Science Data, 8*(2) (2016), 605–649.

Moore, F. C., & Diaz, D. B. (2015). Temperature impacts on economic growth warrant stringent mitigation policy. *Nature Climate Change, 5*(2), 127–131. https://doi.org/10.1038/nclimate2481

National Centers for Environmental Information (NCEI). (n.d.). *Quantifying the ocean carbon sink*. National Oceanic and Atmospheric Administration (NOAA). Retrieved February 24, 2025, from: https://www.ncei.noaa.gov/news/quantifying-ocean-carbon-sink

Reuters. (2023, December 10). *Insight: No global carbon price? Some companies set their own*. Reuters. Accessed November 4, 2024 from: https://www.reuters.com/sustainability/no-global-carbon-price-some-companies-set-their-own-2023-12-10/

The Guardian. (2024, October 14). *Nature's carbon sink collapse raises doubts over global heating models and emissions targets*. The Guardian. Retrieved February 24, 2024, from: https://www.theguardian.com/environment/2024/oct/14/nature-carbon-sink-collapse-global-heating-models-emissions-targets-evidence-aoe

World Bank. (2024). *Guidance note on shadow price of carbon in economic analysis (English)*. World Bank Group. Accessed November 4, 2024, from http://documents.worldbank.org/curated/en/099553203142424068/IDU1c94753bb1819e14c781831215580060675b1

Putting a Price on Carbon

Introduction

Building on Chapter 10's exploration of green technologies and their financial strategies as well as the valuation of these projects in the previous chapter, we turn now to examine carbon markets as a critical tool for mitigating greenhouse gas emissions and financing sustainable innovation. Carbon markets address the market failure identified by Nicholas Stern, where the external costs of greenhouse gas emissions are not borne by those responsible, leading to insufficient incentives for reduction (Stern, 2007). By placing a price on carbon, these markets encourage firms to adopt sustainable practices, invest in green technologies, and integrate emissions management into their operational and financial strategies.

The chapter focuses on two primary categories of carbon markets: compliance systems like the EU Emissions Trading System (EU ETS) and voluntary initiatives supporting emissions-reducing projects. These mechanisms are analyzed for their ability to influence corporate decision-making, including investment planning, cost management, and risk mitigation. For example, compliance markets impose mandatory obligations with market-driven flexibility, while voluntary markets allow firms to align offset purchases with broader sustainability goals.

This chapter also considers the broader implications of carbon pricing, including its role in addressing corporate financial risks and opportunities. It explores how firms balance operational costs, regulatory compliance, and the need to innovate in response to evolving carbon market dynamics. By connecting these elements, the chapter provides a comprehensive view of how carbon markets serve as a bridge between policy goals and corporate strategies in the transition to a low-carbon economy.

S. Dow and Y. Shi, *Corporate Finance Under Climate Crisis*,
https://doi.org/10.1007/978-3-031-83487-5_12

12.1 The Role of Carbon Markets in Corporate Finance

Carbon prices are shaped by regulatory frameworks, energy market dynamics, macroeconomic trends, and financial market behavior. These factors collectively determine the supply and demand for emissions allowances, the tradable permits central to cap-and-trade systems worldwide. While carbon pricing serves as a critical tool for incentivizing decarbonization, it also introduces price volatility, complicating financial planning and investment strategies within the corporate sector.

Carbon markets play a pivotal role in corporate finance by addressing key challenges faced by businesses in a transitioning economy. For carbon-intensive industries, these markets provide a structured way to internalize the cost of emissions, aligning financial decision-making with climate goals. This can help offset the higher cost of capital often associated with carbon-intensive industries, which are viewed as riskier by investors due to regulatory and reputational concerns.

For green industries and emissions-intensive sectors investing in low-carbon projects, participation in carbon markets can improve access to financing. By monetizing emissions reductions through the sale of allowances or credits, companies can generate additional revenue to fund their sustainability initiatives. Moreover, consistent carbon pricing offers a predictable regulatory framework, enabling firms to plan long-term capital expenditures with greater confidence. For example, sectors such as steel and cement can justify substantial investments in low-carbon technologies, such as hydrogen-based production, when sustained carbon price signals are anticipated.

However, price volatility in carbon markets remains a significant barrier, particularly for long-term green infrastructure projects. Carbon prices are subject to various influences, including regulatory shifts, economic fluctuations, and geopolitical disruptions, which can lead to sudden price swings. This complicates financial planning, increases project risk, and deters investment, especially for projects with extended payback periods, such as renewable energy installations or hydrogen production facilities (Wu et al., 2024).

12.2 Carbon Pricing Mechanisms

As of April 2023, there were **seventy-three carbon pricing mechanisms** operating worldwide covering approximately **23% of global emissions**. These mechanisms include carbon taxes and compliance carbon markets (CCMs) but do not count voluntary carbon markets (VCMs) nor hybrid markets that combine elements of both compliance and voluntary markets. Figure 12.1 illustrates the growth in implementation of these mechanisms since 1990 when Finland became the first country to impose a carbon tax. While it might seem that globally we are well on the way to pricing emissions, there is in fact tremendous need for price evolution in these systems. The World Bank reports that in 2023, despite having added five more mechanisms due to expansion of existing programs (from 68 to 73) the carbon

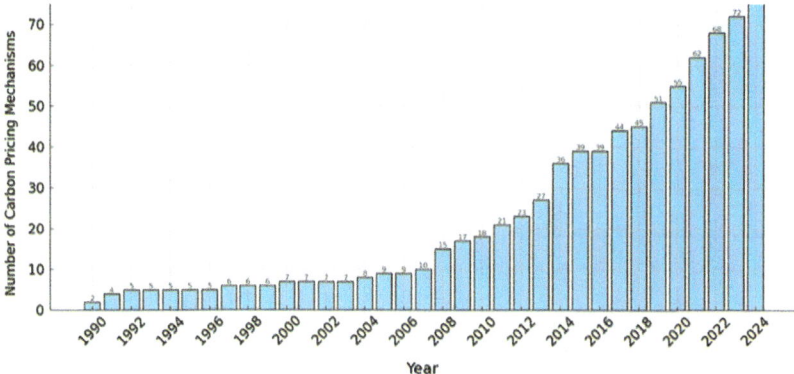

Fig. 12.1 Number of Carbon pricing mechanisms in operation worldwide (1990–2024) (*Source* World Bank [2024])

price associated with these mechanisms is far below what is required by 2030 to incentivize meaningful GHG reductions. Less than 5% of taxes and CCMs had carbon prices consistent with carbon neutrality by 2050. If we look at the pace of climate degradation, the cost of carbon management via carbon markets or taxes must increase.

Carbon Taxes

While this chapter focuses on carbon markets as a primary tool for reducing greenhouse gas emissions, it is important to note the role of carbon taxes as an alternative approach. A carbon tax places a direct price on each ton of carbon dioxide or other greenhouse gas emitted, requiring companies and individuals to pay based on their actual emissions. The economic logic is straightforward: a higher tax incentivizes emissions reductions through efficiency improvements, innovation, or substitution with low-carbon alternatives. Figure 12.2 illustrates carbon tax rates around the world.

The effectiveness of a carbon tax is closely tied to the tax rate. For example, in 2024, carbon tax rates ranged from less than $1.00 per ton of CO_2 in Poland to $167.00 per ton in Uruguay (see Fig. 12.2). While Uruguay's high tax rate provides a strong economic incentive for firms to reduce emissions and invest in mitigation technologies, Poland's negligible tax creates little motivation for such actions.

Carbon Markets

Carbon markets are mechanisms designed to reduce greenhouse gas (GHG) emissions by introducing economic incentives. These markets operate under two main

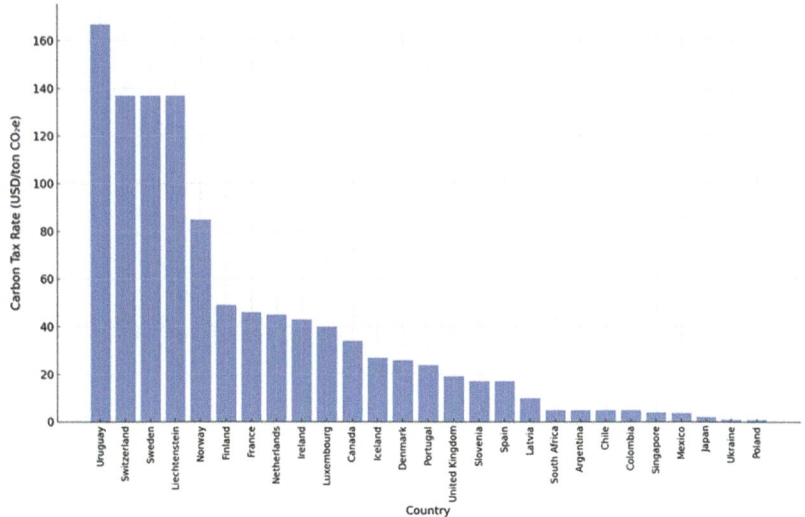

Fig. 12.2 Carbon Tax Rates Worldwide by Country (as of April 2024) (*Source* World Bank [2024])

categories: compliance carbon markets, where participation is legally mandated, and voluntary markets, where firms or entities engage for non-regulatory reasons. Compliance carbon markets play a pivotal role in achieving emissions reduction targets under frameworks like the Paris Agreement. The EU Emissions Trading System (EU ETS) exemplifies a robust and mature compliance carbon market. As we will see as we work through this chapter, prices vary considerably across time and between compliance carbon markets and voluntary carbon markets. Figure 12.3 displays the geographic distribution of compliance carbon markets and the percentage of carbon market transactions covered by each as of 2022.

Fig. 12.3 Distribution of Carbon Market Size Worldwide (*Source* World Bank [2024])

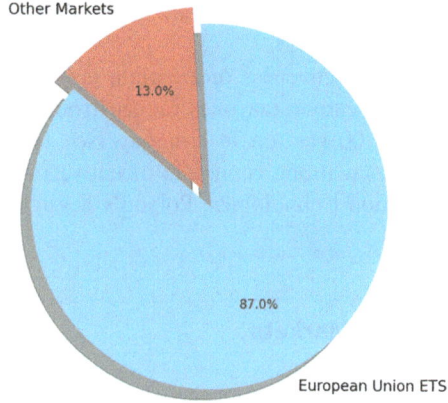

While Fig. 12.3 underscores the dominance of mature markets like the EU ETS, significant momentum is building in middle-income economies. The expansion of carbon pricing in middle-income countries is vital for increasing global emissions coverage, currently at 24% of global GHG emissions. Successful implementation in these regions could add approximately 3% to global coverage, moving closer to the Global Carbon Pricing Challenge target of 60% by 2030 (World Bank, 2024).

To illustrate, countries like Brazil and India are moving toward implementation of their own Emissions Trading Systems. Brazil has proposed legislation to establish an emissions trading system (ETS) as part of its broader climate agenda. The proposed system is expected to target energy-intensive industries, supporting Brazil's efforts to meet its Nationally Determined Contributions (NDCs) under the Paris Agreement. In India, the government is evolving its Perform, Achieve, and Trade (PAT) program into an intensity-based ETS aimed at reducing emissions in industries such as steel and cement. This initiative aligns with India's commitment to achieving net-zero emissions by 2070.

12.3   Sectoral Expansion of Carbon Pricing Mechanisms

Carbon pricing is extending its geographic coverage, but as we see in this section, it is also expanding into previously underregulated sectors. Traditionally focused on the power and industrial sectors, mechanisms like emissions trading systems (ETSs) and carbon taxes are increasingly being applied to other areas, including maritime transport, aviation, waste management, and building emissions. This broader application reflects the growing recognition that achieving global climate goals requires addressing emissions across all parts of the economy.

In **maritime transport**, the European Union's inclusion of shipping emissions within its ETS in 2024 is a landmark development. This move obliges shipping companies to pay for their greenhouse gas emissions, creating incentives for cleaner fuels and more efficient operational practices. Other jurisdictions such as Japan and Singapore exploring similar measures.

The **aviation secto**r is also experiencing increased regulation through carbon pricing. The Carbon Offsetting and Reduction Scheme for International Aviation (CORSIA), established by the International Civil Aviation Organization (ICAO), requires airlines to offset emissions exceeding 2019 levels for international flights. Additionally, the EU ETS, which already covers aviation, is being strengthened through the gradual removal of free allowances for airlines, further promoting the adoption of cleaner technologies and offsetting practices.

Waste management is another area where carbon pricing mechanisms are making inroads. Methane emissions from landfills, a major contributor to climate change, are now subject to pricing under systems like California's cap-and-trade program and New Zealand's ETS. By attaching a cost to emissions, these policies incentivize better waste handling practices, such as recycling and methane capture, helping to mitigate emissions from this historically underregulated sector.

Similarly, the **building and heating sectors** are being targeted by carbon pricing in countries like Germany, where a national pricing system for fuel use in heating has been introduced. By creating financial incentives to adopt energy-efficient technologies and renewable heating solutions, these policies are driving the transition toward more sustainable practices in residential and commercial buildings.

12.4   Carbon Border Adjustment Mechanism: A Bridge Across Geographies and Sectors

The **Carbon Border Adjustment Mechanism (CBAM)** exemplifies the growing geographic and sectoral reach of carbon pricing mechanisms. It bridges the gap between highly regulated domestic markets, like the European Union (EU), and regions with less stringent or absent carbon pricing policies. By imposing a carbon price on imports, CBAM seeks to address the issue of carbon leakage—where industries relocate production to jurisdictions with lower regulatory standards—and levels the playing field for domestic producers. At the same time, CBAM targets carbon-intensive sectors, ensuring that industries such as cement, steel, aluminum, and fertilizers begin to decarbonize regardless of where they operate.

CBAM integrates geography and sectors by linking global trade flows with decarbonization efforts. The mechanism requires importers to account for the embedded carbon emissions in products brought into the EU, effectively expanding the reach of the EU's Emissions Trading System (ETS) beyond its borders. The initial phase, which began in October 2023, focuses on monitoring and reporting emissions, laying the groundwork for full implementation in 2026. At that stage, importers will need to purchase CBAM certificates equivalent to the carbon price faced by domestic producers under the ETS.

This dual focus on geography and sectors ensures that decarbonization pressures are not fragmented across countries. However, CBAM also highlights the risks of uneven implementation. If mechanisms like CBAM proliferate independently across jurisdictions, they could create a patchwork of regulations that harms the competitiveness of industries in highly regulated markets while providing loopholes for less regulated ones. This underscores the importance of global coordination to align carbon pricing efforts and avoid fragmented systems that impose unnecessary costs on industries and consumers.

The international implications of CBAM extend beyond its immediate application to EU imports. Major trading partners, including countries like India, Türkiye, and Indonesia, are responding by developing their own carbon pricing mechanisms to minimize the costs of compliance with CBAM and to retain revenue domestically rather than paying it to the EU. Similarly, advanced economies such as Canada, Japan, and the United Kingdom are exploring their own border adjustment mechanisms, inspired by the EU's approach. These efforts suggest that CBAM could catalyze a broader alignment of carbon pricing policies, driving both geographic and sectoral integration on a global scale.

At the sectoral level, CBAM's initial scope targets some of the most carbon-intensive industries, such as steel and cement, but the EU has signaled its intention to expand this coverage to include additional sectors in the future. By incorporating emissions from internationally traded goods, CBAM directly incentivizes cleaner production processes across global supply chains. This influence is particularly significant for industries operating in countries without established carbon pricing frameworks, as they face growing pressure to align with the EU's standards to remain competitive in the European market.

For developing economies, CBAM presents both challenges and opportunities. On the one hand, it increases the regulatory burden on exporters, particularly those with limited capacity to monitor and report emissions. On the other hand, it could encourage investments in low-carbon technologies and attract financial support from international initiatives aimed at helping these economies adapt to stricter climate policies. To balance these dynamics, the EU and other CBAM adopters will need to ensure that such mechanisms do not disproportionately disadvantage developing countries or create new trade barriers.

Summing Up

CBAM represents a significant step in the evolution of carbon pricing, demonstrating how geographic and sectoral dimensions can converge to create a unified approach to decarbonization. By extending the reach of carbon pricing across borders and industries, CBAM underscores the potential for innovative mechanisms to drive global climate action. However, its success will depend on the extent to which it fosters cooperation rather than fragmentation, ensuring that industries worldwide can transition toward a low-carbon future without undue competitive distortions.

CBAM is deeply tied to the EU's broader climate policy framework, anchored by the **European Union Emissions Trading System (EU ETS)**. The EU ETS, as a "pure" compliance market, provides the foundation for CBAM and demonstrates how structured carbon pricing mechanisms can drive decarbonization across geographies and sectors.

12.5   The European Union Emissions Trading System (EU ETS)

As we saw in Fig. 12.3 the EU ETS is the world's largest and most influential compliance carbon market, covering approximately 40% of the EU's greenhouse gas emissions. Launched in 2005, it operates as a cap-and-trade system, setting a fixed limit on emissions that decreases annually. Firms within the regulated sectors must either reduce their emissions or purchase allowances (EUAs) to comply. The flexibility of trading allowances ensures that emissions reductions are achieved cost-effectively, minimizing economic disruptions. The system also permits the banking of allowances, enabling firms to save unused allowances for future use, which encourages early investments in emissions reductions. The enforceable accountability provided by the EU ETS ensures that firms are bound to reduce their emissions, driving significant environmental progress while maintaining economic stability.

How the EU ETS Works

Allowances

Under the EU ETS, a cap is set on the total amount of certain GHGs that can be emitted by regulated entities. This cap is reduced annually to align with emissions reduction targets. EU Allowances (EUAs)—each representing the right to emit one ton of CO_2 equivalent, are acquired either through free allocation or through a government auction. Free allocations of allowances are distributed to some sectors, particularly to those sectors that would experience deterioration in their competitive landscape if the costs of emissions were too high. Otherwise, most EUAs are distributed through government auctions where prices are determined by supply and demand reflecting real-time market conditions. Once distributed, EUAs can be traded in the secondary market. Firms with excess allowance may sell them, while those needing additional allowances purchase them to meet compliance obligations. Firms can also bank their allowances, meaning that the firm can save unused allowances for future years if they are not needed immediately. Permitting banking of allowances encourages early emissions reductions. However, it is a one-sided bank as borrowing is not permitted.

Compliance Process

Regulated firms must measure and report their emissions annually, verified by independent auditors. This process is known as: Monitoring, Reporting, and Verification (MRV). By April of the following year, firms must surrender EUAs equivalent to their emissions. Shortfalls require firms to acquire allowances from the market, while surpluses can be banked or sold.

Price Volatility in the EU ETS

As we noted at the beginning of this chapter, carbon pricing provides firms with a number of benefits that could reduce their cost of capital and encourage mitigation of GHG emissions through green investment. However, in order for carbon markets to perform their function of delivering reliable price signals to market participants carbon prices need to be stable. The evolution of EUA prices, however, has been anything but stable. EUA prices have gone through significant highs and lows, at one point in 2007, (during Phase 1 of the EU ETS), being essentially worthless.

Fast forward to the present day, and the EUA market has matured significantly. Compared to the first phase of the EU ETS (2005–2007), prices now reflect a tighter cap, improved market mechanisms, and increasing demand from compliance entities and financial participants. However, volatility persists, driven by a mix of regulatory changes, economic shocks, and speculative trading. According to the European Central Bank (ECB), the price of EUAs more than tripled in just two years, rising from approximately €25 per ton in mid-2020 to over €90 per ton by early 2022 (European Central Bank, 2022). This dramatic price movement was fueled by expectations of stricter caps under the European Green Deal and disruptions caused by the Russia-Ukraine conflict, which reshaped energy markets and drove up demand for allowances.

Price Drivers of EUAs

The price of EUAs is influenced by a complex interplay of economic, regulatory, and market-based factors. Understanding these drivers is critical for both policymakers and market participants, as EUA prices directly affect the cost of compliance and shape investment decisions toward decarbonization.

Regulatory Frameworks

Carbon pricing systems are fundamentally policy-driven, with regulatory changes acting as a primary determinant of price levels. Key factors include the setting of emissions caps, the allocation of allowances (free or auctioned), and mechanisms designed to stabilize the market. For instance, many systems include tools to address allowance oversupply, such as automatic adjustments to the allowance pool. These measures aim to prevent prolonged price collapses while maintaining sufficient price signals to incentivize emissions reductions. However, regulatory uncertainty—such as unexpected policy delays or sudden changes to cap levels—can trigger sharp fluctuations in allowance prices, creating challenges for market participants.

Energy Market Dynamics

Fluctuations in energy prices strongly influence carbon prices due to their impact on emissions. In many systems, power producers' choice of fuel—such as coal or natural gas—depends on the relative costs of these fuels. Higher-carbon fuels like coal become more attractive when lower carbon options, such as natural gas, are expensive, leading to increased emissions and higher demand for allowances. This relationship underscores the importance of the **fuel-switching threshold**, where the relative economics of fuels dictate the emissions intensity of energy production (Gas Exporting Countries Forum [GECF], 2024). Unanticipated changes in temperature, but not the level of temperature itself, will also affect carbon prices. (Batten et al., 2021).

Macroeconomic Activity

The broader economic environment significantly affects carbon prices. During periods of economic growth, industrial activity and energy consumption typically increase, leading to higher emissions and greater demand for allowances. Conversely, economic slowdowns reduce industrial output and emissions, dampening prices. These dynamics were evident during the 2008 global financial crisis and the early stages of the COVID-19 pandemic, both of which saw sharp reductions in carbon prices as economic activity contracted. As economies recover, however, demand for allowances often rebounds, particularly when coupled with tighter regulatory targets (Hintermann, 2010).

Geopolitical and Market Sentiment

Geopolitical events can add layers of complexity to carbon markets, disrupting energy supplies and altering emissions patterns. For example, conflicts affecting natural gas or oil supplies can prompt shifts in energy production, influencing the demand for allowances. At times, geopolitical uncertainty may decouple carbon prices from traditional drivers like energy costs, as seen when allowance prices fell sharply during the early stages of the Russia-Ukraine conflict despite rising energy prices (*The Guardian*, 2022).

Financial Market Dynamics

Financial institutions play an increasingly active role in carbon markets, treating allowances as tradeable financial assets. Speculative trading by hedge funds, investment banks, and proprietary trading firms increases market liquidity but can also amplify short-term price movements. These players react to regulatory announcements, economic data, and market sentiment, driving volatility that complicates planning for firms reliant on stable carbon price signals. While speculative activity often improves price discovery, its impact on short-term price volatility remains a concern for market stability (European Central Bank [ECB], 2022).

The volatility of EUA prices presents challenges for the corporate sector. Regulated companies, such as power producers and industrial firms, often turn to hedging strategies to manage these risks. Futures trading is particularly important for compliance firms which use it to hedge against price volatility. For example, utilities and industrial firms may purchase EUA futures to lock in prices for future allowance needs, stabilizing compliance costs and mitigating financial risk. However, the widespread use of futures also reflects market sentiment about long-term trends, including anticipated regulatory changes or shifts in energy policy. This dual role of futures trading—as a risk management tool for compliance entities and a speculative instrument for financial participants—underscores its centrality in shaping EUA price dynamics.

Summing Up

Carbon pricing mechanisms, such as taxes and compliance markets, are central to global efforts to internalize the cost of greenhouse gas emissions and drive decarbonization. Carbon taxes provide clear price signals by setting a fixed cost per ton of CO_2, while compliance markets like emissions trading systems (ETSs) leverage market forces to achieve emissions reductions efficiently. Both approaches have demonstrated their ability to influence corporate behavior and support the transition to a low-carbon economy.

However, compliance mechanisms are not the only tools available. Beyond regulated systems, voluntary carbon markets have emerged as a complementary avenue, enabling firms and individuals to take proactive measures in offsetting emissions. These markets, driven by private sector initiatives, play an important role in bridging gaps left by formal policies and fostering innovation in emissions reduction projects. The next section provides an overview of the growing significance and challenges of voluntary carbon markets in the broader carbon pricing landscape.

12.6 Voluntary Carbon Markets: Definition, Market Scale, and Strategic Importance

Voluntary Carbon Markets (VCMs) enable private entities, including corporations, non-governmental organizations, and individuals, to offset their greenhouse gas emissions through the purchase of carbon credits. These credits represent verified reductions or removals of emissions achieved through projects such as reforestation, renewable energy development, and methane capture. Unlike compliance carbon markets, which are regulated by government mandates, VCMs operate outside legal frameworks, driven by voluntary commitments such as corporate sustainability goals, net-zero pledges, and stakeholder expectations. This decentralized structure allows VCMs to address emissions in regions or sectors not covered by compliance markets, extending the global reach of decarbonization efforts.

Carbon credits within VCMs are created and verified according to rigorous standards set by private registries and independent bodies. These mechanisms ensure the credibility of the offsets by addressing key factors such as additionality (reductions that would not occur without the project), permanence (long-term sequestration of carbon), and transparency. These safeguards are critical to maintaining the legitimacy of VCMs and ensuring that they deliver measurable environmental benefits.

Market Size and Growth Potential

The global VCM is growing rapidly. In 2021, transactions reached approximately USD 2 billion, quadrupling from the previous year due to an increase in corporate net-zero commitments and societal pressure for climate action. Projections indicate that VCMs could expand to USD 50–100 billion annually by 2030, reflecting the scale of investment required to meet the goals of the Paris Agreement and effectively contribute to global climate mitigation efforts. This ambitious growth highlights the need for enhanced market mechanisms, increased demand for high-quality credits, and robust regulatory frameworks to ensure the integrity and scalability of carbon offset projects (World Bank, 2024).

However, the size of VCMs remains modest compared to compliance carbon markets (CCMs). For instance, revenues from compliance markets exceeded USD 100 billion in 2023, demonstrating the larger scale and more structured nature of regulated markets (World Bank, 2024). This comparison underscores the importance of developing VCMs to complement compliance systems and broaden the global carbon pricing ecosystem.

Strategic Role in Corporate Finance

VCMs play a critical role in corporate finance, offering firms a flexible mechanism for managing emissions while navigating financial and reputational risks. By participating in VCMs, companies can offset emissions that are difficult or expensive to eliminate internally, signaling their commitment to sustainability and enhancing their appeal to ESG-focused investors. Studies indicate that firms actively engaged in purchasing high-quality offsets benefit from improved creditworthiness and lower borrowing costs, as lenders view them as better prepared for the transition to a low-carbon economy (Taskforce on Scaling Voluntary Carbon Markets, 2021).

In addition to enhancing access to capital, VCMs allow firms to manage regulatory risks proactively. By purchasing offsets, companies can prepare for stricter emissions regulations or potential carbon taxes, effectively future-proofing their operations. This approach positions firms as leaders in climate action while reducing potential liabilities. However, over-reliance on offsets without addressing internal emissions reductions risks accusations of greenwashing, which can damage stakeholder trust and corporate reputation.

From a financial perspective, firms must carefully manage the costs associated with VCM participation. The price of offsets varies significantly based on project type and quality. Nature-based solutions, such as reforestation, often command premium prices due to their co-benefits, including biodiversity preservation and community development. Companies must evaluate these costs against the reputational and strategic advantages of offsetting emissions, particularly in hard-to-abate sectors.

Challenges and Credibility Concerns

Despite their potential, VCMs face challenges that threaten their credibility and effectiveness. The lack of standardized governance across VCMs has led to disparities in credit quality, with some projects failing to deliver measurable or additional emissions reductions. Instances of low-quality or fraudulent offsets have eroded trust in the market, emphasizing the need for stricter oversight and alignment with compliance market standards (World Bank, 2024).

Price disparities further complicate the market. High demand for nature-based solutions often drives up prices, while industrial offsets may be less expensive but lack the co-benefits associated with ecosystem preservation. This variation forces firms to weigh cost against impact, a process that can strain resources and expertise.

VCMs are also susceptible to volatility and speculation, particularly as the market grows. The rapid expansion of transactions has introduced risks, including the dominance of large firms prioritizing high-profile projects and the limited accessibility of reliable offsets for smaller entities. Addressing these challenges will require improved transparency, enhanced verification mechanisms, and the adoption of digital platforms to streamline transactions and improve efficiency.

Broader Role in Climate Action

Beyond their financial implications, VCMs contribute to sustainable development goals by funding emissions reduction projects in developing countries. These initiatives deliver co-benefits, such as job creation, community resilience, and biodiversity conservation, alongside carbon mitigation. For example, forest conservation programs like REDD + (Reducing Emissions from Deforestation and Forest Degradation) exemplify how VCMs can address both environmental and social objectives (Taskforce on Scaling Voluntary Carbon Markets, 2021).

VCMs are poised to play an increasingly prominent role in bridging the gap between voluntary climate commitments and mandatory regulatory frameworks. Their ability to extend emissions reductions into unregulated sectors makes them a valuable complement to compliance markets. However, their continued growth and relevance will depend on addressing quality concerns, improving governance, and ensuring alignment with broader climate goals.

Summing Up: Balancing Flexibility and Credibility

Voluntary Carbon Markets represent a flexible yet essential tool in the global carbon pricing ecosystem. Their ability to address emissions outside regulatory frameworks makes them a strategic complement to compliance markets. However, the effectiveness of VCMs hinges on their integration into broader sustainability frameworks that prioritize internal emissions reductions and ensure transparency and credibility. Firms that balance the use of offsets with substantive decarbonization efforts are better positioned to manage financial and reputational risks while demonstrating leadership in climate action.

12.7   Carbon Pricing: A Corporate Finance Perspective

Carbon pricing is central to climate policy, aiming to internalize the external costs of greenhouse gas emissions and incentivize corporate decarbonization. From a corporate perspective, understanding the various mechanisms—carbon taxes, compliance markets, voluntary markets, and indirect pricing signals—is essential for strategic planning and financial risk management. Each mechanism presents distinct challenges and opportunities, shaping firms' financial strategies and operational decisions.

Carbon Taxes

Carbon taxes place a fixed price on emissions, directly charging firms per ton of CO_2-equivalent emitted. The simplicity of this mechanism provides a clear financial incentive for companies to invest in cleaner technologies, improve energy efficiency, and transition to lower carbon alternatives. For instance, as of 2024, tax rates range from under $1 per ton in Poland to $167 per ton in Uruguay, creating vastly different cost pressures depending on jurisdiction.

From a corporate finance perspective, the predictability of carbon taxes is a significant advantage. Firms can incorporate fixed tax rates into financial models, enabling accurate budgeting and investment planning. However, high tax rates may pose competitive disadvantages, particularly for firms in emissions-intensive sectors that compete with companies operating in regions without comparable carbon costs. For these firms, carbon taxes can be both an operational challenge and a driver of innovation, encouraging investments in technologies that reduce long-term emissions costs.

Compliance Carbon Markets

Compliance carbon markets, such as the EU ETS, establish a cap-and-trade system where emissions allowances are allocated or auctioned and traded among firms. This mechanism provides flexibility, allowing companies that exceed their emissions limits to purchase allowances while enabling those with surplus allowances to generate revenue through sales.

For corporations, compliance markets create opportunities to manage emissions dynamically. The market-driven nature of these systems rewards efficiency and innovation while providing mechanisms for cost optimization. However, the inherent price volatility in markets like the EU ETS presents a challenge for financial planning. Allowance prices can fluctuate significantly due to regulatory changes, energy market dynamics, and geopolitical events. Firms often rely on hedging strategies, such as futures contracts, to mitigate these uncertainties, which adds complexity to financial management.

Participation in compliance markets can also enhance a firm's reputation and access to capital, particularly as sustainability becomes a key criterion for ESG-focused investors. However, managing exposure to volatile compliance costs requires sophisticated financial planning and operational strategies.

Voluntary Carbon Markets (VCMs)

Voluntary Carbon Markets allow firms to purchase carbon offsets to neutralize their emissions. Unlike compliance markets, participation in VCMs is driven by corporate sustainability goals rather than legal requirements. VCMs provide firms with the flexibility to support diverse emissions reduction projects, such as renewable energy deployment, reforestation, and carbon capture initiatives, aligning their actions with broader environmental commitments.

VCMs offer strategic advantages, including the ability to preempt future regulations and signal sustainability leadership to stakeholders. However, reliance on offsets without addressing internal emissions reductions poses reputational risks. Accusations of greenwashing can erode trust if offsets lack verifiable additionality or permanence. Furthermore, the voluntary nature of these markets leads to less predictable pricing, complicating cost management for firms seeking certainty in their carbon strategies.

Indirect Carbon Price Signals

Indirect carbon pricing mechanisms, such as fuel excise taxes, fossil fuel subsidy reforms, and differentiated VAT rates, embed carbon costs into broader economic systems. These policies indirectly influence corporate behavior by altering the cost structure of high-emissions products or incentivizing low-carbon technologies. For example, phasing out fossil fuel subsidies effectively raises the cost of carbon-intensive energy sources, encouraging firms to transition to renewables. Similarly, tax exemptions or reductions for renewable energy technologies lower adoption costs, incentivizing investments.

For corporations, indirect signals complement formal pricing mechanisms by embedding emissions considerations into economic systems. However, the fragmented and inconsistent nature of these policies across jurisdictions can complicate long-term planning. Firms must navigate shifting regulatory environments and align their strategies with often unpredictable policy changes.

Comparing Mechanisms

From a corporate finance perspective, each mechanism presents unique attributes that influence firms' strategic decisions:

- **Predictability**: Carbon taxes provide fixed costs, enabling accurate financial planning, while compliance and voluntary markets introduce variability that requires risk management strategies.
- **Flexibility**: Compliance and voluntary markets allow firms to trade allowances or offsets, offering operational adaptability. In contrast, carbon taxes provide no trading mechanism but offer stability.
- **Incentive Structures**: Carbon taxes directly penalize emissions, creating a straightforward financial disincentive. Compliance and voluntary markets, by contrast, reward efficiency and innovation through trading mechanisms.
- **Market Dynamics**: Taxes provide cost stability but lack market-driven incentives. Compliance and voluntary markets foster dynamic pricing and opportunities for revenue generation but are subject to significant price volatility.

Summing Up

From the corporate finance perspective, understanding the nuances of each carbon pricing mechanism is critical for effective financial and operational planning. Firms must weigh the predictability and simplicity of carbon taxes against the flexibility and innovation potential of compliance and voluntary markets. Additionally, indirect pricing signals play a supportive role, embedding carbon costs into broader economic systems. By leveraging these mechanisms strategically, companies can navigate the transition to a low-carbon economy while managing risks and seizing opportunities.

Conclusion

Carbon markets are pivotal mechanisms in the global effort to reduce greenhouse gas emissions, integrating environmental goals with economic incentives. By internalizing the costs of emissions, these markets compel firms to adopt sustainable practices, innovate, and align operations with decarbonization goals. Compliance systems, such as the EU Emissions Trading System, showcase the power of market-based mechanisms to achieve cost-effective emissions reductions while fostering flexibility and efficiency. Voluntary carbon markets, though less structured, offer firms a pathway to meet stakeholder expectations and contribute to climate goals, provided challenges around offset quality and transparency are addressed.

From a corporate finance perspective, carbon pricing fundamentally reshapes financial planning and decision-making. The volatility of allowance prices, coupled with regulatory uncertainty, poses risks but also creates opportunities for firms that proactively engage with these mechanisms. Investing in emissions-reducing technologies, leveraging high-quality offsets, and integrating carbon costs into financial models can enhance competitiveness, improve access to capital, and bolster reputation.

As global climate action intensifies, the importance of carbon markets will continue to grow. Their success hinges on robust design, consistent implementation, and widespread participation. For firms, balancing immediate compliance obligations with long-term sustainability strategies will be key. By embedding carbon markets into their broader financial and operational frameworks, companies can position themselves to thrive in a carbon-constrained economy while making meaningful contributions to the fight against climate change.

Key Takeaways

1. **Carbon Pricing Mechanisms**:
   - Carbon pricing is a crucial tool for addressing greenhouse gas emissions by internalizing the external costs of pollution.
   - Mechanisms include **carbon taxes**, **compliance carbon markets (e.g., EU ETS)**, and **voluntary carbon markets (VCMs)**, each offering unique advantages and challenges.
2. **Compliance Carbon Markets**:
   - Systems like the **EU ETS** impose caps on emissions, requiring firms to acquire allowances to cover their emissions or invest in emissions-reducing technologies.
   - The EU ETS has evolved through phases, introducing reforms like banking allowances and the Market Stability Reserve (MSR) to address price volatility and oversupply.
   - Price drivers include energy costs, economic conditions, regulatory changes, and geopolitical events, such as the Russia-Ukraine war.
   - Compliance markets incentivize innovation and efficiency while exposing firms to price volatility, necessitating strategies like hedging to manage financial risks.
3. **Voluntary Carbon Markets (VCMs)**:
   - VCMs allow firms to purchase carbon offsets to meet sustainability goals beyond regulatory requirements.
   - Projects include renewable energy, reforestation, and carbon capture, requiring standards like additionality, permanence, and verification to ensure credibility.

- VCMs offer flexibility and reputational benefits but face challenges like greenwashing, price variability, and quality assurance.

4. **Indirect Carbon Price Signals**:
   - Policies like fuel excise taxes, fossil fuel subsidies, and VAT differentials indirectly influence carbon pricing by altering the economics of high-emission fuels and products.
   - These measures complement direct pricing mechanisms and provide additional levers for emissions reduction.

5. **Corporate Finance Implications**:
   - Carbon markets reshape corporate financial strategies, linking emissions management with budgeting, investment planning, and cost of capital.
   - Firms participating in carbon markets can enhance creditworthiness and investor confidence, particularly if they align with ESG principles.
   - Volatility in carbon pricing, especially in compliance markets, requires robust financial planning and risk management strategies, such as futures trading and hedging.

6. **Challenges and Opportunities**:
   - Both compliance and voluntary markets face issues like price volatility, regulatory uncertainty, and credibility of offsets.
   - Despite challenges, these markets drive innovation, support the transition to low-carbon technologies, and create opportunities for firms to enhance competitiveness and sustainability.

7. **Future Outlook**:
   - Carbon markets will play an increasingly central role in global climate strategies, with growing pressure for robust design, transparent standards, and widespread participation.
   - Firms that proactively engage with these mechanisms, balancing immediate compliance with long-term investments in sustainability, will be better positioned to thrive in a carbon-constrained economy.

Questions

1. Define and differentiate between compliance carbon markets and voluntary carbon markets. Provide examples of each.
2. Discuss the role of carbon taxes in incentivizing emissions reductions. How do they compare to carbon markets in terms of predictability and flexibility for firms?
3. Explain the concept of additionality in voluntary carbon markets. Why is it essential for the credibility of carbon offsets?

4. Identify and analyze the primary factors driving price volatility in the EU Emissions Trading System (EU ETS).
5. How do indirect carbon price signals, such as fuel excise taxes and fossil fuel subsidies, complement direct pricing mechanisms?
6. Discuss the financial implications of participating in carbon markets for firms in high-emitting sectors. Include examples of hedging strategies.
7. Evaluate the challenges faced by voluntary carbon markets in ensuring the quality of offsets and avoiding greenwashing.
8. What are the corporate finance implications of carbon price volatility for budgeting and investment decisions?
9. Compare the impact of regulatory uncertainty on compliance markets and voluntary markets.
10. How can firms balance short-term compliance costs with long-term sustainability investments in the context of carbon markets?

References

European Central Bank. (2022). Speculative trading and carbon prices in the EU ETS. Economic Bulletin, Focus Issue 6. Retrieved November 4, 2024, from https://www.ecb.europa.eu/press/economic-bulletin/focus/2022/html/ecb.ebbox202203_06~ca1e9ea13e.en.html

European Central Bank. (2022, May 25). Russia-Ukraine war increases financial stability risks, ECB Financial Stability Review finds. Retrieved November 4, 2024, from https://www.ecb.europa.eu/press/pr/date/2022/html/ecb.pr220525~fa1be4764d.en.html

Gas Exporting Countries Forum. (2024). How carbon pricing shapes coal-to-gas transition. GECF Expert Commentary. Retrieved February 24, 2025, from https://www.gecf.org/_resources/files/events/gecf-expert-commentary-how-carbon-pricing-shapes-coal-to--gas-transition/gecf-ec-2024-coal-to-gas-switching-final.pdf

Hintermann, B. (2010). Allowance price drivers in the first phase of the EU ETS. *Journal of Environmental Economics and Management, 59*(1), 43–56. https://doi.org/10.1016/j.jeem.2009.07.002

Stern, N. (2007). *The economics of climate change: The Stern review.* Cambridge University Press.

Taskforce on Scaling Voluntary Carbon Markets. (2021). *Final report.* Institute of International Finance. Retrieved November 4, 2024, from https://www.iif.com

The Guardian. (2022, March 2). EU carbon permit prices crash after Russian invasion of Ukraine. Retrieved November 4, 2024 from https://www.theguardian.com/environment/2022/mar/02/eu-carbon-permit-prices-crash-after-russian-invasion-of-ukraine

World Bank. (2024). State and Trends of Carbon Pricing 2024. World Bank. Accessed November 4, 2024, from: http://hdl.handle.net/10986/41544

Wu, Y., Liu, X., & Tang, C. (2024). Carbon market and corporate financing behavior—From the perspective of constraints and demand. *Economic Analysis and Policy, 81,* 873–889. https://doi.org/10.1016/j.eap.2024.01.006

Carbon, Crisis, and Strategy: ExxonMobil and Shell Face the Future

13

Foreword

This case examines the climate strategies of Royal Dutch Shell and Exxon-Mobil as they navigate the escalating climate crisis. With the physical and financial risks of climate change intensifying, major oil companies must develop strategies that ensure both sustainability and profitability. While both firms have pledged to achieve net-zero emissions by 2050, their approaches differ significantly.

As two of the most exposed public companies to climate transition risk, Shell and ExxonMobil face mounting regulatory pressures and the risk of rising carbon costs that could erode profitability. The critical question is whether their proposed solutions preserve or destroy shareholder value. To evaluate this, you will assess each firm's worth under various climate scenarios using the Discounted Cash Flow (DCF) approach. This requires:

- Forecasting Free Cash Flow to Firm (FCFF) and terminal value.
- Estimating the cost of capital and how it may evolve.
- Considering the impact of different climate scenarios on future cash flows.

Fundamentally, valuing these firms is akin to a comprehensive capital budgeting exercise, consolidating the impact of their past, present, and planned projects into today's market value. This case study requires an analysis of energy market dynamics, including:

- Fossil fuel demand projections and the feasibility of green energy investments.
- The economic and regulatory implications of transition risk.
- The financial viability of
- Carbon Capture, Utilization, and Storage (CCUS).

© The Author(s), under exclusive license to Springer Nature Switzerland AG 2025
S. Dow and Y. Shi, *Corporate Finance Under Climate Crisis*,
https://doi.org/10.1007/978-3-031-83487-5_13

Your challenge is to value Shell and ExxonMobil under different climate scenarios and assess whether their respective strategies position them for long-term financial sustainability.

Introduction

The Stern Review on the Economics of Climate Change famously declared that climate change represents "the greatest and widest-ranging market failure ever seen" (Stern, 2007). This market failure stems from the fact that greenhouse gas emissions impose significant externalities—costs borne not by those who emit but by society as a whole. For decades, the fossil fuel sector has been a central contributor to these emissions. Major oil and gas companies, including Shell and ExxonMobil, have long understood the environmental risks associated with their operations. Internal research, media investigations, and public policy debates reveal a troubling gap between what these companies knew and the actions they took.

For decades, industry insiders were aware of the links between fossil fuels and climate change. By the late 1970s, ExxonMobil's internal scientists had accurately modeled the relationship between CO_2 emissions and global warming, predicting the trajectory of climate change with striking precision. Similarly, Shell's internal forecasts in the 1990s warned of the societal and economic risks posed by unmitigated fossil fuel use. Yet, these insights were not reflected in their public narratives. Instead, companies prioritized casting doubt on climate science, lobbying against regulations, and delaying meaningful action (Supran et al., 2023).

13.1   The Fossil Fuel Sector and GHG Emissions

Fossil fuels remain the single largest source of global greenhouse gas emissions, accounting for over 75% of CO_2 emissions in 2022 (IEA, 2021). As scientific evidence mounted, public and investor pressure on companies like Shell and ExxonMobil grew. In response, both companies have articulated strategies to address climate change, but their approaches diverge significantly.

Shell has embraced a broader net-zero emissions (NZE) pledge, committing to become a net-zero energy company by 2050. This commitment involves scaling up investments in renewable energy, such as wind and solar, as well as green hydrogen and electric vehicle infrastructure (Shell, 2024). Shell's strategy signals a pivot from its traditional oil and gas focus toward becoming a diversified energy company.

In contrast, ExxonMobil's strategy centers on Carbon Capture, Utilization, and Storage (CCUS). While acknowledging the need for emissions reductions, Exxon-Mobil positions CCUS as a pragmatic solution that allows continued fossil fuel production while mitigating its environmental impact. ExxonMobil argues that CCUS technologies are essential to meeting global energy demand while aligning with decarbonization goals (ExxonMobil, 2024).

These contrasting approaches highlight the tension between transitioning entirely away from fossil fuels and finding ways to make their continued use more sustainable. However, the underlying question remains: Are these strategies sufficiently aligned with the global urgency of achieving NZE, or are they primarily designed to preserve existing business models?

13.2 The Tragedy of the Horizon

Mark Carney, the former Governor of the Bank of England, coined the term "tragedy of the horizon" to describe the temporal disconnect between the long-term risks of climate change and the short-term decision-making horizons of businesses, investors, and policymakers (Carney, 2015). For fossil fuel companies, this tragedy manifests in the prioritization of immediate profitability over long-term sustainability.

The fossil fuel sector is uniquely affected by the tragedy of the horizon because the impacts of climate change—stranded assets, regulatory restrictions, and declining demand for fossil fuels—are projected to intensify over the coming decades. Yet, many companies continue to invest in oil and gas production, effectively following a "burn now" philosophy that seeks to maximize profits before regulatory and market conditions render their reserves less valuable.

Critics argue that this approach leverages the temporal gap inherent in the tragedy of the horizon. By focusing on short-term returns, companies may exacerbate transition risks, including the likelihood of stranded assets and increased regulatory scrutiny. For instance, ExxonMobil's continued investment in fossil fuel projects suggests an effort to extract maximum value from its reserves, even as it promotes CCUS as a solution. Similarly, while Shell has made significant investments in renewable energy, its overall capital allocation remains heavily weighted toward oil and gas exploration and production.

13.3 Bridging the Gap

The divergence between Shell and ExxonMobil's strategies underscores the complexity of addressing transition risk. Shell's investments in green energy represent a bet on the rapid decarbonization of global energy markets, but these investments carry significant financial and operational risks. ExxonMobil's focus on CCUS reflects a more cautious approach, but it is contingent on the development of supportive policies, such as carbon pricing and government subsidies.

As these companies navigate their respective paths, they face a shared challenge: balancing short-term financial performance with the long-term need for decarbonization. The question is not only whether their strategies will preserve market value but also whether they will align with the realities of a rapidly warming world.

13.4   The Growing Threat of Stranded Assets

Stranded assets arise when once-valuable resources, such as oil reserves or coal mines, become economically unfeasible due to external pressures. The combination of climate policies, technological advancements, and shifting consumer preferences will likely render substantial portions of the fossil fuel industry's asset base obsolete within the coming decades. Stranded assets represent not only a financial risk but also a strategic challenge, as companies must decide whether to reinvest in traditional resources or pivot toward low-carbon technologies.

13.5   Curious Investment Behavior in the Fossil Fuel Sector

Despite the increasing likelihood of asset stranding, fossil fuel companies continue to invest heavily in new extraction projects. ExxonMobil, for example, announced a major oil discovery off the coast of Guyana, adding over 10 billion oil-equivalent barrels to its reserves (ExxonMobil, 2021). BP is moving forward with the Seagull gas field in the North Sea, which is expected to produce 50,000 barrels of oil equivalent per day (BP, 2023). Similarly, Shell has greenlit its Whale deepwater oil project in the Gulf of Mexico, which aims to produce 100,000 barrels of oil equivalent daily (Shell, 2021). These investments reflect a persistent reliance on fossil fuels even as the global energy transition accelerates.

Such investment decisions raise questions about corporate strategy in the face of mounting risks. One academic explanation comes from the study "Burn Now or Never? Climate Change Exposure and Investment of Fossil Fuel Firms," which suggests that companies are engaging in a "burn now" strategy (Adolfsen et al., 2024). This approach seeks to maximize short-term profits from fossil fuel assets before they are rendered unprofitable by future regulatory measures or market shifts. In essence, these firms are hedging against a future where stringent climate policies curtail their operations. Is this another strategic alternative pursued by both ExxonMobil and Shell?

13.6   Lessons from Other Sectors

The risk of stranded assets is not confined to fossil fuels. Industries like utilities and automotive have already faced significant disruptions. For example, the
European electricity sector experienced €129 billion in asset impairments between
2010 and 2015 as renewable energy technologies disrupted traditional coal and
gas power generation. This shift, fueled by rapid improvements in wind and solar
technology, offers a cautionary tale for fossil fuel firms. Once the paradigm shifts,
the transition from old to new technologies can be swift and unforgiving, leaving
incumbents with devalued assets and declining market share.

13.7   Shell and ExxonMobil: Facing Stranded Assets

For Shell and ExxonMobil, the threat of stranded assets necessitates a strategic
response. Shell's investments in renewable energy reflect an acknowledgment of
these risks, signaling a willingness to adapt to the changing landscape. Exxon-
Mobil, on the other hand, emphasizes Carbon Capture, Utilization, and Storage
(CCUS) as a means to extend the life of its fossil fuel operations while mitigating
emissions. These divergent strategies illustrate the broader industry debate: Should
companies transition away from fossil fuels entirely or find ways to decarbonize
their existing operations?

   Ultimately, the success or failure of these approaches will depend on external
factors such as policy implementation, technological advancements, and market
dynamics. What remains clear is that the risk of stranded assets will continue
to loom large over the fossil fuel industry, shaping its investment decisions and
long-term viability.

13.8   Investor and Stakeholder Expectations

The fossil fuel sector is under increasing scrutiny from investors and stakeholders who expect companies to address the risks and opportunities posed by climate
change. These expectations are articulated through various mechanisms, including shareholder resolutions, activist campaigns, ESG (Environmental, Social, and
Governance) frameworks, and even legal cases. As the transition to a low-carbon
economy accelerates, oil and gas companies must navigate growing demands for
transparency, accountability, and measurable progress toward decarbonization.

   Institutional investors, such as BlackRock and Vanguard, have prioritized ESG
metrics in evaluating corporate performance. For example, BlackRock CEO Larry
Fink has repeatedly called for companies to articulate how their business models
align with a net-zero economy, signaling a broader shift in investor priorities. This
pressure has led to an increase in shareholder resolutions targeting fossil fuel companies. Resolutions often call for disclosures on climate risk, alignment with the
Paris Agreement, or even the adoption of binding emissions reduction targets.

ExxonMobil has felt the sting of activist investors. In 2021, a small activist hedge fund, Engine No. 1, successfully secured three board seats at ExxonMobil by rallying institutional investors to demand a stronger focus on climate strategy. This marked a watershed moment in investor activism, emphasizing that even established industry giants must heed shareholder demands for sustainable business practices (Fink, 2022; Reuters, 2021).

13.9   Legal Cases as a Tool for Accountability

Stakeholder expectations are also increasingly articulated through legal cases, as communities, governments, and NGOs hold fossil fuel companies accountable for their contributions to climate change. In one landmark case, the Dutch court ordered Shell to reduce its emissions by 45% by 2030, stating that the company's current climate policies were insufficient to meet its human rights obligations. Although Shell successfully appealed the ruling, the case highlights how legal mechanisms are shaping corporate climate accountability. The number of lawsuits is also on the rise. Shell and ExxonMobil have the dubious distinction of leading the pack in the number of lawsuits in the past twenty years, as reported by Zero Carbon Analytics (Zero Carbon Analytics, n.d.). Oil and gas giants that fail to align with climate goals of investors and other stakeholders may have significant financial and reputational consequences.

13.10   Shifts in Consumer and Community Expectations

Beyond institutional investors, local communities and consumers are also playing a pivotal role in reigning in oil and gas companies. Increasing awareness of climate risks has fueled boycotts of high-emission companies and driven demand for greener alternatives. Communities most affected by fossil fuel extraction and climate impacts are leading grassroots campaigns, advocating for stricter regulations, and demanding reparations for damages caused by industrial activities.

For Shell and ExxonMobil, responding to these articulated expectations is crucial for retaining investor confidence and mitigating reputational risks. Shell has proactively engaged with stakeholders by setting ambitious net-zero targets and increasing investments in renewable energy. However, critics argue that the pace and scale of these changes remain inadequate. ExxonMobil, on the other hand, has faced criticism for its slower transition efforts but has begun to respond to stakeholder pressures by expanding its CCUS projects and disclosing more information on climate-related risks.

13.11   Transition Plans for Shell

Shell plc has committed to transforming into a net-zero emissions energy business by 2050, aligning with the goals of the Paris Agreement to limit global temperature rise to 1.5 °C above pre-industrial levels. This ambition encompasses reducing emissions from its operations (Scope 1 and 2) and the energy products it sells (Scope 3), which constitute over 90% of its total reported emissions. What does this mean for firm value (its own "business as usual" framework)?

Capital Expenditure (Capex) in Green Investments

Shell's financial commitment to renewable energy has evolved over recent years:

2021: The company invested $2.4 billion in its Renewables and Energy Solutions segment, with $1.8 billion allocated to low-carbon energy solutions.
2022: Shell increased its investment to $3.5 billion in the same segment, dedicating $2.9 billion to low-carbon energy solutions.
2023: The company invested $5.6 billion in low-carbon energy solutions, accounting for more than 23% of its total capital spending.

13.12   Green Investments: Challenges and Risks for the Oil and Gas Sector

While green investments offer an avenue for diversification and alignment with global decarbonization goals, they are not a panacea for the challenges faced by the oil and gas sector. Transitioning into the green energy space introduces a new set of financial risks and operational challenges that companies must navigate carefully.

Financial Risks

High Upfront Capital Requirements
Investments in green energy, such as renewable power plants or hydrogen production, often require significant upfront costs. These projects may strain balance sheets, particularly for companies with declining cash flows from fossil fuel operations.

Uncertain Returns
Green energy markets are influenced by policy changes, subsidies, and evolving technologies, which can lead to volatility in returns. For example, the phase-out of subsidies for wind or solar in certain markets could materially affect project economics.

Market Competition

Entrants from outside the oil and gas sector, including technology firms and utilities, bring different expertise and cost structures, increasing competition. Oil and gas firms may find it challenging to establish a competitive edge in these domains.

Stranded Asset Risks

Transitioning to green investments may leave traditional assets stranded or underutilized, leading to write-downs and impairments on fossil fuel-based infrastructure.

13.13   Operational Challenges

Technology and Expertise Gap

The oil and gas sector's core expertise often does not align directly with renewable energy operations, which may require different technological skills, supply chain management, and operational strategies.

Intermittency and Storage

Renewable energy projects, particularly wind and solar, face issues related to intermittency and the need for large-scale energy storage solutions. Managing these operational challenges demands innovation and robust infrastructure investments.

Geographic Mismatch

Many oil and gas operations are concentrated in regions unsuitable for renewable energy production due to lack of sun, wind, or regulatory support. Shifting operations to optimal locations may incur additional costs.

Regulatory Compliance and Carbon Accounting

Green energy projects often come with complex regulatory frameworks and requirements for transparent carbon accounting. Failing to meet these standards could expose firms to fines, reputational damage, or loss of market access.

13.14   ExxonMobil's Climate Strategy

ExxonMobil's strategic approach centers on balancing its traditional fossil fuel operations with significant investments in Carbon Capture, Utilization, and Storage (CCUS) technologies, all while navigating a complex policy landscape.

ExxonMobil positions itself as a leader in CCUS, a technology aimed at capturing carbon dioxide emissions from industrial sources and either reusing or storing them to prevent atmospheric release. The company has committed over $15 billion through 2027 to lower greenhouse gas emissions, with a substantial portion allocated to CCUS initiatives. This investment underscores ExxonMobil's strategy to mitigate emissions from its operations and those of its industrial partners.

13.15   Investments in Fossil Fuels Alongside CCUS

Despite its CCUS endeavors, ExxonMobil continues to invest heavily in fossil fuel exploration and production. In 2023, the company announced plans to increase oil and gas production in the Permian Basin, aiming for a 25% boost by 2025. This dual investment strategy reflects ExxonMobil's belief in the ongoing demand for fossil fuels, even as it develops technologies to reduce associated emissions.

13.16   ExxonMobil's Strategy: CCUS and Fossil Fuels

Opportunities

Leveraging Existing Infrastructure: ExxonMobil's CCUS projects capitalize on its expertise in large-scale industrial operations, providing a pathway to decarbonize without abandoning core operations.

Immediate Emissions Reductions: CCUS offers near-term solutions for emissions-intensive industries, addressing global demand for mitigation strategies.

Potential First-Mover Advantage: Heavy investment in CCUS may position ExxonMobil as a leader in carbon management technologies.

13.17   Risks

Dependence on Policy Support: CCUS is heavily reliant on incentives like tax credit and carbon pricing to be economically viable.

Prolonged Fossil Fuel Dependence: Continued investment in oil and gas risks asset stranding and misalignment with NZE goals.

Market and Reputational Risks: Stakeholders increasingly demand reductions in Scope 3 emissions, where CCUS has limited impact compared to renewable transitions.

13.18   The Biggest Risk of CCUS for ExxonMobil

That's a beautiful list of risks and opportunities of the CCUS adventure, but the single most important risk is CCUS a hallucination? This is certainly the view of Fatih Birol, Executive Director of the IEA. Current investment in carbon capture (Carbon Capture, Utilization, and Storage, Carbon Capture and Storage; and Direct Air Capture) for all users, including heavy industry and oil and gas is about $4 billion per year. If all emissions from the oil and gas sector were to be covered, an annual investment of $4 trillion would be required. The scale should give you pause: $4 trillion is $4000 billion. Or as Mr. Birol summed it up in an interview with the Guardian during COP28 (Harvey, 2023): "It's a fantasy. It is an illusion."

13.19   ExxonMobil and Shell: Are They Aligned with NZE?

As we've seen, Shell and ExxonMobil have adopted markedly different strategies to navigate the energy transition. Table 13.1 compares their climate strategies side by side and suggests the extent to which their actions line up with their NZE intentions.

It might seem that Shell is on the right track with its emphasis on green technology. Industry wide, oil and gas companies are devoting only 2.5% of their total (Harvey, 2023) capital expenditure to green alternatives. To make a dent, the sector has to get much busier. Shell made record profits in 2022 and announced in early 2023 that it was scaling back its green investments (Carbon Brief, 2023). In other

Table 13.1   Comparison of Shell and ExxonMobil's NZE strategies

Aspect	Shell	ExxonMobil
Core Strategy	Transition to renewable energy; diversify energy portfolio	Maintain fossil fuel production; focus on CCUS for emission reductions
Target Scope	Scope 1, 2, and 3 emissions	Primarily Scope 1 and 2 emissions
Key Investments	$5.6 billion in low-carbon solutions (2023)	$15 billion in CCUS by 2027
Alignment with NZE	Strong alignment with Paris Agreement	Weak alignment; relies heavily on fossil fuels
Criticism	High costs, operational challenges in renewables	CCUS scalability and cost feasibility questioned
Policy Dependence	Moderately dependent on subsidies for renewables	Heavily reliant on tax credits and carbon pricing
Market Strategy	Diversification into biofuels, green hydrogen, EV charging	Continued investment in oil and gas alongside CCUS
Public Perception	Positive due to visible renewable projects	Mixed; criticized as a stalling tactic

words, having a good year was not going to move the needle on their commitment to renewable energy.

As far as ExxonMobil is concerned, their future seems to hinge on a technology they can't deliver on.

Shell and ExxonMobil: Climate Strategy Side by Side. Table 13.1 compares the NZE strategies of both companies.

13.20   ExxonMobil and Shell: What Will the Future Look Like?

ExxonMobil' future and Shell's future to a large extent is going to depend on old-fashioned economics: what are forecast revenues and costs? What is the market for oil? What will happen to our costs? In other words: what are the market forces at work? To answer these questions, they have to engage in robust scenario development.

13.21   Role of Energy Transition Scenarios

Both companies could make good use of scenario analysis to see where the future lies in the transition to a low-carbon economy. In fact, making wide use of resources from the IPCC, the IEA, and the NGFS will help both companies and the sector in general prepare for what's coming next. These scenarios are instrumental in providing insights into the potential outcomes of various energy strategies and informing decisions that align with climate goals and energy security objectives. Clearly, projections offered by these different scenario brands will play a central role in determining firm value.

The value of Shell and ExxonMobil is going to vary under different scenarios and over different time horizons. Barring some kind of miracle—an unknown unknown—the implications of varying scenarios will reveal a variety of pathways. Scenario analysis, by framing the future, will help them gauge where their market value is going to land.

To ascertain if their climate strategies will be value-creating or value-destroying, ExxonMobil and Shell need to look at a number of factors that thorough scenario analysis will facilitate.

1. Economic Growth and Industrial Activity

Oil and gas demand is strongly tied to economic growth, particularly in emerging markets. Higher GDP growth leads to increased industrial activity, transportation, and energy consumption, driving up energy demand. Conversely, during economic recessions, demand tends to decline.

2. Energy Transition Policies

Government policies aimed at transitioning to renewable energy sources, including carbon pricing, subsidies for renewable energy, and bans on internal combustion engine vehicles, can reduce oil and gas demand. Some countries implementing aggressive climate policies may phase out fossil fuels faster.

3. Technological Advances

Advances in energy storage, battery efficiency, and renewable energy generation technologies reduce reliance on oil and gas. For example, improvements in wind and solar technologies are directly displacing fossil fuels in power generation (IRENA, 2022).

4. Consumer Behavior and Preferences

Increasing awareness of climate change and preference for greener energy sources influence demand. If consumers shift toward public transportation, EVs, or renewable-powered homes, oil and gas demand diminishes.

5. EV Penetration

The adoption rate of EVs has a direct inverse relationship with oil demand. High EV penetration, especially in transportation-heavy markets like China and the United States, reduces oil demand (BNEF, 2023).

6. Sectoral Shifts

Structural shifts in sectors such as transportation, industrial production, and chemicals significantly impact demand. In the aviation sector, the adoption of Sustainable Aviation Fuel will reduce reliance on conventional fuel. In heavy industry, the switch to green hydrogen will reduce demand for oil.

7. Urbanization and Demographics

In emerging markets, rapid urbanization can either increase or decrease demand depending on infrastructure development. For instance, cities investing in mass transit systems reduce dependency on oil, whereas urban sprawl may increase demand.

8. Geopolitical and Supply-Side Factors

Geopolitical stability, trade policies, and OPEC+production decisions shape demand. Supply disruptions or embargoes often lead to temporary shifts in consumption patterns (BP, 2023).

13.22   Current and Projected Demand for Oil

All of the factors work together to determine the demand for oil. The global energy landscape is undergoing a significant transformation, with shifting demands for oil, gas, and renewable energy sources. Energy transition scenarios play a crucial role in understanding and guiding this evolution.

As of 2024, global oil demand has reached approximately 102 million barrels per day (bpd). Projections indicate a decline to around 80 million bpd by 2035 in a net-zero emissions scenario, and to 100 million bpd under current policy trajectories. This anticipated decrease is attributed to the rise of renewable energy sources and advancements in energy efficiency.

Both companies wonder how alternative energy will affect their fortunes. Both have a watchful eye on the evolution of electric vehicles, (EVs)—the more EVs the less their sales to the auto sector!

13.23   Current and Projected Demand for Natural Gas

Natural gas demand is expected to experience modest growth in the near term, with a peak around 2040 at approximately 4,700 billion cubic meters (bcm), followed by a gradual decline through 2050. This trend reflects the increasing integration of renewable energy and the implementation of energy efficiency measures.

13.24   Current and Projected Demand for Renewable Energy

Renewable energy sources, particularly solar and wind, are projected to see substantial growth. In Europe, renewables could account for up to 87% of the energy supply by 2050. Globally, renewable power generation capacity is expected to rise from 4,250 gigawatts (GW) today to nearly 10,000 GW by 2030, covering the growth in global electricity demand and reducing reliance on coal-fired generation.

13.25   Firm Value

What is the value of the firm in Climate Crisis: that's the big question! First of all, you are not going to arrive at a single estimate…the value of the firm is going to depend on which climate future occurs. However, within each scenario are wide ranging assumptions about what could happen.

1. Revenue Projections: Fossil Fuels vs. Renewable Energy vs. CCUS Markets

Shell's revenue outlook is increasingly tied to its renewable energy investments, such as wind, solar, and green hydrogen. However, fossil fuels still contribute the majority of its revenues, with oil and gas making up over 60% of its sales in 2023. Shell's renewable energy projects have the potential for growth, but they rely on market expansion and supportive policies, which introduce uncertainty.

ExxonMobil:

ExxonMobil's revenue remains predominantly linked to fossil fuels, with investments in CCUS serving as a smaller, complementary revenue stream. CCUS has significant growth potential as industrial emitters seek decarbonization solutions, but its scalability depends heavily on carbon pricing and regulatory incentives.

Comparative Insight:

While Shell's renewables strategy offers long-term growth potential, it faces volatility during the transition phase. ExxonMobil's revenue reliance on fossil fuels may ensure near-term stability but risks stagnation as global energy demand shifts toward low-carbon solutions.

2. Cost Structures: CapEx, O&M, and R&D in Transition Technologies

Shell:

Shell's capital expenditures (CapEx) on renewables have risen substantially, reaching $5.6 billion in 2023, representing over 23% of total CapEx. However, renewable projects require higher upfront investment and operational expenditures (O&M) compared to fossil fuel operations. Additionally, R&D in technologies like green hydrogen adds to cost burdens, potentially straining cash flows in the short term.

ExxonMobil:

ExxonMobil has maintained higher CapEx in fossil fuel operations, with significant investments in new oil fields like Guyana. Its CCUS investments are also capital-intensive, particularly in scaling large-scale storage facilities. However, CCUS may have lower O&M costs compared to renewable energy, provided regulatory frameworks remain favorable.

Comparative Insight:

Shell's diversification into renewables introduces financial strain due to high CapEx and R&D demands. ExxonMobil, while more conservative in its energy mix, may face stranded asset risks if its fossil fuel-heavy investments are devalued.

3. Risk Adjustments: Transition Risks, Regulatory Risks, and Market Uncertainties

Shell is more exposed to regulatory and policy risks due to its reliance on subsidies and incentives for renewable energy projects. Additionally, transition risks such as consumer shifts toward renewables could accelerate faster than anticipated, creating market uncertainties. However, its alignment with NZE goals may mitigate long-term reputational risks.

ExxonMobil faces significant transition risks, including regulatory challenges like carbon pricing and litigation over its environmental impact. Market uncertainties tied to CCUS adoption and the longevity of fossil fuel demand further compound risks. However, its conservative approach may shield it from the volatility associated with nascent renewable markets in the short term.

Comparative Insight:

Maybe Shell's strategy carries more regulatory risk due to its dependence on favorable policies, whereas ExxonMobil faces greater market risk due to potential misalignment with global decarbonization trends. On the other hand, ExxonMobil's reliance on CCUS hinges on higher carbon prices to make them worthwhile (in terms of savings)…so it's not very clear. Think it over!

4. Discount Rates and Cost of Capital: Incorporating Transition Risk into WACC

Shell's weighted average cost of capital (WACC) is impacted by the perceived riskiness of its renewables portfolio. Investors may demand higher returns to compensate for the uncertainties associated with new markets and technologies. However, successful execution of its green strategy could lower WACC over time as the company demonstrates resilience. The argument for or against renewables impact on WACC depends on the required return on these projects. It could be lower if investors prefer green projects to carbon-intensive projects. It could be higher if the demand for uncertainty compensation wins over.

ExxonMobil:

ExxonMobil's WACC could reflect the historic relatively lower risk of its fossil fuel operations, but this seems to be increasing as investors factor in transition risks and potential stranded assets. Moreover, the heavy reliance on CCUS requires stable policies, and uncertainty in carbon pricing could further elevate its cost of capital.

Comparative Insight:

Shell might face higher WACC in the short term due to the perceived risks of renewables, while ExxonMobil's WACC could increase if fossil fuel investments lose value amid transition risks (Table 13.2).

Conclusion

Have ExxonMobil and Shell been engaging in the heavy lifting required by scenario analyses, given the current situation: Shell and ExxonMobil both committing to NZE but frankly not doing much to get there?

Could both strategies be consistent with a "Burn now or Never?" Both companies are aware that rolling back the clock entirely is not going to be an option as physical climate risk escalates around the globe.

Which strategy could prove the most valuable? The values of Shell and ExxonMobil are shaped by their distinct approaches to the energy transition. Shell's focus on renewables aligns better with long-term NZE goals but

Table 13.2 Comparative analysis of factors influencing firm value

Factor	Shell	ExxonMobil	Comparative insight
Revenue Projections	– Revenue tied to renewables (wind, solar, green hydrogen) – Fossil fuels still make up 60% of revenues (2023) – Growth potential relies on market expansion and supportive policies	– Revenue predominantly linked to fossil fuels – CCUS serves as a smaller complementary stream – Growth in CCUS depends on carbon pricing and incentives	– Shell's renewables strategy offers long-term potential but is volatile during transition – ExxonMobil's fossil fuel focus provides short-term stability but risks stagnation
Cost Structures	– High CapEx on renewables ($5.6 billion in 2023) – Renewables have high O&M and R&D costs (e.g., green hydrogen) – Financial strain in the short term	– High CapEx in fossil fuels (e.g., Guyana oil field) – CCUS investments are capital-intensive but may have lower O&M costs – Dependent on favorable regulations	– Shell's renewables diversification introduces financial strain – ExxonMobil risks stranded assets if fossil fuels are devalued
Risk Adjustments	– Greater regulatory and policy risks due to dependence on subsidies – Transition risks from consumer shifts to renewables – NZE alignment mitigates long-term reputational risks	– Faces significant regulatory risks (e.g., carbon pricing) – Market uncertainties tied to CCUS adoption and fossil fuel longevity – Conservative approach shields short-term volatility	– Shell faces higher regulatory risks due to reliance on policies – ExxonMobil's strategy depends on higher carbon prices, adding market risk
Discount Rates and WACC	– Short-term higher WACC due to renewables' perceived risk – Successful execution may lower WACC by demonstrating resilience – WACC depends on investor sentiment (green vs. carbon-intensive projects)	– Historically lower WACC for fossil fuels but increasing with transition risks – CCUS dependence on stable policies adds uncertainty to WACC	– Shell faces higher WACC initially due to renewables – ExxonMobil's WACC may increase with stranded asset risks and transition uncertainties

introduces higher short-term costs and risks. ExxonMobil's reliance on fossil fuels and CCUS offers stability now but risks misalignment with future energy market dynamics, which could impact its valuation as the global transition accelerates. Balancing these factors will be critical for investors assessing the long-term viability of each company.

References

Adolfsen, J. F., Heissel, M., Manu, A.-S., & Vinci, F. (2024, June). *Burn now or never? Climate change exposure and investment of fossil fuel firms* (ECB Working Paper No. 2024/2945). Available at SSRN: https://ssrn.com/abstract=4863762 or https://doi.org/10.2139/ssrn.4863762

BP. (2023, November 8). *Production begins from BP-operated Seagull field in North Sea.* BP. Accessed November 4, 2024, from: https://www.bp.com/en/global/corporate/news-and-ins ights/press-releases/production-begins-from-bp-operated-seagull-field-in-north-sea.html

Carbon Brief. (2023, March 9). *Shell hits the brakes on growing renewables unit after record 2022 profit.* Accessed November 4, 2024 at https://www.carbonbrief.org/daily-brief/shell-hits-the-brakes-on-growing-renewables-unit-after-record-2022-profit/

Carney, M. (2015). *Breaking the tragedy of the horizon—Climate change and financial stability. Speech given at Lloyd's of London.* Retrieved November 4, 2024, from https://www.bankof england.co.uk, https://www.bankofengland.co.uk/speech/2015/breaking-the-tragedy-of-the-horizon-climate-change-and-financial-stability

ExxonMobil. (2021, October 7). *ExxonMobil increases Stabroek resource estimate to approximately 10 billion barrels.* Accessed November 4, 2021, at: https://corporate.exxonmobil.com/news/news-releases/2021/1007_exxonmobil-increases-stabroek-resource-estimate-to-approx imately-10-billion-barrels

ExxonMobil. (2024). *Advancing climate solutions.* Retrieved November 4, 2024 from: https://cor porate.exxonmobil.com/sustainability-and-reports/advancing-climate-solutions

Fink, L. (2022). *BlackRock CEO Larry Fink's annual letter to CEOs.* Retrieved November 4, 2024, from: https://www.blackrock.com/corporate/investor-relations/larry-fink-ceo-letter#: ~:text=Their%20time%20horizons%20can%20span,our%20clients%20are%20invested%20in

Harvey, F. (2023, December 2). Oil and gas firms must convert to renewables or face decline, says IEA chief. *The Guardian.* Accessed November 4, 2024 from https://www.theguardian.com/environment/2023/dec/02/oil-and-gas-firms-must-convert-to-renewables-or-face-decline-says-iea-chief

International Energy Agency (IEA). (2021). *Net zero by 2050: A roadmap for the global energy sector.* IEA.

Reuters. (2021, June 29). *Little Engine No.1 beat Exxon with just $12.5 mln-Sources.* Retrieved November 4, 2021, from: https://www.reuters.com/business/little-engine-no-1-beat-exxon-with-just-125-mln-sources-2021-06-29/

Shell. (2021, July 25). *Shell invests in the whale development in the Gulf of Mexico.* Accessed November 4, 2021, at: https://www.shell.com/news-and-insights/newsroom/news-and-media-releases/2021/shell-invests-in-the-whale-development-in-the-gulf-of-mexico.html#:~:text=Shell%20Offshore%20Inc.%2C%20a%20subsidiary,topsides%20from%20our%20Vito%20project

Shell. (2024). *Energy Transition Strategy 2024.* Retrieved November 4, 2024, from https://www.shell.com/sustainability/our-climate-target/shell-energy-transition-strategy.html#:~:text=In%20summary%2C%20these%20beliefs%20are,underpin%20the%20future%20energy%20system

Stern, N. (2007). *The economics of climate change: The Stern review.* Cambridge University Press.

Supran, G., Rahmstorf, S., & Oreskes, N. (2023). Assessing ExxonMobil's global warming projections. *Science, 379*(6628).

Zero Carbon Analytics. (n.d.). *Website.* Accessed November 4, 2024, from: https://zerocarbon-analytics.org/

Part IV

Governance and the Future of Finance

Corporate Governance Under Climate Change: Ensuring Strategic Cohesion from Today to 2100

14

Introduction

The increasing urgency of climate change demands a fundamental shift in corporate governance practices. This chapter delves into how corporate governance has evolved to address climate-related challenges, highlighting the pivotal role of robust governance frameworks in balancing immediate financial performance with long-term sustainability goals. As businesses navigate this complex landscape, it becomes imperative to integrate environmental, social, and financial considerations into core strategies, fostering a culture of accountability, stakeholder engagement, and continuous improvement. Through a critical examination of current practices, regulatory developments, and innovative approaches, this chapter aims to provide a comprehensive understanding of the governance reforms necessary to achieve strategic cohesion in the face of escalating climate risks.

14.1 The Evolving Role of Governance in Addressing Climate Change Challenges

Corporate governance has evolved significantly over the past century, moving from informal oversight to more structured frameworks aimed at maximizing shareholder value. The primary model that emerged in the twentieth century was shareholder-centric, fashioned by Milton Friedman's assertion that the sole responsibility of business is to increase its profits (Friedman, 1970). This model became entrenched, focusing predominantly on short-term financial gains.

However, this focus on short-term financial gains exposed significant vulnerabilities within corporate structures. These vulnerabilities became glaringly apparent during the 2008 financial crisis. The crisis marked a significant crossroads for corporate governance. Governance failures, particularly in risk management, board oversight, and executive compensation, contributed to the collapse of major financial institutions like Lehman Brothers and Bear Stearns. These institutions engaged in risky lending practices and complex financial products without adequate oversight or understanding, leading to massive financial losses and global economic instability (Financial Crisis Inquiry Commission, 2011). The crisis underscored the critical need for robust governance frameworks to prevent such systemic risks (Kirkpatrick, 2009).

With the introduction of the Dodd-Frank Act in 2010, efforts to restore stability and trust in the financial system were set in motion. Concurrently, there was a growing recognition of the importance of Environmental, Social, and Governance (ESG) factors, leading to a paradigm shift in corporate governance that began to prioritize broader stakeholder interests. The shift toward integrating ESG criteria into corporate strategies gains further momentum with the Paris Agreement of 2015. The landmark accord not only galvanized global efforts to combat climate change but also underscored the necessity for companies to adopt sustainable practices as a core component of their strategic planning. This shift reflects a broader understanding that long-term corporate success is inextricably linked to its environmental impact. It marked a tipping point for corporate governance concerning environmental issues.

The climate crisis presents ongoing, escalating risks that affect every aspect of business operations, necessitating robust governance frameworks. Unlike past financial crises, this persistent threat requires comprehensive approaches integrating environmental, social, and financial considerations. Corporations face another pivotal governance crossroads, where balancing immediate financial performance with long-term sustainability goals is paramount. Investors and stakeholders increasingly demand proactive management of climate-related risks, recognizing that companies addressing environmental concerns are better positioned for long-term success. As ESG investing expands, greater corporate responsibility and transparency are expected, underscoring the need for strong governance practices.

Internationally, the Task Force on Climate-related Financial Disclosures (TCFD) continues to influence global standards, encouraging companies to disclose how they identify, assess, and manage climate-related risks and opportunities (Task Force on Climate-Related Financial Disclosures, 2017). The formation of the International Sustainability Standards Board (ISSB) aims to harmonize sustainability reporting standards globally, providing a unified framework for companies to follow.[1] In late 2023, the IFRS Foundation took over the monitoring of the progress of companies' climate-related disclosures.

However, managing corporate strategies under climate risk presents significant difficulties. Climate change introduces complex risks that span physical impacts, regulatory changes, market shifts, and reputational concerns. Many corporations are not adequately equipped to manage these challenges due to a lack of

---

[1] IFRS Foundation History, 2021, https://www.ifrs.org/about-us/who-we-are/#history, accessed December 3, 2024.

comprehensive risk assessment frameworks, insufficient integration of climate considerations into strategic planning, and limited expertise on climate issues within their boards and management teams. Studies have shown that only a minority of companies have fully embedded climate risk into their governance structures (Khanna & Cespa, 2018; Morkoetter et al., 2015). While awareness of climate risks is increasing, the translation of this awareness into actionable strategies remains limited. This gap underscores the governance crossroads corporations face, necessitating a shift toward more robust governance practices.

To meet stakeholder expectations and manage corporate strategies to ensure long-term resilience and firm value, companies must embed strong governance practices that prioritize sustainability. This involves setting clear, science-based targets for reducing carbon emissions, enhancing sustainability reporting, and engaging in transparent communication about environmental efforts. Boards must be proactive in risk management, ensuring that environmental considerations are embedded in strategic planning and operational decision-making.

Effective governance in the context of climate change also means fostering a culture of accountability, stakeholder engagement, and continuous improvement. Environmental performance should be regularly monitored and reported, and companies should collaborate with a broad range of stakeholders to ensure that strategies are robust across varying time frames, from surviving today's competition to thriving decades from now.

14.2   Current Challenges in Corporate Governance

Corporate governance is at a critical juncture, where the need to address climate change imposes new demands and challenges on corporate structures and decision-making processes. The evolving landscape of environmental risks requires a reassessment of traditional governance practices to ensure long-term sustainability and resilience. This section explores the challenges that corporations face in balancing immediate financial performance with long-term climate goals, the limitations of current ESG practices, the lack of sustainability expertise on boards, and the misalignment of incentive structures.

First and foremost is the difficulty of balancing immediate financial performance with long-term climate goals. Financial markets have the inertia to prioritize short-term gains, pressuring companies to deliver immediate returns to shareholders. This short-termism can conflict with the long-term investments and strategies required for meaningful climate action. Companies may find it difficult to justify significant expenditures on sustainability initiatives when the financial benefits are not immediately apparent (*Financial Times*, 2019). Furthermore, policy ambiguity regarding climate regulations exacerbates this challenge. Inconsistent or unclear policy signals can create uncertainty, making it harder for companies to commit to long-term climate strategies. This tension between short-term financial imperatives and long-term environmental responsibilities requires robust governance to act ambidextrously and navigate effectively.

The lack of clear, standardized climate-related performance metrics presents another significant challenge. While Environmental, Social, and Governance (ESG) practices have gained traction, they often suffer from inconsistencies and lack of rigor. Current ESG metrics can be broad and sometimes fail to capture specific climate-related impacts for adequate financial analysis. Alex Edmans introduces the concept of "rational sustainability," which advocates for integrating sustainability into business practices in a way that aligns with long-term shareholder value (Edmans, 2024). Rational sustainability emphasizes creating long-term value through evidence-based, outcome-focused strategies rather than merely adhering to ESG labels. However, the current state of ESG practices frequently falls short of this ideal, as companies may engage in superficial sustainability efforts without substantial impact. The absence of robust metrics makes it difficult for stakeholders to assess corporate performance on climate issues accurately, hindering accountability and progress.

Board diversity and the presence of sustainability expertise are critical for effective climate-related governance, yet many boards lack these essential elements. Research indicates that diverse boards are more likely to consider a broader range of perspectives and are better equipped to address complex issues like climate change (Jiang et al., 2019). However, many corporate boards still lack adequate representation from individuals with expertise in sustainability or environmental science. This gap can lead to insufficient understanding and prioritization of climate risks and opportunities. Without knowledgeable voices on the board, companies may struggle to develop and implement effective climate strategies (Deloitte, 2020).

Incentive structures within corporations often remain misaligned with the goals of effective climate strategies. Executive compensation and performance bonuses are typically tied to short-term financial metrics rather than long-term sustainability achievements. This misalignment can discourage executives from pursuing ambitious climate goals that might entail upfront costs or longer payoff periods (Ioannou et al., 2019; Li et al, 2020). Aligning incentives with climate performance is crucial for driving meaningful action. For instance, incorporating climate-related targets into executive compensation packages can ensure that corporate leaders are motivated to prioritize sustainability alongside financial performance.

These challenges—balancing immediate financial performance with long-term goals, lack of clear metrics, insufficient board expertise, and misaligned incentives—do not exist in isolation. Instead, they reinforce one another, creating a cycle that perpetuates the status quo. The pressure for short-term financial performance discourages investment in sustainability, while the lack of clear metrics and board expertise further diminishes the emphasis on climate goals. Misaligned incentives then perpetuate these issues by failing to reward long-term, sustainable decision-making. Together, these interconnected challenges create a formidable barrier to the robust governance needed to navigate the climate crisis effectively. Breaking this cycle requires comprehensive governance reforms that address each of these issues holistically, fostering an environment where sustainable practices can thrive and drive long-term corporate success.

14.3 Climate Stewardship and Effective Governance Practices

There is no silver bullet to break the cycle of reinforcing governance challenges outlined in the previous section. To begin, comprehensive governance reforms call for climate and environmental stewardship, a critical aspect of modern corporate governance that aims to integrate a company's responsibility to manage its environmental impact sustainably. Effective governance practices are built around the converging and widely adopted principles of climate and environmental stewardship advocated by the following organizations, which constitute an ever-strong institutional environment for business-society relationship, corporate reporting, climate-related corporate strategy and risk management, policy advocacy and cross-sector collaboration, and stakeholder engagement.

- Global Reporting Initiative (GRI)
- United Nations Principles for Responsible Investment (UNPRI)
- Science-Based Targets Initiative (SBTi)
- Task Force on Climate-related Financial Disclosures (TCFD)
- Task Force on Nature-related Financial Disclosures (TNFD)
- A Roadmap for the Global Energy Sector by International Energy Agency (IEA)
- Circular Economy Principles by Ellen MacArthur Foundation

Effective governance practices in turn are essential to ensure that climate and environmental stewardship is not just a peripheral concern but a core element of corporate strategy and operations. This section explores the key governance practices adopted by leading companies to support climate and environmental stewardship, enhancing both corporate sustainability and long-term business success.

Incorporating environmental sustainability into the core corporate strategy ensures that environmental goals are aligned with business objectives. Companies like Unilever and IKEA have integrated sustainability into their business models, resulting in significant environmental and economic benefits. For instance, Unilever's Sustainable Living Plan aims to decouple its growth from environmental impact while increasing its positive social impact.

Strong board oversight is crucial for ensuring that environmental goals are prioritized and met. Boards should include directors with expertise in environmental issues and sustainability. Regular environmental performance reviews and accountability mechanisms, such as sustainability committees, help ensure that environmental goals are integrated into overall governance. Companies like Microsoft have established board committees dedicated to sustainability oversight.

Transparent reporting on environmental performance and sustainability initiatives is essential for accountability and stakeholder engagement. Adopting reporting frameworks such as the Global Reporting Initiative (GRI) and aligning with the Task Force on Climate-related Financial Disclosures (TCFD) recommendations can enhance transparency. For example, Apple provides comprehensive environmental progress reports detailing their initiatives and performance against set targets.

Engaging with stakeholders, including investors, customers, employees, and communities, is crucial for understanding and addressing their environmental concerns. Companies should engage in regular dialogue with stakeholders to gather input and feedback on sustainability practices. Effective engagement strategies include sustainability reports, public consultations, and collaborative projects. Patagonia, for example, actively involves stakeholders in its environmental initiatives and communicates its sustainability efforts transparently.

Identifying and managing environmental risks as part of the overall risk management framework is essential for long-term sustainability. Companies should conduct regular environmental risk assessments and integrate these findings into their strategic planning and decision-making processes. Tools such as scenario analysis and stress testing can help companies anticipate and mitigate environmental risks. The TCFD provides guidelines for incorporating climate-related risks into financial disclosures, helping companies to better manage these risks.

Aligning executive and employee incentives with environmental performance encourages the adoption of sustainable practices. Companies can link a portion of executive compensation to the achievement of specific environmental targets. For instance, linking bonuses to reductions in carbon emissions or improvements in energy efficiency ensures that leadership is incentivized to prioritize sustainability. Companies like Danone have incorporated environmental targets into their executive compensation packages to drive performance in this area.

14.4   Governance Reform, Director Agency, and Board Diversity

One key success factor for comprehensive governance reforms, arguably the most important one, is board director agency, characterized by intentionality, forethought, self-reactiveness, and self-reflectiveness. Directors who exhibit high levels of personal agency are more likely to proactively set and pursue ambitious goals, anticipate and plan for future challenges, adjust strategies in response to new information, and critically evaluate the outcomes of their decisions (Bandura, 2001). Personal agency enhancement for sustainability is a multifaceted process that involves belief in one's ability to make a difference, alignment of personal value with sustainability goals, expertise in sustainability, and access to enabling tangible and intangible resources such as information, technologies, social support, policy, and incentives (Lozano, 2018; Tur-Porcar et al., 2018).

Studies found there are serious knowledge and experience gaps in the boardroom in climate-related knowledge and expertise that may impact ability to drive future change (Heidrick & Struggles, 2021; Iliev & Roth, 2023). While 75% of boards believe climate change is crucial for their companies' success, 69% of boards do not require climate change knowledge for board membership, and 74% do not prioritize climate change in executive performance metrics. On the other hand, a board that gains sustainability expertise increases a firm's overall sustainability performance by over 7%, and such increases come from improving sustainability practices.

Sustainability expertise may be acquired through recruitment of new directors or education of existing ones. Given the importance and urgency of the reforms emanated by the climate crisis, both approaches should be adopted. This concerns the related issue of board director tenure. The average tenure of board directors in Fortune 100 companies is approximately 8.4 years. This figure is consistent with the broader trends observed across the S&P 500, where the average tenure has been reported to be around 7.8 years. Long-serving directors may struggle to break free from entrenched norms without the infusion of fresh viewpoints and climate-related expertise that newer directors bring. Extended tenures can provide stability and deep organizational knowledge, but they also risk perpetuating the status quo and resisting necessary reforms. This entrenchment is particularly problematic when addressing dynamic challenges like climate change, where adaptability and innovative governance reforms are essential. Directors with long tenures may be less inclined to support transformative changes, preferring familiar strategies that align with their historical experiences, thus reinforcing existing practices that may be inadequate for new environmental realities.[2]

Diversity in board composition, including varying tenures and expertise, is crucial for breaking the cycle of reinforcing challenges and the perpetuation of the status quo. A diverse board can blend the benefits of experience with innovative thinking, creating a dynamic environment that is more responsive to the need for governance reforms. Studies have shown that diverse boards are better equipped to address complex issues, as they incorporate a wider range of perspectives and are more likely to challenge conventional wisdom (Forbes, 2020; Harvard Business Review, 2021). By ensuring a mix of tenures and fostering an inclusive culture that values different viewpoints, boards can enhance their collective behavior, driving more effective and forward-thinking governance practices. This diversity is essential for addressing the significant challenges posed by climate change, leading to more robust and sustainable corporate strategies.

In conclusion, climate and environmental stewardship is a critical component of effective corporate governance. By integrating sustainability into their core strategies, ensuring robust board oversight, maintaining transparency, engaging stakeholders, incorporating sustainability into risk management, and aligning incentives with environmental performance, companies can manage their environmental impact effectively. Despite the challenges, the benefits of strong

---

[2] 2023 Spencer Stuart Board Index, https://www.spencerstuart.com/-/media/2023/september/usbi/2023_us_spencer_stuart_board_index.pdf?sc_trk=BDB9A48933CA433C9DDD7D4E85D62A38, accessed December 3, 2024.

environmental stewardship far outweigh the costs, leading to enhanced reputation, operational efficiency, and investor confidence. Corporate governance reform is necessary to enable climate and environmental stewardship, and board diversity and improved board director agency is key to kick start and accelerate the governance reform.

14.5   Case Study: ExxonMobil vs. CalPERS

ExxonMobil is a global leader in the oil and gas industry, known for its vast operations and considerable influence within the sector. The company's governance structure includes a board of directors composed of individuals with extensive experience in business, finance, and energy. However, this board has historically faced criticism for lacking sufficient expertise in environmental sustainability and climate science. Despite recent efforts to enhance its climate policies, including investments in carbon capture and storage (CCS) and support for carbon pricing, ExxonMobil's strategies have been viewed by many as insufficient to address the growing climate crisis comprehensively.

The California Public Employees' Retirement System (CalPERS) is one of the largest public pension funds in the United States, managing around $469 billion in assets. As a major institutional investor, CalPERS holds significant sway over the companies in which it invests. Known for its advocacy for strong corporate governance and sustainability, CalPERS has pushed for greater transparency and action on climate-related risks, making it a formidable force in demanding climate accountability from companies like ExxonMobil.

The conflict between ExxonMobil and CalPERS centers on climate risk management and governance. CalPERS, along with other institutional investors, has consistently urged ExxonMobil to adopt more stringent climate risk assessments and disclosures. Specific demands included setting more ambitious greenhouse gas reduction targets and enhancing transparency regarding climate-related financial risks.

In early 2024, ExxonMobil filed a lawsuit against two shareholder groups, Arjuna Capital and Follow This, who had submitted climate-related proposals. This legal action prompted significant backlash from major investors, including the California Public Employees' Retirement System (CalPERS), which announced its intention to vote against all ExxonMobil board members in the May 2024 annual meeting.[3] This case underscores the escalating tensions between corporate management and shareholders over climate governance.

Despite withdrawal of the proposal by the activist investors, ExxonMobil pressed ahead, hoping to win a ruling that would set a precedent for such cases. Thus, the conflict escalated, highlighting a deep divide over climate strategies

---

[3] Calpers to Vote Against Exxon Board Members, Citing Lawsuit Against Arjuna and Follow This, Wall Street Journal, May 20, 2024.

and governance. ExxonMobil's management maintained that the proposals were impractical and detrimental to shareholder value, while critics argued that the company's aggressive legal tactics undermined shareholder democracy and effective climate governance.[4]

ExxonMobil's actions in response to CalPERS and other shareholders reveal significant inconsistencies in its climate-related strategies. While the company has publicly committed to addressing climate change, its resistance to adopting more rigorous climate risk management and disclosure practices suggests a reluctance to embrace necessary governance reforms fully. This reluctance undermines the credibility of ExxonMobil's environmental stewardship and raises questions about its long-term strategic vision.

ExxonMobil's strategy appears heavily focused on short-term financial performance, often at the expense of long-term sustainability imperatives. This approach is increasingly at odds with the expectations of shareholders and regulators, who emphasize the importance of robust climate risk management for long-term financial stability.

The key takeaways from this case study include the necessity for corporations to align their governance structures with comprehensive climate risk management and the critical role of transparency in climate-related disclosures. For other companies, the ExxonMobil vs. CalPERS conflict underscores the importance of engaging constructively with shareholders on climate issues and integrating sustainability into core business strategies. Companies that fail to do so risk facing similar conflicts, potential litigation, and damage to their reputation and investor relationships.

14.6   Climate-Related Innovations in Corporate Governance

The most noticeable innovation is the integration of climate risk into financial reporting. The Task Force on Climate-related Financial Disclosures (TCFD) has developed guidelines that encourage companies to disclose how climate-related risks and opportunities impact their financial statements. This includes the analysis of both physical risks (such as those from extreme weather events) and transition risks (such as policy changes and shifts in market demand) related to climate change (TCFD, 2017). Incorporating these risks into financial disclosures helps investors better understand the long-term viability of a company's strategy in the context of climate change.

Companies are increasingly adopting science-based targets to align their strategies with the goals of the Paris Agreement. The Science-Based Targets initiative (SBTi) provides a framework for companies to set greenhouse gas reduction targets that are consistent with keeping global warming below 2 °C above pre-industrial

---

[4] ExxonMobil rediscovers its swagger. Economist, May 29, 2024.

levels (Science-based Targets Initiative, 2023). This approach ensures that corporate actions are scientifically grounded and contribute meaningfully to global climate goals.

Enhanced stakeholder engagement is another innovation in climate governance. Companies are moving toward more inclusive practices that involve shareholders, employees, customers, and communities in their sustainability initiatives. This can include regular dialogues, sustainability committees, and collaborative projects aimed at addressing climate-related issues. Engaging stakeholders helps companies align their strategies with societal expectations and enhances transparency and accountability.

The use of digital technologies, such as artificial intelligence (AI) and blockchain, is transforming climate governance. AI can optimize energy use and predict environmental impacts, while blockchain can ensure transparency in carbon trading and supply chain management.[5] These technologies enable more precise tracking of emissions and verification of sustainability claims, thereby enhancing the credibility of corporate sustainability efforts.

Linking executive compensation to climate performance metrics is another innovative approach. Companies like Danone and Shell have begun tying bonuses and other incentives to the achievement of specific environmental targets, such as reductions in carbon emissions or improvements in energy efficiency (Danone, 2021; Shell, 2021). This alignment ensures that senior management is directly accountable for the company's environmental performance, fostering a culture of sustainability at the highest levels of leadership.

The development of green financial products, such as carbon credits and offsets, green bonds, sustainability-linked loans, and insurance products associated with nature-based solutions, represents a growing trend in climate finance.[6,7] These financial instruments are designed to fund projects that have positive environmental impacts, such as renewable energy installations, energy efficiency upgrades, or emission reduction. They provide companies with the financial resources needed to invest in sustainable practices while also offering investors opportunities to support environmentally beneficial projects.

Establishing board-level sustainability committees is becoming a frequent practice among leading companies. These committees are tasked with overseeing the company's sustainability strategy, ensuring that environmental considerations are

---

[5] PwC-Microsoft, "How AI can enable a sustainable future." https://www.pwc.co.uk/sustainability-climate-change/assets/pdf/how-ai-can-enable-a-sustainable-future.pdf, accessed December 3, 2024.

[6] International Finance Corporation. (2021). "Green Bond Market." https://www.ifc.org/content/dam/ifc/doc/mgrt/202206-emerging-market-green-bonds-report-2021-vf-2.pdf, accessed December 3, 2024.

[7] ICMA Group, "Green Bond Principles—2022." https://www.icmagroup.org/sustainable-finance/the-principles-guidelines-andhandbooks/green-bond-principles-gbp/, accessed December 3, 2024.

integrated into business decisions. They provide a structured approach to monitoring and advancing climate-related initiatives, thereby strengthening governance frameworks.

These innovations in corporate governance reflect a broader shift toward integrating climate considerations into core business practices. By adopting these strategies, companies can enhance their resilience to climate risks, meet stakeholder expectations, and contribute to global sustainability efforts. The continuous evolution of these practices highlights the dynamic nature of corporate governance in addressing the pressing challenge of climate change.

14.7   Future Directions in Climate Governance

As the world faces the mounting impacts of climate change, the corporate sector is increasingly under pressure to evolve and innovate in its governance practices. This section explores the future directions in climate governance, highlighting the integration of advanced technologies, evolving regulatory and policy frameworks, and the crucial role of corporate culture and leadership in driving sustainable business practices. By examining these areas, we can better understand how companies can navigate the complexities of climate risks, meet stakeholder expectations, and secure long-term resilience and success.

Climate and digital technologies. As the urgency of climate change becomes increasingly apparent, new trends and technologies are emerging to address this global challenge. Among the most significant trends are the advancements in renewable energy technologies, such as solar and wind power, which continue to become more efficient and cost-effective. Innovations in energy storage, like advanced battery technologies, are crucial for overcoming the intermittency of renewable energy sources, making them more reliable and widespread. The rise of hydrogen as a clean energy source, especially green hydrogen produced using renewable energy, is gaining momentum as a potential game-changer in sectors that are difficult to decarbonize, such as heavy industry.[8]

Carbon capture and storage (CCS) technology is another critical area of development. The oil majors are investing heavily in CCS to mitigate CO_2 emissions and comply with stricter environmental regulations. Unlike CCS or other Emission Mitigation Technologies (EMT), Negative Emission Technologies (NET) offer a way to actively remove CO_2 from the atmosphere and reverse some of the existing atmospheric CO_2 accumulation. NET has become increasingly prominent due to scientific assessments indicating that simply reducing emissions will not be enough to avoid severe climate impacts.

---

[8] International Energy Agency. (2021). "Global Hydrogen Review 2021." https://www.iea.org/reports/global-hydrogen-review-2021, accessed December 3, 2024.

Digital technologies, including artificial intelligence (AI) and blockchain, are also playing a pivotal role. AI can optimize energy use and improve the efficiency of renewable energy systems, while blockchain technology is being used to enhance transparency and accountability in carbon trading and improve supply chain and environmental data management. These technologies facilitate more accurate tracking of emissions and verification of corporate sustainability claims.

Emerging climate technologies are critical for corporate governance reforms as they enhance sustainability, ensure compliance with regulations, and meet stakeholder expectations. Advancements in renewable energy technologies, such as solar, wind, and advanced battery storage, make energy systems more reliable and cost-effective, supporting companies' transitions to sustainable practices. The rise of green hydrogen and carbon capture and storage (CCS) technologies demonstrates corporate commitment to reducing and reversing carbon footprints, aligning with long-term climate goals. Digital technologies like AI and blockchain enhance transparency and accountability in carbon trading and supply chain management, fostering accurate environmental reporting and building stakeholder trust (World Economic Forum, 2020). Integrating these technologies into corporate strategies requires innovative governance frameworks, promoting a culture of continuous improvement and resilience against climate risks.

Political and legal environments. Regulatory and policy frameworks are evolving rapidly to drive corporate climate governance. The European Union's Corporate Sustainability Reporting Directive (CSRD), effective from January 2023, mandates comprehensive sustainability reporting for a broad set of large companies, as well as listed SMEs operating within the EU. This directive aims to enhance transparency and accountability in corporate sustainability practices, influencing global governance standards.[9] The United States is also seeing significant developments. The Securities and Exchange Commission (SEC) took a monumental step on March 6, 2024 toward integrating climate risk into the fabric of public company reporting with the adoption of the "Enhancement and Standardization of Climate-Related Disclosures for Investors" final rule (the Final Rule), addressing growing investor demand for transparent, reliable, and comparable climate-related information.

However, the Final Rule is facing extraordinary legal challenges across the United States and the ideological spectrum, with critics arguing that it exceeds the SEC's authority by overstepping into environmental regulation rather than sticking to its core mandate of ensuring fair, orderly, and efficient markets.[10] They argue that the SEC's rules may violate the Administrative Procedure Act (APA)

---

[9] New rules on corporate sustainability reporting: The CSRD Directive, https://finance.ec.europa.eu/capital-markets-union-and-financial-markets/company-reporting-and-auditing/company-reporting/corporate-sustainability-reporting_en, accessed December 3, 2024.

[10] Long-Awaited SEC Climate Disclosure Rule Draws Legal Challenges Across the Ideological Sepctrum, Thompson Coburn LLP, https://www.thompsoncoburn.com/insights/publications/item/2024-04-09/long-awaited-sec-climate-disclosure-rule-draws-legal-challenges-across-the-ideological-spectrum, accessed December 3, 2024.

by being arbitrary and capricious, lacking substantial evidence that enhanced disclosures will benefit corporate performance and shareholder value. Additionally, there are First Amendment concerns, with opponents claiming that mandatory disclosures could be seen as compelled speech, forcing companies to adopt the SEC's policy views on climate change.

Financial industry groups, such as the Bank Policy Institute and the Financial Services Forum, have criticized the proposed requirements for being overly broad and costly to implement, particularly the Scope 3 emissions disclosures, which they argue are complex and could lead to regulatory redundancy (American Banker, 2024; DLA Piper, 2024). These critics of the Final Rule also contend that such mandates would impose undue burdens on American energy producers, potentially limiting capital access and increasing energy prices. Additionally, a coalition of 16 Republican attorneys general, backed by the American Legislative Exchange Council, has challenged the SEC's authority, arguing that the rules represent an overreach of ESG policies that could unfairly favor "green" companies over traditional energy firms.[11] The new administration under President Donald Trump will almost certainly slow down the implementation of the Final Rule, if not reverting the hard-fought shift toward integrating climate risk into financial reporting.

Legal and procedural challenges have also been significant against the European Union's Corporate Sustainability Reporting Directive (CSRD). Business groups argue that the extensive reporting requirements, which cover a broad range of sustainability issues beyond climate change, are overly burdensome and costly, especially for SMEs. They contend that the CSRD's comprehensive nature, which includes detailed disclosures on the entire value chain, imposes significant administrative and financial burdens that could be detrimental to business operations and competitiveness. Legal pushbacks against the CSRD primarily focus on the scope and complexity of the mandated disclosures. Critics argue that the requirement to report on value chain impacts is impractical and poses significant challenges in data collection and verification. There are also concerns about the potential for increased litigation and reputational risks due to mandatory disclosures of negative environmental impacts. Additionally, some argue that the directive's broad definition of materiality, which includes environmental and societal impacts, could lead to inconsistent and subjective reporting, complicating compliance, and enforcement.[12]

The political and legal environment under climate change is fraught with uncertainties, as evidenced by the contentious debates over regulatory mandates like the SEC's climate disclosure rules and the EU's Corporate Sustainability Reporting

---

[11] Parallel ESG Investigations: Reading the Congressional Tea Leaves, Brownstein Client Alert, May 15, 2023, https://www.bhfs.com/insights/alerts-articles/2023/parallel-esg-investigations-reading-the-congressional-tea-leaves, accessed December 3, 2024.

[12] Materiality: The Word that Launched a Thousand Debates, posted on Harvard Law School Forum on Corporate Governance, https://corpgov.law.harvard.edu/2021/05/14/materiality-the-word-that-launched-a-thousand-debates/, accessed December 3, 2024.

Directive (CSRD). These uncertainties stem from varying political agendas, legal challenges, and stakeholder resistance, which can create an unpredictable operating landscape for corporations. Corporate governance reform is essential in this context, as it equips companies with robust frameworks for risk management, compliance, and strategic adaptation. By embracing governance reforms, corporations can foster a culture of continuous learning and agility, enabling them to better navigate regulatory changes, anticipate legal challenges, and respond effectively to stakeholder demands. This proactive approach not only mitigates risks but also positions companies to capitalize on opportunities arising from the transition to a sustainable economy, ensuring long-term resilience and competitiveness.

The role of corporate culture and leadership. Corporate culture and leadership are fundamental to effective climate governance. Companies with a strong sustainability culture are more likely to integrate environmental considerations into their core strategies and operations. However, it is essential to ensure that these commitments are genuine and not merely superficial efforts at greenwashing. Companies should set measurable sustainability goals and report their progress transparently to build credibility and trust with stakeholders.

Cultural changes within corporations are crucial. To cultivate a robust sustainability culture, companies should integrate sustainability metrics into performance evaluations, offer training programs, and encourage employee-led sustainability initiatives. This approach ensures that sustainability is embedded at all organizational levels, fostering a genuine commitment to environmental stewardship.

Inclusive and diverse boards are increasingly recognized as vital for addressing complex sustainability challenges and making informed decisions. To achieve this diversity, companies should set diversity targets, implement unbiased recruitment processes, and provide ongoing education on environmental and social governance issues. These steps will help overcome resistance and ensure that diverse perspectives are represented in decision-making.

One notable trend is the rise of Chief Sustainability Officers (CSOs) in the C-suite, highlighting the importance of sustainability in corporate leadership. Companies like Microsoft and Danone have appointed CSOs to ensure that sustainability is embedded at the highest levels of decision-making. CSOs play a critical role in influencing corporate strategy, driving cross-departmental collaboration, and advocating for sustainability initiatives. It is essential that CSOs have direct access to the board to effectively champion these efforts.

Despite these advancements, resistance to cultural change remains a significant challenge. Many organizations may face internal pushback from stakeholders accustomed to traditional business practices and skeptical of new sustainability initiatives. Overcoming this resistance requires strong leadership and clear communication of the long-term benefits of sustainability. Engaging stakeholders through inclusive dialogue, demonstrating quick wins, and highlighting successful case studies can help build support for climate-related governance reforms. Additionally, aligning sustainability goals with financial performance and demonstrating their impact on the company's bottom line can further mitigate resistance.

Corporate culture and leadership play a critical role in effective climate governance. By promoting authentic sustainability practices, cultivating a robust sustainability culture, achieving board diversity, and empowering CSOs, companies can navigate the complexities of climate governance and drive meaningful change. These efforts will enhance corporate resilience, meet stakeholder expectations, and ensure long-term sustainability.

Conclusion

As the world stands at the crossroads of unprecedented environmental challenges, corporate governance emerges as a cornerstone of resilience and sustainability. This chapter has underscored the transformative power of robust governance frameworks in addressing the multifaceted risks posed by climate change. Companies can no longer afford to view sustainability as a peripheral concern; it must be embedded in the core of corporate strategies, ensuring that immediate financial objectives align seamlessly with long-term environmental stewardship.

The evolution of corporate governance—from profit-centric models to frameworks that integrate Environmental, Social, and Governance (ESG) considerations—marks a pivotal shift in how businesses operate. However, this transition is fraught with challenges, from balancing short-term market pressures against long-term climate imperatives to addressing gaps in board expertise and aligning incentive structures with sustainability goals. Overcoming these hurdles requires a holistic approach, combining visionary leadership, stakeholder engagement, and innovative governance practices.

The case study of ExxonMobil and CalPERS vividly illustrates the tensions between traditional governance models and the demands of modern climate accountability. It serves as a cautionary tale, emphasizing the need for transparency, stakeholder alignment, and the integration of climate risks into corporate decision-making.

Looking ahead, the future of corporate governance will be shaped by the rapid advancements in renewable energy, digital technologies, and regulatory frameworks. Companies that proactively adopt these innovations, cultivate diverse and inclusive leadership, and foster a culture of genuine sustainability will not only mitigate risks but also unlock new opportunities for growth and value creation.

In closing, the integration of climate and environmental stewardship into governance is not merely an ethical imperative; it is a strategic necessity. Businesses that rise to this challenge will be better positioned to navigate the complexities of the climate crisis, earning the trust of stakeholders and securing their place in a sustainable future. By embracing governance reforms and prioritizing sustainability, corporations can transform potential vulnerabilities into enduring strengths, charting a path toward resilience, innovation, and shared prosperity.

Key Takeaways

- **Evolution of Corporate Governance**
  - Corporate governance has significantly evolved from informal oversight to structured frameworks focusing on shareholder value, particularly influenced by Milton Friedman's profit-centric model.
  - The 2008 financial crisis highlighted the critical need for robust governance frameworks to manage systemic risks and led to reforms like the Dodd-Frank Act and a shift toward integrating ESG criteria into corporate strategies.
- **Current Governance Crossroads:**
  - The climate crisis presents ongoing, escalating risks requiring comprehensive governance approaches that integrate environmental, social, and financial considerations.
  - Unlike past financial crises, climate change demands a holistic approach to governance, emphasizing proactive risk management and long-term sustainability.
- **Challenges in Corporate Governance:**
  - Balancing short-term financial performance with long-term climate goals is challenging due to market pressures and policy ambiguities.
  - ESG practices often lack clear, standardized metrics, making it difficult to assess corporate performance on climate issues accurately.
  - Many boards lack sustainability expertise, hindering effective climate-related governance.
  - Misaligned incentive structures discourage long-term climate strategies.
- **Climate Stewardship and Governance Practices:**
  - Effective governance practices include integrating environmental sustainability into core strategies, robust board oversight, transparent reporting, stakeholder engagement, risk management, and aligning incentives with environmental performance.
  - Companies like Unilever, Microsoft, and Danone exemplify these practices through their sustainability initiatives and governance structures.
- **Governance Reform, Director Agency, and Board Diversity:**
  - Board director agency, characterized by intentionality, forethought, self-reactiveness, and self-reflectiveness, is crucial for driving governance reforms.
  - Sustainability expertise on boards can significantly enhance corporate sustainability performance.
  - Board diversity, including varying tenures and expertise, is essential for addressing the complex challenges of climate change and breaking the cycle of reinforcing governance challenges.
- **Case Study: ExxonMobil vs. CalPERS:**
  - The conflict between ExxonMobil and CalPERS highlights the need for rigorous climate risk management and transparency.
  - ExxonMobil's resistance to adopting comprehensive climate strategies and disclosures underscores the importance of aligning governance structures with climate risk management.

- **Climate-Related Innovations in Corporate Governance**:
  - Innovations include integrating climate risk into financial reporting, adopting science-based targets, enhancing stakeholder engagement, leveraging digital technologies like AI and blockchain, linking executive compensation to climate performance, developing green financial products, and establishing board-level sustainability committees.
- **Future Directions in Climate Governance**:
  - Emerging trends and technologies, such as advancements in renewable energy, energy storage, hydrogen, and digital technologies, are crucial for enhancing corporate sustainability.
  - Regulatory and policy frameworks, such as the EU's Green Deal and the SEC's proposed climate risk disclosures, are evolving to drive corporate climate governance.
  - Corporate culture and leadership, including the rise of CSOs and fostering inclusive and diverse boards, play a fundamental role in effective climate governance.

Questions

1. Explain the evolution of corporate governance from profit-centric models to ESG integration. What factors drove this transition?
2. What are the key challenges companies face when balancing short-term financial performance with long-term climate goals?
3. Discuss the role of board diversity and sustainability expertise in effective climate governance.
4. How do science-based targets and digital technologies enhance corporate climate governance? Provide examples.
5. Using the ExxonMobil vs. CalPERS case study, analyze the importance of aligning governance structures with climate accountability.
6. What future trends in technology, regulatory frameworks, and corporate culture will shape climate governance?
7. How can companies align executive compensation with climate performance metrics? Why is this alignment critical?
8. Describe the significance of integrating sustainability into risk management frameworks. How does this improve corporate resilience?

References

American Bank. (2024). *Bank groups blast SEC climate rule despite Scope 3 concessions*, March 7, 2024. https://www.americanbanker.com/news/bank-groups-blast-sec-climate-rule-despite-scope-3-concessions

Bandura, A. (2001). Social cognitive theory: An agentic perspective. *Annual Review of Psychology*, *52*, 1–26. https://doi.org/10.1146/annurev.psych.52.1.1

Danone. (2021). *One planet. One health*. Danone. Accessed November 4, 2024 from: https://www.danone.com/impact/planet.html

Deloitte. (2020). *Board diversity: A progress report*. Deloitte Insights. Accessed November 4, 2024 from: https://www2.deloitte.com/us/en/pages/center-for-board-effectiveness/articles/missing-pieces-report-board-diversity.html

DLA Piper. (2024). *California climate disclosure laws: Considerations for compliance*, January 11, 2024. https://www.dlapiper.com/en/insights/publications/2024/01/california-climate-disclosure-laws-considerations-forcompliance

Edmans, A. (2024). Rational sustainability. *Journal of Applied Corporate Finance*, forthcoming. Accessed November 4, 2024 from: https://doi.org/10.1111/jacf.12609

Financial Crisis Inquiry Commission. (2011). The Financial Crisis Inquiry Report.

Financial Times. (2019). Short-termism, shareholder primacy, and securities markets law: Reexamining the relationship. *Financial Times*.

Forbes. (2020). *Board diversity and performance*. Forbes. https://www.forbes.com/

Friedman, M. (1970). The social responsibility of business is to increase its profits. *The New York Times Magazine*. https://www.nytimes.com/1970/09/13/archives/a-friedman-doctrine-the-social-responsibility-of-business-is-to.html

Harvard Business Review. (2021). The board's role in sustainability. Harvard Business Review. https://hbr.org/

Heidrick & Struggles. (2021). *Changing the climate in the boardroom*. https://www.heidrick.com/-/media/heidrickcom/publications-and-reports/changing-the-climate-in-the-boardroom.pdf

Iliev, P., & Roth, L. (2023). Director expertise and corporate sustainability. *Review of Finance*, *27*(6), 2085–2123. https://doi.org/10.1093/rof/rfad012

Ioannou, I., et al. (2019). Incentive pay and corporate climate change targets. *Strategic Management Journal*, *40*(7), 1097–1118. https://doi.org/10.1002/smj.3005

Jiang, C., et al. (2019). Corporate environmental responsibility and firm performance: Does board diversity matter? *Journal of Business Ethics*, *155*(2), 497–511. https://doi.org/10.1007/s10551-017-3505-0

Khanna, V., & Cespa, G. (2018). Climate change and corporate governance: Evidence from weather shocks. *Journal of Finance*, *73*(4), 2059–2096. https://doi.org/10.1111/jofi.12628

Kirkpatrick, G. (2009). *The corporate governance lessons from the financial crisis*. OECD.

Li, J., et al. (2020). The impact of executive compensation on corporate social responsibility. *Journal of Business Ethics*, *163*(1), 69–88. https://doi.org/10.1007/s10551-018-3928-5

Lozano, R. (2018). Scrutinizing sustainability change and its institutionalization in organizations. *Frontiers in Sustainability*.

Morkoetter, S., et al. (2015). Corporate governance and climate change: The banking sector. *Journal of Business Finance & Accounting*, *42*(9–10), 1247–1273. https://doi.org/10.1111/jbfa.12145

Science Based Targets Initiative. (2023). *Science-based targets for corporates*. https://sciencebasedtargets.org/

Shell. (2021). *Shell's climate target*. Shell. https://www.shell.com/sustainability/

Smyth, J. (2024, May 28). Exxon vs CalPERS: Inside the clash over shareholder rights and fiduciary duty. *Financial Times*.

Task Force on Climate-related Financial Disclosures (TCFD). (2017). *Recommendations of the Task Force on Climate-related Financial Disclosures*. Financial Stability Board. https://www.fsb-tcfd.org/publications/

Tur-Porcar, A., Roig-Tierno, N., & Llorca Mestre, A. (2018). Factors affecting entrepreneurship and business sustainability. Sustainability, *10*(3), 1–15. https://doi.org/10.3390/su10020452

The End and Beginning: The Shift Toward Stakeholder Capitalism

Introduction

In an era marked by tumults in the global political and economic landscape, it is useful to step back from the unrelenting push toward a stakeholder-centric corporate model and explore how it has reached this point in corporate history along with the evolving paradigms in capitalism. Tracing the journey from the emerging world of shareholder centrism seventy or so years ago to the nuanced realms of stakeholderism, this chapter delves into the intricate interplay between corporate strategies, governance, and financial frameworks amid this shift. The objectives are twofold. First is to examine the brief history of modern capitalism and its profound manifestations of the evolution in human and corporate behavior, societal expectation, and technology. Based on the lessons learned, we intend to project how the burgeoning role of technology, particularly Artificial General Intelligence, may shape a sustainable future for humanity and the planet.

15.1 The Rise of Shareholder Capitalism: Corporate Goliaths and the Shareholder Crown

The Industrial Revolution, a scene straight out of a Dickens novel, marked the initial rise of capitalism. This era, bursting with steam engines and smoky factories, witnessed the birth of a system where production and market expansion were the beating heart. Government intervention was as scarce as hen's teeth in this laissez-faire playground. "The chief business of the American people is business,"

S. Dow and Y. Shi, *Corporate Finance Under Climate Crisis*,
https://doi.org/10.1007/978-3-031-83487-5_15

Calvin Coolidge might have quipped,[1] capturing the ethos of the age with unerring precision.

Businesses mushroomed, and with them, the ideology of free market capitalism gained unbridled momentum. This was a time when markets reigned supreme, orchestrated by the Invisible Hand, a term Adam Smith originally mentioned in his 1759 work Theory of Moral Sentiments, but became known from his main work The Wealth of Nations in 1776. It implied the unintended greater social impacts brought about by individuals acting in their own self-interests.[2]

The Birth of Corporate Goliaths

Then came the post-World War II era, a period marked by a robust economic boom that reshaped the global economic landscape. Corporations rose like Goliaths, becoming central to economies and growing in size and influence. These corporations required significant capital, more than individual proprietors or partnerships could provide, leading to the proliferation of joint-stock companies where multiple investors (shareholders) could pool resources, while their personal assets were protected from corporate debts or legal liabilities since corporations were granted legal personhood, with rights and responsibilities separate from their owners.

This limited liability encouraged investment by incentivizing more people to invest in corporations. This period of boom and expansion saw an ever-expanding base of investors whose interests needed to be managed. This era was akin to an economic symphony, with corporations and shareholders as the lead players in a grand capitalist concerto that resonated globally.

Foundational Theories: Agency Problem and Market Primacy

Enter Eugene Fama and Michael Jensen, two prominent economists who helped lay the theoretical foundation for shareholder supremacy, about fifty years after Henry Ford lost his case in the Michigan Supreme Court to retain considerable earnings from the Ford Motor Company against the wishes of the Dodge Brothers, minority shareholders of the company.[3] Jensen's best-known work is the 1976 paper he co-authored with William H. Meckling, *Theory of the Firm: Managerial Behavior, Agency Costs and Ownership Structure* (Jensen & Meckling, 1976). One of the most widely cited economics papers of the last 40 years, it painted the public corporation as an ownerless entity, made up of only contractual relationships between principals (shareholders) and agents (managers), with potential conflicts

---

[1] Source: "The Press Under a Free Government: Address to the American Society of Newspaper Editors," on January 17, 1925, in Foundations of the Republic, University Press of the Pacific, 2004, ISBN: 9781410215987.

[2] Source: https://en.wikipedia.org/wiki/Invisible_hand, accessed December 3, 2024.

[3] Dodge v. Ford Motor Co, 204 Mich. 459, 170 N.W. 668 (Mich. 1919).

when their interests diverged and incentives not perfectly aligned. They unveiled the complexity in corporate dynamics and called it the "agency problem."

Fama's Efficient Market Hypothesis (EMH) portrayed markets as oracles; they are "informationally efficient," meaning that stock prices at any given time fully reflect all available information and collective wisdom (Fama, 1970). This hypothesis underpins the argument that the market is the best judge of a company's value, which in turn supports the idea that maximizing shareholder value (as reflected in stock prices) should be a primary goal of corporations. Stock markets became the ultimate resource allocator and the perfect arbiter for solving the agency problem. Here, the concept of shareholder primacy crystallized, emphasizing both market primacy and the maximization of shareholder wealth.

Milton Friedman and the Era of Shareholder Value

"There is one and only one social responsibility of business—to use its resources and engage in activities designed to increase its profits," Milton Friedman, Fama's fellow Nobel Prize winner, pronounced in his New York Times op-ed on September 13, 1970, setting the tone for decades of corporate America (Friedman, 1970). Their theories became the gospel of the corporate world, guiding strategies and actions.

In the real world, these influential theories translated into corporate policies that placed shareholder value on a pedestal. The focus of corporate governance shifted on performance-based executive compensation as a mechanism to align the interests of managers with shareholders. This has been a key feature of shareholderism, where executive rewards are often tied to stock performance, theoretically incentivizing management to prioritize shareholder value. Practices such as share buybacks and strategic mergers were justified as tributes to the altar of shareholder interests.

15.2   The Emergence of Stakeholderism: A New Melody in the Corporate Symphony

Challenging the Status Quo: Shifts in Academic Thought

As the twentieth century approached its twilight, a profound change began to unfurl in the academic world (Freeman & Reed, 1983), heralding the rise of stakeholderism. This was not merely a subtle shift in business philosophy but a revolutionary change in how corporations viewed their role in society. No longer were the interests of shareholders the sole beacon guiding corporate decisions; a broader, more inclusive perspective began to take root, considering the welfare of employees, customers, suppliers, communities, and the environment alongside the traditional focus on shareholders.

The Efficient Market Hypothesis (EMH) suggested that markets efficiently incorporate all available information into stock prices. In a stakeholder-centric approach, this hypothesis was challenged to consider how well markets reflect the interests and values of broader stakeholders, not just financial performance, certainly not short-term maximal profits that appeal to numerous stockholders and drive stock prices from time to time. Stakeholder capitalism emphasizes long-term value creation over short-term profit maximization. This shift challenges the traditional shareholder-centric view by arguing that sustainable long-term value for shareholders is achieved by balancing the interests of all stakeholders.

Traditionally, agency theory primarily focused on the relationship between shareholders (principals) and managers (agents). Stakeholder capitalism expanded this framework by considering multiple principals (all stakeholders) with diverse and sometimes conflicting interests. This complexity required a rethinking of how neglecting non-shareholder stakeholders might cost the corporation and how these new agency conflicts, existing or potential, were identified and managed. For example, focusing solely on shareholder interests, especially short-term ones, at the expense of environmental sustainability could lead to long-term costs for the company, such as reputational damage or regulatory penalties.

Adapting these theories that debuted stakeholder capitalism involved broadening the scope of who constitutes a principal, redefining agency costs, focusing on long-term value, and redesigning corporate governance structures to be more inclusive of diverse stakeholder interests. It was by no means an easy call for change. Some argued that the shift toward stakeholders put company directors on the spot to make near-impossible trade-offs and its best-case scenario would be just enlightened shareholderism (Bebchuk & Tallarita, 2020).

Expanding the Definition of Corporate Responsibility and the Rise of ESG

Nevertheless, the impact of this shifting focus was profound. Companies started to integrate corporate social responsibility (CSR) and sustainability into their core strategies. Business models were reimagined to balance economic objectives with social and environmental considerations. This change was not just about adhering to regulatory requirements or mitigating risks; it was about redefining the purpose of business in society. Companies began to recognize that their long-term success was intrinsically linked to the well-being of the communities they served, the health of the environment, and the fairness of their practices.

The increasing importance of environmental, social, and governance (ESG) factors challenged the traditional emphasis on financial metrics for assessing corporate performance. But integrating ESG metrics into the evaluation of managerial performance and aligning executive compensation with both financial and non-financial goals still had a long way to go in solving the new agency problem.

With the emerging stakeholderism, the role of corporate leaders evolved. They were no longer just stewards of shareholder wealth but guardians of a broader set of interests. This required a new set of skills and a deeper understanding of the complex web of relationships that sustain a business. Leaders had to engage with a diverse range of stakeholders, understand their concerns and aspirations, and find ways to align them with the company's objectives.

The shift toward stakeholderism also had implications for corporate governance. Traditional governance structures, designed to serve shareholders' interests, were re-evaluated. New models of governance emerged, emphasizing transparency, accountability, and inclusivity. Boards of directors expanded their focus to include environmental, social, and governance issues. Voluntary and mandatory reporting and disclosure practices evolved to provide a more comprehensive view of a company's performance, encompassing both financial and non-financial metrics.

This evolution in the corporate world, characterized by a significant shift in focus from shareholders to a broader spectrum of stakeholders, marks the Great Shift toward stakeholderism. It is not just a change in business strategy, organizational routines, or performance management; it represents a paradigm shift in the very essence of corporate identity and purpose.

15.3 What Really Caused the Great Shift from Shareholder Primacy?

What sparked the decade-long departure from the relentless pursuit of shareholder profits to a broader, more inclusive vision of corporate responsibility? The answer lies in the seismic changes reshaping our world, technological revolutions, a globalized marketplace, and a rising tide of social and environmental awareness. As climate crises intensify, inequality deepens, and younger generations demand ethical action, businesses are no longer insulated from societal expectations. This transformation isn't just a fleeting trend; it's a fundamental rethinking of what it means to succeed in business. Why did the old model falter, and what forces ignited this revolution? The Great Shift invites us to explore the catalysts behind this new corporate ethos, one where companies thrive by serving the many, not just the few.

Globalization and the Information Revolution

The late 20th and early twenty-first centuries brought globalization and the information revolution. Advances in technology and the rise of social media made information more accessible, increasing public awareness of corporate practices and their impacts on society and the environment and demand for greater transparency. Rising concerns about climate change, social inequality, and ethical governance led to a re-evaluation and growing expectation for businesses to act responsibly and ethically, addressing social and environmental issues alongside profit-making.

Societal and Cultural Pressures: Climate Change and Inequality

Worldwide population growth and global economic development prompted concerns over the depletion of natural resources that have forced businesses to consider the long-term sustainability of their operations. Consumers were increasingly favoring companies that demonstrate ethical practices and sustainability, influencing market trends. There was a growing desire among employees to work for organizations that aligned with their values, impacting talent attraction and retention.

Companies that considered the broader implications of their actions were often better positioned to mitigate risks. There was a growing recognition that sustainable practices could lead to long-term profitability and resilience. In some regions, legal and regulatory changes were pushing companies to consider the wider impact of their decisions. For example, global agreements and standards (like the Paris Agreement and the United Nations Sustainable Development Goals) were also guiding businesses toward a stakeholder-oriented approach. Investors were increasingly considering environmental, social, and governance (ESG) factors in their investment decisions. Some shareholders themselves often advocated for companies to adopt more sustainable and responsible business practices.

Business leaders who recognized the importance of sustainability and ethical responsibility were driving change within their organizations and industries. Companies were aware that their reputation and brand value were enhanced by being socially responsible and environmentally conscious.

The Role of Generational Values in Shaping Business Strategies

Although the world was never flat as Thomas Freidman's short-lived claim (Friedman, 2005), companies operated in an increasingly interconnected world and they had to be responsive to a diverse range of cultural and societal norms. Younger generations, like Millennials and Gen Z, who were more value-driven in their consumption and employment choices, were influencing corporate strategies. The Great Shift was a response to a complex mix of environmental, social, economic, and cultural factors. It reflected a broader understanding that long-term business success was intricately linked to the well-being of all stakeholders, not just shareholders.

This era of stakeholderism marked a radical departure from the Friedman-esque mantra of shareholder primacy. Naomi Klein, in her provocative work *This Changes Everything: Capitalism vs. The Climate,* articulated the urgent need for businesses to re-evaluate their role in society, particularly in the context of environmental sustainability. "Our economic model is at war with life on earth," Klein argued, underscoring the dire consequences of unfettered capitalism that prioritizes profit over planetary health (Klein, 2014). Her work shed light on the interconnectedness of economic systems, environmental crises, and social inequities, urging businesses to adopt more holistic and sustainable approaches.

Simultaneously, R. Edward Freeman's theory of stakeholderism, elaborated in his seminal book *Strategic Management: A Stakeholder Approach*, provided a theoretical foundation for this new business ethos. Freeman challenged the conventional wisdom of shareholder primacy, positing that businesses should consider the interests of all stakeholders in their decision-making processes. "The primary responsibility of management is not just to the shareholders but to the stakeholders," Freeman asserted, expanding the corporate responsibility beyond the narrow confines of profit maximization. His work was a clarion call for a more ethical and sustainable approach to business, one that recognized the interdependence of all stakeholders in the corporate ecosystem.

15.4 Why Does the Great Shift Matter?

The Great Shift toward stakeholderism and sustainable business practices is profoundly connected to the future plight of humanity, as it encompasses environmental, social, and economic aspects that are crucial for the long-term survival and well-being of human societies.

The shift emphasizes reducing environmental impacts and mitigating climate change, which are essential for preventing catastrophic environmental changes (e.g., extreme weather events, sea-level rise, and loss of biodiversity) that could threaten human survival, because environmental degradation and climate change have direct and indirect impacts on public health, including increased risks of diseases, poor air quality, and shortage of food and clean water. Sustainable practices help in the preservation of ecosystems, crucial for maintaining biodiversity and natural resources that humans rely on.

The stakeholder approach focuses on unfair labor practices, community stress, and economic inequalities, leading to more frequent and severe economic crises, financial instability, and social unrest for large sections of the global population. Emphasizing health and safety in business practices directly impacts the well-being of employees and communities, which is fundamental for a thriving human society.

Sustainable economic models aim for resilience and long-term prosperity, as opposed to short-term gains that may lead to economic crises. Sustainable practices can drive innovation and create new economic opportunities and jobs, vital for human prosperity. A continued focus on short-term gains and shareholder primacy could erode public trust in institutions and diminish social cohesion.

The shift recognizes that many challenges, such as climate change and inequality, are global and require international cooperation, which is vital for the collective future of humanity. It fosters a greater understanding and exchange between diverse cultures and societies, which is essential for global peace and cooperation. As environmental and economic stresses could lead to increased conflict, political instability, and forced migration, posing challenges to global security and cooperation, without collaborative efforts to address these global challenges, international tensions could escalate, hindering effective responses to shared problems.

The shift can drive innovation in sustainable technologies, which are crucial for solving many of the challenges faced by humanity. Emphasizing science-based approaches in business and policy decisions is critical for addressing complex global issues effectively. Failing to integrate sustainability into technological and scientific development could result in missed opportunities to address critical challenges and improve the quality of life.

The Great Shift is not just an economic or business paradigm shift; it is a transformational approach that intertwines with the very fabric of human survival, well-being, and prosperity. By aligning business practices with the broader needs of the planet and its inhabitants, this shift plays a crucial role in shaping a sustainable and equitable future for humanity.

15.5 How Challenging Is the Great Shift?

The journey toward stakeholderism is not without its challenges. Balancing the diverse and sometimes conflicting interests of various stakeholders is a complex task. It requires a nuanced understanding of the trade-offs involved and a commitment to ethical decision-making. Moreover, this shift demands a cultural change within organizations, with values like empathy, collaboration, and social responsibility becoming as important as traditional business acumen.

The regulatory environment remains uncertain, but the force of its push is increasing. Sara Olsen, an impact management expert, maps the fast-changing landscape of impact measurement and management[4] and observes a notable change from the laissez-faire impact investment market flushed with the Dotcom slush money to the challenging parallel pursuits of stakeholder impact and fiduciary duty to shareholders (2007–2015), which she calls the purgatory stage of impact management. And recently, the landscape is visibly shifting toward an increasingly forceful regulatory phase. It is only a matter of time when clear ESG reporting standards will emerge, mandates for non-financial or integrated financial and non-financial disclosure will be mandated, sustainability regulations enforcement mechanisms will be established, and impacts will promulgate and incentives (e.g., tax benefits, subsidies, or preferential treatment) for leading companies committed to ESG principles or penalties for the laggards will ensue.

Adding to the uncertainty is the fact that, in many jurisdictions, legal structures and corporate laws are centered around protecting shareholder interests. Adapting these laws to accommodate a broader range of stakeholders is a complex legal challenge. There is a lack of consistency in how stakeholder interests are defined and protected across different countries and industries, complicating multinational operations.

---

[4] Source: Impact Measurement and Management: A Brief History, a talk hosted by Impact Entrepreneur, 2023. https://impactentrepreneur.com/event/navigating-impact-measurement-management-imm-with-sara-olsen/, accessed December 4, 2024.

Unlike shareholders, whose primary interest is typically financial return, stakeholders (like employees, customers, communities, and the environment) have varied and sometimes conflicting interests. Quantifying non-financial interests such as environmental impact or community welfare is complex, and balancing these diverse needs with financial objectives can be challenging. Integrating stakeholder interests may involve increased costs or reduced profitability in the short term, which can be a concern for businesses and their shareholders. Shifting focus can lead to tension with traditional investors who prioritize financial returns, potentially affecting investment and stock performance.

Companies adopting a stakeholder approach may face competitive pressures from firms that continue to prioritize short-term shareholder returns. Creating a level playing field where all companies in an industry consider stakeholder interests is challenging, especially in highly competitive markets. Corporate strategies need to be reimagined and financial models recalibrated.

Corporate governance faces a critical test. Traditional corporate governance models are designed around maximizing shareholder value. Adapting these models to incorporate stakeholder interests requires significant changes in corporate strategy and decision-making processes. Shareholderism often emphasizes short-term profits, while stakeholderism leans toward long-term sustainability and ethical conduct. This shift can create tension in strategic planning and performance evaluation.

Moving toward a stakeholder model requires a cultural shift within the organization, which can be resistant to change, especially in well-established corporations. Employees at all levels need to understand and embrace the stakeholder approach, requiring significant training and development efforts.

But within these challenges lie opportunities—for innovation, for ethical leadership, for creating a balance between profit and planet. This is the new frontier, the next chapter in capitalism's narrative. The Great Shift is not just about corporations and markets; it is a story about humanity, our planet, and our future. It is a narrative of transformation and hope, a journey toward a world where business is about people, planet, and prosperity. This is our story, unfolding in the vibrant tapestry of human endeavor, shaping humanity's future.

15.6   Strategies for the Great Shift

In the grand narrative of corporate evolution, the shift from shareholderism to stakeholderism marks a profound transformation, changing not just how businesses operate but also how they view their place in the world. This coincides with the dynamic and ever-evolving world of corporate sustainability, the journey of a business from its nascent understanding of environmental and social responsibility to a deeply ingrained culture of sustainability unfolds in stages, each marked by specific initiatives and milestones. This evolution can be viewed as a captivating narrative, one that unfolds over time, transforming the very core of a corporation's identity and operations while forging and treading a path that others might follow.

It is a story of responsibility, innovation, and a commitment to a future where businesses thrive in harmony with the environment and society.

Laying the Foundation for the Great Shift

The immediate actions focus on laying the foundation by acknowledging the Great Shift. In the initial phase, corporations find themselves at a crossroads, acknowledging the need to transition from a narrow focus on shareholder value to a broader stakeholder-oriented approach. This phase is characterized by a sense of urgency and reflection, as businesses re-evaluate their core values and strategies to align with the principles of stakeholderism.

The initial phase is about laying a robust foundation for sustainability, a period characterized by an air of anticipation and the promise of transformation. It is a time for defining clear and tangible sustainability objectives that resonate not just with the corporate mission but also echo the expectations of a diverse array of stakeholders.

Imagine a company setting a goal to significantly reduce its carbon footprint or to achieve a certain percentage of renewable energy use. The vibrancy of this stage lies in its potential; it is where ideas take shape and intentions are set. The establishment of a governance structure dedicated to overseeing sustainability efforts marks a commitment, a declaration of the company's dedication to this path.

In other words, understanding the implications of stakeholderism for corporate governance and operations is the focus. A good understanding can be achieved only through a strong tie between the leadership of the company and dedicated cross-functional teams that assess the impact of this shift on various aspects of the business.

The Key Performance Indicators:

- The clarity and ambition reflected in the sustainability goals set.
- The robustness and functionality of the sustainability governance structure.
- The visible integration of high-level sustainability considerations in daily corporate decisions.

Redefining the Corporate Mission

Within the first few years, the plot thickens. The emphasis is on laying the groundwork for this transformative journey. Corporations begin to redefine their mission and objectives to reflect a stakeholder-centric approach, considering the interests of customers, employees, suppliers, communities, and the environment, alongside those of shareholders. They begin to translate their sustainability rhetoric into action.

This period is marked by the adoption of recognized sustainability reporting frameworks, providing a language and structure to communicate efforts and progress transparently. In parallel, clear communication channels with stakeholders are established to understand their needs and expectations and provide greater transparencies on achievements and challenges.

The training of employees in sustainability practices is akin to sowing seeds in fertile ground; it is an investment in the future, ensuring that the workforce is not just aware but also engaged in this transformative journey. The company begins to see the first fruits of its labor as sustainability metrics start to show measurable improvements.

The Key Performance Indicators:

• The extent of adherence to and compliance with global sustainability reporting standards.
• The percentage of the workforce trained and actively engaged in sustainability practices.
• Tangible improvements in environmental and social metrics, mirroring the company's commitment to its goals.

Shifts in Corporate Policies and Practice

The mid-term phase, spanning three to five years, is where the integration of stakeholderism into all facets of the business becomes more pronounced. This phase sees significant shifts in corporate policies and practices to ensure that the interests of all stakeholders are considered in decision-making processes.

Investments in advanced technology for data collection and analysis become central to this stage, enabling companies to monitor and refine their sustainability initiatives with precision. The development and launch of sustainable products and services during this period signify a company's commitment to innovation and its willingness to pioneer new paths in sustainability.

Key Developments include revamping supply chain management to ensure ethical and sustainable practices and implementing employee welfare programs and fostering a culture of inclusivity and diversity.

The Key Performance Indicators:

• The degree to which sustainability is integrated across various business units.
• The effectiveness and accuracy of technological systems in monitoring and reporting sustainability metrics.
• The innovation quotient: the number and impact of sustainable products and services introduced in the market.

The Sustainability Culture

In the long-term, spanning over five years, stakeholderism becomes deeply institutionalized within the corporate fabric. It is no longer seen as an add-on but as an intrinsic part of the business ethos. Sustainability becomes more than a strategy; it transforms into a culture. This is a period of maturation and influence, where the company not only practices stakeholderism internally but also advocates for it within their industries and influences their peers and partners to adopt similar practices.

The long-term phase is about solidifying the gains, ensuring that sustainability is not just a trend but a legacy. The company's sustainability rankings and stakeholder perceptions become indicators of its success and influence. Innovations in this phase are not just about products or services but also about pioneering new ways of doing business that are environmentally conscious and socially responsible, with which the company leads industry-wide initiatives focused on sustainability and ethical practices and is recognized as a benchmark for stakeholder-centric operations and sustainable practices.

The Key Performance Indicators:

- The company's standing and reputation in sustainability rankings.
- The breadth and depth of sustainable practices across the supply chain.
- The number of sustainable innovations and patents, signifying the company's role as a leader in sustainability.

The Modern Corporation in the Modern World

Sustainability is a dynamic field, and businesses should regularly review and adapt their strategies in response to new challenges and opportunities. Throughout this journey, the shift from shareholderism to stakeholderism redefines the very essence of what it means to be a corporation in the modern world. It is a narrative of change and responsibility, of a business world that is increasingly aware of its impact on society and the environment, striving to create value not just for shareholders but for all stakeholders. This transformative journey is not just about adopting new strategies or practices; it is about embracing a new philosophy that prioritizes the well-being of the planet and its people alongside economic prosperity.

Future Pathways

How do we know if the Great Shift is successful? A change that does not have sweeping and long-lasting effects should not be called a Great Shift. The Great Shift toward stakeholderism and sustainable business practices is a transformative and ongoing process. Success or failure depends on interacting factors and is measured by numerous indicators. Whether this shift will ultimately be successful is contingent on a comprehensive and sustained effort across various sectors and stakeholders. To gauge its success or failure, we need to consider several criteria:

Environmental and Social Impact

- Reduction in Carbon Footprint: A significant indicator is the reduction of greenhouse gas emissions by businesses and industries.
- Positive Social Impact: Success can also be measured by improvements in social factors like community development, employee well-being, and reduced inequality.

Economics

- Sustainable Profitability: The ability of businesses to maintain profitability while adhering to sustainable practices.
- Investment Trends: A shift in investment toward sustainable and ethical companies could signal a change in priorities for investors.

Policy and Governance Changes

- Regulatory Compliance and Reforms: The adoption and implementation of policies that enforce sustainable practices.
- Corporate Governance: Changes in corporate governance to incorporate stakeholder interests as a key component of business strategy.

Cultural and Behavioral Changes

- Consumer Behavior: Changes in consumer preferences toward sustainable products and services.
- Corporate Culture Shifts: A transformation in corporate cultures that genuinely values sustainability and stakeholder welfare.

Technological Advancements

- Innovation in Sustainable Technologies: The development and adoption of technologies that support sustainable business operations.
- Integration of Sustainability in Business Operations: Using technology to integrate sustainability into the core of business operations.

Global and Local Integration

- Global Collaboration: Effective international cooperation in addressing global sustainability challenges.
- Local Engagement: Active engagement and support from local communities and stakeholders.

Long-term Resilience and Adaptability

- Resilience to Economic and Environmental Shocks: The ability of businesses to withstand and adapt to economic and environmental changes.
- Flexibility in Business Models: The adaptability of businesses to evolving sustainability challenges and stakeholder needs.

The success of the Great Shift will be a cumulative effect of changes in environmental, social, economic, technological, and governance dimensions. It is a dynamic process, and its success may not be immediately visible but rather assessed over the long-term horizon.

15.7   The Future of the Great Shift: Three Thought Experiments

As we navigate through the dynamic landscape of the Great Shift toward stakeholderism, we now embark on an intriguing trio of thought experiments. Each experiment serves as a vivid tableau, illustrating the profound impact of this shift. First, we explore the nuanced interplay between human behavioral archetypes and the Great Shift, revealing how our intrinsic and societal tendencies are reshaped. Next, we delve into the contrasting terrains of economic and political systems, observing their adaptation to this new paradigm. Finally, we peer into the future, contemplating the role and influence of Artificial General Intelligence in a stakeholder-centric world. These thought experiments offer a rich tapestry of perspectives, enhancing our understanding of the transformative power of the Great Shift.

The Great Shift and Human Behavioral Constraints

In the tangled journey of human evolution, our species has donned various behavioral cloaks, each reflecting the predominant ethos of its era.[5] From the biologically driven *Homo Sapiens* to the rationality-embracing *Homo Rationis*, these behavioral adaptations have been instrumental in shaping societies and cultures across centuries. This thought experiment, framed in Table 15.1, delves into the evolution of these behavioral archetypes. It explores the transition from the era of industrialization, marked by a shift toward *Homo Rationis* and the rise of capitalism, to the post-industrial age, where the limitations of market supremacy and rational self-interest become increasingly apparent. As we stand at the crossroads

---

[5] Source: Wikipedia, Name for human species—List of binomial names, https://en.wikipedia.org/wiki/Names_for_the_human_species#, accessed December 3, 2024.

of history, the emerging challenge is not to reject the rationality of *Homo Rationis* entirely but to evolve into *Homo Synthesizer*—a harmonizer of societal norms with rational, logical thought, striving for a sophisticated balance that integrates cultural understanding with rational strategy.

Take for example ESG investing, Alex Edmans reaffirms that the long-established *raison d'être* for corporations is improving long-term shareholder value and sees ESG factors as implementation supplements, rather than a monumental

Table 15.1 Human behaviors and constraints

Human Binomina	Behavioral Characteristics	Constraints and Controls
Homo Sapiens	Biologically and evolutionarily driven, emphasizing natural aspects	Biological needs and instincts, natural environmental factors
Homo Tribalis	Acts according to norms and values of tribal society, community-oriented	Tribal customs, traditions, and communal decisions
Homo Socius	Influenced by societal norms and structures, societal expectations	Societal norms, civic norms, social expectations, cultural influences
Homo Ludens	Engages in play, games, and recreational activities for enjoyment	Cultural values around leisure, societal attitudes toward play
Homo Politicus	Involved in political actions and decision-making	Political systems, laws, governance structures
Homo Individualis	Emphasizes individual identity and achievement, values personal freedom and autonomy	Cultural norms of individualism, personal aspirations, societal attitudes toward individual success
Homo Environmentalis	Focused on environmental conservation and sustainability	Environmental policies, ecological awareness, resource concerns
Homo Rationis	Believes social norms and structures follow a super-human, rational law, leading to an invisible hand and market dynamics	Rational principles, perceived natural laws of economics and society, faith in progress and innovation
Homo Economicus	Rational, self-interested, utility maximizer in economic contexts	Market forces, economic incentives, cost–benefit analysis
Homo Faber	Human as a maker and user of tools, emphasizing creativity and innovation	Availability of resources, technological advancement
Homo Synthesis	Human as a harmonizer and synthesizer of societal norms with a rational, logical approach to decision-making, emphasizes a sophisticated balance where social and cultural influences are critically assessed and integrated with a rational worldview	Cultural and societal norms, as well as rational thinking and logical analysis, societal expectations weighed against rational principles, seeking effective, logical solutions within specific social contexts, making decisions and actions that reflect a blend of cultural understanding and rational strategy

Note Binomial names in italic are invented by the author

shift. Recognizing the pressure of voluntary ESG disclosures for firms to internal-ize externalities, especially under an ineffective regulatory environment, he argues that it is crucial for firms to be transparent about the sacrifices in shareholder returns to pursue ESG-focused strategies (Edmans, 2023). Shareholders are ratio-nal, but they may simultaneously be willing to accept lower financial returns for societal benefits like reduced carbon emissions.

The journey through the various incarnations of human behavior, from Homo Sapiens to Homo Synthesizer, is not just an academic exercise but a mirror reflect-ing our collective evolution. As we contemplate the future, Homo Synthesizer emerges as an ideal, embodying the synthesis of cultural wisdom and rational anal-ysis. This archetype represents an evolved human consciousness that appreciates the complexity of societal structures and the necessity of logical decision-making, all while remaining deeply rooted in the cultural and social fabrics that define our existence. Embracing this synthesis means moving beyond the constraints that once defined us, acknowledging the multifaceted nature of human experience, and forging a path forward that is as informed by our shared history as it is by our aspirations for a balanced and sustainable future.

Can the Great Shift Succeed Without Capitalism?

The question of whether humanity can survive and sustain itself without capitalism is a deeply complex and theoretical one, involving numerous social, economic, political, and environmental considerations. It is important to note that the answer depends on the alternative systems considered, how they are implemented, and the adaptability of societies to change. Here are some considerations:

- Capitalism, in its various forms, has been dominant in much of the world, but there are also examples of other economic systems, like socialism, mixed economies, and traditional economies, each with their unique impacts on societal well-being. Non-capitalist systems may focus more on equitable dis-tribution of resources, which could address issues like poverty and inequality but might face challenges in efficiency and innovation.
- Critics of capitalism often point to its focus on continuous growth and consumption as unsustainable in the long term, especially in terms of envi-ronmental impact. Alternative economic systems might prioritize sustainability over growth. Different economic systems may have varying approaches to resource management, affecting environmental outcomes.
- Alternative systems to capitalism might place a greater emphasis on social wel-fare, healthcare, and education, potentially leading to more equitable societies. The success of any economic system depends significantly on cultural values and societal norms, which vary widely across the world.
- The survival and sustainability of humanity under different economic systems would depend heavily on the form of governance and the extent of democratic participation and human rights. In an increasingly interconnected world, the

shift away from capitalism would require significant international cooperation and coordination.

• Capitalism is often credited with driving innovation due to its competitive nature. Alternative systems would need to find different mechanisms to encourage technological and scientific advancement.

• Various non-capitalist societies have existed throughout history, offering lessons on the advantages and challenges of alternative systems. The future sustainability of humanity might depend more on how principles like equity, environmental stewardship, and innovation are integrated into any economic system, rather than the system itself.

While it is speculative to assert definitively whether humanity can survive and sustain without capitalism, the question underscores the importance of considering diverse economic models and their implications for society, environment, and the global future. It also highlights the need for adaptability, innovative thinking, and global cooperation in addressing the complex challenges facing humanity.

Integrating principles like equity, environmental stewardship, and innovation into different economic systems requires tailored approaches that respect the inherent characteristics and goals of each system.

Table 15.2 provides a comparative overview of how the principles of equity, environmental stewardship, and innovation can be integrated into different economic systems, highlighting the unique approaches and mechanisms that two contrasting economic systems, market capitalism and centralized socialism, might employ.

There are areas for common strategies under both systems. Participation in international agreements and adherence to global standards related to equity, environmental protection, and innovation is one of them. It is also important to communicate about the established mechanisms for monitoring the effectiveness of policies and practices and adapt them as necessary to meet changing circumstances and new challenges. Engagement with various stakeholders, including communities, businesses, and non-governmental organizations, to ensure policies and practices that are effective and inclusive is equally important across the systems.

Integrating principles like equity, environmental stewardship, and innovation into different economic systems requires a mix of regulatory measures, incentives, state initiatives, and market mechanisms. Both market capitalist and centralized socialist systems have unique strengths and challenges in adopting these principles, and a nuanced approach tailored to each system's characteristics is crucial.

On the other hand, integration can become politicized in numerous ways, as these principles often intersect with broader political ideologies, interests, and values. The ways in which this line of thinking is politicized can vary significantly between market capitalist and centralized socialist systems, as well as within different political contexts.

In market capitalist systems, political debates on regulation and free market persist. Left-leaning political groups may advocate for stronger government regulations to ensure equity and environmental protection, viewing state intervention

Table 15.2 Stakeholderism under contrasting economic systems

Aspect	Market Capitalist Economic System	Centralized Socialist Economic System
Regulatory Framework	• Implement and enforce regulations mandating equitable practices and environmental protection (e.g., fair labor laws, environmental standards) • Utilize tax incentives and subsidies for sustainability	• Implement comprehensive state policies for equitable resource distribution and environmental protection • Establish state-funded programs for equitable access
Environmental Solution Development	• Promote market-based environmental solutions like cap-and-trade systems • Support green technology development through subsidies and tax breaks	• Use central planning for sustainable practices and environmental conservation projects • Implement national programs for renewable energy and sustainable agriculture
Investment in Innovation	• Facilitate collaborations between government and private sector for social and environmental challenges	• Invest in state-run research institutions focusing on social and environmental innovations • Promote international collaboration for sustainability
Innovation Ecosystem Development	• Foster innovation through investment in R&D, education, and technological infrastructure • Create innovation hubs and incubators for social and environmental issues	• Emphasize education and public awareness about equity, environmental stewardship, and innovation • Integrate these principles into educational curricula
Socially Responsible Investment	• Incentivize investments in socially and environmentally responsible companies through ESG investing • Enforce strict disclosure requirements for corporate impacts	• Employ controlled market mechanisms to incentivize environmental stewardship and social equity • Use state-controlled funds to support sustainable projects

as necessary to correct market failures. Right-leaning groups often emphasize free market solutions, arguing that too much regulation stifles economic growth and innovation. Business and industry interests may resist environmental regulations that they perceive as costly or restrictive, leading to political lobbying and influence. Environmental groups and activists often counter this by pushing for stricter regulations and policies, leading to political clashes. Discussions around social equity can become politicized, with some viewing welfare programs and equitable resource distribution as essential for social justice, while others see them as an overreach of government power or a deterrent to personal responsibility and economic efficiency.

In centralized socialist systems, the balance between state control and individual freedoms can become a contentious political issue, with debates on how much control the state should exert over economic and social policies. Policies promoting equity and environmental stewardship might be criticized if they are perceived as limiting personal freedoms or economic opportunities. The effectiveness of central planning in driving innovation and sustainability can be a point of political contention, with some arguing for the efficiency of centralized decision-making and others advocating for the inclusion of market mechanisms. In socialist systems, how to engage with global capitalist markets while maintaining socialist principles can be politically divisive, especially in areas like trade, environmental agreements, and technology sharing.

There are common political themes across both systems. The balance between addressing global challenges like climate change and prioritizing national interests can be a source of political tension. Politicization can arise from the perceived interests of different stakeholder groups (workers, businesses, environmentalists, marginalized communities), with political parties often aligning with specific groups. The interpretation of what constitutes equity and sustainability can vary widely, often influenced by underlying political ideologies.

The integration of equity, environmental stewardship, and innovation into economic systems is frequently influenced by political ideologies, interests, and power dynamics. This politicization can manifest in debates over the role of government, the balance of state control and market freedom, and the prioritization of different societal goals and stakeholder interests.

The politicization of principles like equity, environmental stewardship, and innovation can present several challenges to corporate finance, especially as companies navigate varying political landscapes, regulatory environments, and stakeholder expectations. Here are some of the key challenges:

Regulatory Uncertainty and Compliance Costs. Political shifts can lead to changes in regulations related to environmental, social, and governance issues (ESG), impacting corporate strategies and operating costs. Adapting to new regulations or standards can be costly, requiring investments in innovative technologies, training, or operational changes.

Investment and Financing Dynamics. Increasing focus on sustainable and equitable practices can shape investor expectations, influencing funding availability and terms. Politicization can lead to changes in capital markets, with some investors prioritizing ESG factors, while others may resist such trends, affecting stock prices and investment flows.

Risk Management. Companies must contend with market risks and political risks, including potential policy changes and public sentiment shifts. Balancing long-term sustainability goals with short-term financial performance can be challenging, especially in politically volatile environments.

Strategic Planning and Forecasting. Political debates and changes can create uncertainty, making it difficult for companies to plan and forecast effectively. Corporations may need to adjust their strategies to align with political trends and stakeholder expectations, which can be a complex and dynamic process.

Corporate Reputation and Brand Value. Companies must navigate public sentiment and political rhetoric, which can quickly impact their reputation and brand
value. Engaging with a diverse set of stakeholders, including politically active
groups, requires careful consideration and communication strategies.

Global Operations and International Relations. For multinational corporations,
differing political stances and regulations across countries can complicate operations and financial planning. Political decisions on trade can directly impact
corporate finance, affecting supply chains and market access.

The politicization of equity, environmental stewardship, and innovation principles brings a complex set of challenges to corporate finance. Navigating this landscape requires adaptability, proactive risk management, and a keen understanding
of both local and global political dynamics.

Artificial General Intelligence (AGI) and the Great Shift

Technological advancements, particularly information technologies, have had a
major influence on business and economy for the last half century. They will continue to facilitate the shift toward stakeholderism in business, in the following
ways:

Enhanced Data Access, Democratization of Information, and Transparency. The
internet and digital media have democratized access to information, empowering
stakeholders by providing them with the knowledge needed to hold corporations
accountable.

Crowdsourcing and Crowdfunding Platforms. These platforms enable stakeholders to have a direct impact on business decisions and initiatives, fostering a sense
of ownership and participation.

Real-Time Reporting. Technologies like Internet of Things (IoT) and cloud com-
puting enable real-time data collection and reporting on various aspects of business
operations, from carbon footprint to labor practices. This transparency allows
stakeholders to make more informed decisions and holds businesses accountable.

Blockchain for Traceability. Blockchain technology ensures the traceability and
transparency of transactions throughout the value chain and products from origin
to end-user, highlighting ethical sourcing and production practices.

Stakeholder Engagement. Social media and digital platforms have revolution-
ized how businesses communicate with their stakeholders. They allow for direct,
two-way communication and have given stakeholders a more potent voice, pushing
companies toward greater accountability and responsiveness. Virtual engagement
tools like Augmented Reality (AR), Virtual Reality (VR), and various online
collaboration tools enable more interactive and inclusive stakeholder engagement, even across geographical boundaries. Technology enables better monitoring
and management of employee satisfaction and engagement have become crucial
aspects of internal stakeholder management.

Advanced Analytical and Collaboration Capabilities. Big data analytics and
artificial intelligence (AI) allow businesses to analyze vast amounts of data to

understand stakeholder needs and preferences better, predict trends, and make informed decisions that align with stakeholder interests. Smart supply chain management enabled by IoT and AI optimizes supply chain coordination, reducing waste and ensuring efficiency, which is increasingly important to stakeholders concerned about environmental and social impacts. Predictive analytics helps in identifying and mitigating risks related to social and environmental factors, aligning business strategies with stakeholder concerns. The rise of remote work technologies supports a more balanced and flexible work environment, addressing employee well-being and contributing to societal shifts in work-life balance. As AI becomes more integral to business operations, responsible AI development ensuring these systems are ethical and unbiased is key to maintaining stakeholder trust.

In summary, technological advancements have not only facilitated more efficient and effective business operations but have also played a crucial role in driving businesses toward a more stakeholder-centric approach. These technologies enable better transparency, engagement, and responsiveness to stakeholder needs, leading to more sustainable and socially responsible business practices.

Looking ahead, however, should we single out the accelerated progression toward the so-called Artificial General Intelligence (AGI) to examine its potential impact on the economy-scale shift toward stakeholders?

AGI refers to a level of artificial intelligence that can understand, learn, and apply its intelligence to a wide range of problems in a way that is indistinguishable from human intelligence, more precisely put, human collective intelligence. Like its predecessors, the control of fire a million years ago, the invention of wheels 5500 years ago, recent inventions such as steam engine, electricity, and semiconductor, these General Purpose Technologies fundamentally changed how humans live, human societies organize, and their relation to and impact on Planet Earth. It is useful to speculate how AGI can be deeply connected with the Great Shift toward more sustainable, equitable, and stakeholder-focused practices. Here is an exploration of how AGI may intersect with this paradigm shift:

Driving Sustainable Solutions. AGI could be instrumental in addressing complex global challenges like climate change, biodiversity loss, and resource management by analyzing vast amounts of data and simulating potential solutions. It could accelerate innovation in sustainable technologies, helping to create more efficient renewable energy systems, waste reduction technologies, and sustainable agricultural practices. AGI could become a sustainability problem solver at unprecedented scale and speed.

Enhancing Equity and Social Justice. AGI could assist in identifying and addressing social inequalities by analyzing data on economic, health, and social indicators, helping to create more equitable policies and practices in real time. By providing personalized education and healthcare solutions, AGI could contribute to leveling the playing field for underprivileged communities.

Transforming Business Operations and Ethics. AGI could help businesses better understand and respond to the needs of their stakeholders, ensuring decisions are more inclusive and considerate of diverse perspectives. The development of

AGI has already raised significant ethical questions. Its integration into businesses will necessitate a focus on ethical programming and decision-making frameworks, aligning with the principles of the Great Shift.

Impact on Employment and the Economy. AGI could lead to displacement in certain job sectors but also create new opportunities, especially in fields related to AI development, sustainability, and stakeholder engagement. The rise of AGI might necessitate new economic models to address the distribution of wealth and the value of work in an increasingly automated world.

Global Collaboration and Policy. The development and management of AGI will require global collaboration to ensure it is used ethically and beneficially, aligning with the goals of the Great Shift. The integration of AGI into society and economy will need thoughtful governance and policies to ensure it contributes positively to societal goals, including sustainability and equity.

AGI has the potential to be a pivotal force in the Great Shift, offering tools and insights to drive sustainability, equity, and ethical business practices. However, its development and integration must be managed carefully, with a strong focus on ethical considerations and societal impacts, to ensure that its benefits are aligned with the goals of the Great Shift. But how?

In the race toward AGI, there are two distinct paradigms. One is called Effective Accelerationism (with a coder like symbol of e/acc)[6] and the other Superalignment.[7] Effective Accelerationism refers to accelerating AGI development through fast iterations and intense learning under market force and intense competition. Superalignment, although not explicitly opposing the other paradigm, advocates for a cautious and responsible approach toward AGI development and emphasizes the importance of safety and control in AI systems. This involves ensuring that advancements in AI are aligned with ethical considerations and societal benefits.

While Effective Accelerationism entrusts the invisible hand to force AGI development to achieve societal benefits, Superalignment believes in caution, patience, and deliberation by some entities that treat the infantile AGI as loving parents would to their babies. Not any kind of love. Super love. Now, let us explore how these contrasting paradigms could be connected with the Great Shift or the future of capitalism.

Effective acceleration in the context of the Great Shift could involve swiftly implementing sustainable and stakeholder-focused practices within capitalism. This might include facilitating rapid development and adoption of green technologies, accelerating energy transitions to renewable sources, and accelerating evolution of business models to be more inclusive, equitable, and sustainable. This could involve redefining and measuring success beyond profit, to include social and environmental impacts, and quickly integrating the new metrics into corporate decision-making.

---

[6] Source: Wikipedia: Effective Accelerationism, https://en.wikipedia.org/wiki/Effective_accelerationism.

[7] Jan Leike and Ilya Sutzkever, Introducing Superalignment, OpenAI blog, July 5, 2023, https://openai.com/blog/introducing-superalignment.

Superalignment could begin with organizing large-scale consultations to establish guidelines for AGI development including guardrails that developers or users should not surpass. This alignment could foster a more holistic and compassionate approach to business and help develop clarity before AGI deployment. Alignment could focus on institutionalizing the AGI guidelines and guardrails through regulation, governance, and education and organizations commissioned to fulfill the alignment objectives.

Integrating these concepts into the current economic paradigm presents both opportunities and challenges. Effective Accelerationism demands agility and the willingness to embrace rapid change, which can be difficult for established businesses. In tackling urgent global issues, the pursuit of speed and efficiency often comes with elevated risk of potentially overlooking ethical considerations and stakeholder impacts.

Superalignment requires a fundamental shift in corporate, industry or even societal culture, and values, stressing a built-in ethical framework that prioritizes compassion, empathy, and the collective good to ensure AGI decisions and actions are beneficial to humanity and the environment. This is important but seems impractical or even non-democratic, given past human experiences. The impracticality increases with the scale of desired accomplishment. One hope is that AI could be super capable in scaling solutions, without devolving into autocracy or a similar political system.

One could imagine the connection between these contrasting concepts and broader societal and political concepts like individual freedom, market economy, and the spectrum of democracy versus autocracy in several ways. Each concept embodies different priorities and approaches, which can be aligned or contrasted with these societal systems and values. The key challenge, as well as promising opportunity, lies in integrating the strengths of each approach—embracing the speed and innovation of e/acc while ensuring the ethical and altruistic focus of s/align—within the existing societal and political structures.

Conclusion

As the concluding chapter of *Corporate Finance Under Climate Crisis, The Great Shift* brings us full circle, weaving together the themes and challenges explored throughout this book. In Chapter 1, we opened with Nicholas Stern's declaration of climate change as "the greatest market failure the world has ever seen," setting the stage for an urgent examination of how finance must evolve to address the crisis. Here, in *The Great Shift*, we began with Adam Smith's vision of markets governed by the "invisible hand," reflecting on how his foundational ideas resonate—and falter—in the face of today's climate challenges.

This dual perspective encapsulates the journey of this book: from understanding the science of emissions and accounting for carbon to navigating physical and transition risks, pricing those risks in financial markets, and deploying technologies to mitigate greenhouse gas emissions. Along the way, we have uncovered the critical role of corporate governance in steering businesses toward long-term resilience while balancing economic and environmental priorities.

As we return to the beginning, the essence of this journey emerges clearly. Addressing the climate crisis requires not merely incremental change but a rethinking of market and governance systems. The invisible hand must now be guided by principles of accountability and foresight, ensuring that today's financial decisions contribute to a resilient, equitable, and sustainable future. The market, often viewed as a source of innovation and wealth, must also reconcile its contributions to environmental degradation. *The Great Shift* argues for recalibrating markets to balance profit motives with collective responsibility, advocating governance and policy structures that align financial decision-making with long-term sustainability.

By realigning incentives, embracing innovation, and adopting inclusive frameworks, markets can fulfill their potential—not as passive mechanisms, but as purposeful forces driving collective action against the climate crisis. This shift, though monumental, offers an unparalleled opportunity to redefine the role of finance, ensuring a legacy of equity, resilience, and sustainability for generations to come.

Key Takeaways

- **The Rise of Shareholderism and Emergence of Stakeholderism**: Discusses the historical emergence of capitalism during the Industrial Revolution and its evolution into the post-World War II era, highlighting the rise of large corporations and the shareholder primacy model. Explores the shift toward considering broader stakeholder interests, including employees, customers, suppliers, and the environment, alongside shareholders.
- **Causes and Impact of the Shift on Corporate Policies and Practices**: Examines the drivers of the shift toward stakeholderism, such as globalization, information revolution, societal concerns, and legal changes, and its implications for businesses and society. Analyzes how this paradigm shift impacts corporate governance, strategic decision-making, and executive compensation, emphasizing the importance of environmental, social, and governance (ESG) factors.
- **Strategies for the Great Shift**: Outlines the phases of integrating stakeholderism in businesses, from immediate actions to long-term institutionalization, including key performance indicators (KPIs) for each phase.
- *Thought Experiments to Provoke Further Learning (see Appendix for an AI-generated illustration)*
  - **Human Behavioral Constraints**: Presents a thought experiment on human evolution in behavioral archetypes, from Homo Sapiens to Homo Synthesizer, and its relation to the Great Shift in economic paradigms.

- **Planetary Sustainability Under Opposing Economic Models**: A theoretical exploration of whether alternative economic systems to capitalism can achieve sustainability and equity. Discusses how these principles can be integrated into market capitalist and centralized socialist economic systems. Explores the political debates and challenges in integrating these principles into economic systems. Analyzes the challenges and opportunities in corporate finance arising from the politicization of these principles.
- **Artificial General Intelligence and the Great Shift**: Investigates the potential impact of Artificial General Intelligence on the economy and the shift toward stakeholderism. Compares the Effective Acceleration and Super-Love Alignment paradigms in AGI development and their connection to the Great Shift in capitalism. Presents an AI-generated visual representation of the complex themes discussed, symbolizing the transition from shareholderism to stakeholderism and its broader implications.

Questions

1. Define shareholderism and explain its historical emergence during the Industrial Revolution.
2. What role did Eugene Fama and Michael Jensen play in shaping the ideology of shareholder supremacy?
3. How does stakeholderism differ from shareholderism in terms of corporate governance and decision-making?
4. Discuss the significance of ESG (Environmental, Social, and Governance) factors in the shift toward stakeholderism.
5. Identify and explain the key drivers of the shift from shareholderism to stakeholderism in the late 20th and early twenty-first centuries.
6. Compare and contrast the principles of equity, environmental stewardship, and innovation in market capitalist versus centralized socialist systems.
7. Describe the potential impact of Artificial General Intelligence (AGI) on stakeholderism.
8. Discuss the challenges and opportunities in integrating equity, environmental stewardship, and innovation into corporate finance.
9. What are the implications of the Great Shift for corporate strategy and governance?
10. Evaluate the potential success factors for the Great Shift in achieving sustainable and equitable business practices.

Appendix: An Illustration: The Great Shift

This is an AI (DALL-E)-generated illustration representing the interplay of numerous factors affecting humanity's survival and sustainability. This image includes symbolic representations of different economic systems, technological advancements like AGI, cultural diversity, and global challenges such as climate change, environmental degradation, and geopolitical and social instability. These elements are interconnected, displaying complexity and interdependence in the context of the Great Shift toward sustainable and equitable practices.

In the illustration, various elements represent the complex themes and concepts discussed, particularly the shift from shareholderism to stakeholderism and its implications for humanity. Here is a breakdown of key parts of the image and their potential meanings:

1. Economic Systems: Different structures or patterns in the image can represent various economic systems. Look for distinct areas or shapes that might symbolize traditional capitalist structures (potentially rigid, mechanical forms) versus more holistic, organic forms that could represent stakeholder-centric systems.
2. Technological Development (AGI): Elements that appear more futuristic or technologically advanced, resembling circuits, AI imagery, or abstract representations of digital networks, could symbolize the role of AGI and other technological advancements.
3. Cultural Diversity: Diverse human figures, symbols from various cultures, or a range of colors and patterns might be used to represent cultural diversity. This diversity is crucial in the transition to stakeholderism, emphasizing inclusivity and the consideration of diverse perspectives.
4. Global Challenges:
   - Climate Change: This might be represented by elements like the Earth, weather patterns, or symbols indicating temperature changes or natural disasters.
   - Environmental Degradation: Look for imagery that suggests pollution, deforestation, or loss of biodiversity, such as barren landscapes or industrial imagery.
   - Geopolitical and Social Instability: Chaotic or fragmented sections of the image could represent instability, while more harmonious areas might symbolize peace and stability.
5. Transition from Shareholderism to Stakeholderism: The key theme of the transition might be depicted through a visual journey from one part of the image to another. The starting point could be more chaotic, disjointed, or mechanistic, representing the focus on shareholder value, moving toward a more integrated, harmonious design that signifies the integrated approach of stakeholderism.
6. Interconnectedness: Lines or pathways connecting different elements of the image can illustrate the interdependence of these factors. This interconnectedness is central to understanding how changes in one area (like technology or economics) can impact others (like culture or the environment).

References

Bebchuk, L. A., & Tallarita, R. (2020). The illusory promise of stakeholder governance. *Cornell Law Review, 106*, 91–178. https://doi.org/10.2139/ssrn.3544978

Edmans, A. (2023). The end of ESG. *Financial Management, 52*(1), 3–17.

Fama, E. F. (1970). Efficient capital markets: A review of theory and empirical work. *Journal of Finance, 25*(1), 383–417.

Fama, E. F. (1991). Efficient capital markets: II. *Journal of Finance, 46*(5), 1575–1617. https://doi.org/10.1111/j.1540-6261.1991.tb04636.x

Freeman, R. E., & Reed, D. L. (1983). Stockholders and stakeholders: A new perspective on corporate governance. *California Management Review, 25*(3), 88–106. https://doi.org/10.2307/411 65018

Friedman, M. (1970, September 13). The social responsibility of business is to increase its profits. *The New York Times Magazine*. https://www.nytimes.com/1970/09/13/archives/a-friedman-doctrine-the-social-responsibility-of-business-is-to.html

Friedman, T. L. (2005). *The world is flat: A brief history of the twenty-first century*. Farrar, Straus, and Giroux.

Jensen, M. C., & Meckling, W. H. (1976). Theory of the firm: Managerial behavior, agency costs and ownership structure. *Journal of Financial Economics, 3*, 305–360.

Klein, N. (2014). *This changes everything: Capitalism vs. the climate*. Simon & Schuster.

Index

The manufacturer's authorised representative in the EU is Springer
Nature Customer Service Centre GmbH, Europaplatz 3, 69115 Heidelberg,
Germany. If you have any concerns regarding our products, please
contact ProductSafety@springernature.com

Printed and bound by CPI Group (UK) Ltd, Croydon, CR0 4YY
27/04/2026
02097573-0008